Barron's Review Course Series

Let's Review:

Chemistry—
The Physical Setting

Sixth Edition

Albert S. Tarendash, M.S.
Assistant Principal—Supervision (Retired)
Department of Chemistry and Physics
Stuyvesant High School
New York, New York

Chemistry/Physics Faculty (Retired)
The Frisch School
Paramus, New Jersey

BARRON'S

All inquiries should be addressed to:
Barron's Educational Series, Inc.
250 Wireless Boulevard
Hauppauge, New York 11788
www.barronseduc.com

ISBN: 978-1-4380-0959-9

ISSN: 2164-7380

PRINTED IN THE UNITED STATES OF AMERICA

9 8 7 6 5 4 3 2

10%
POST-CONSUMER WASTE
Paper contains a minimum of 10% post-consumer waste (PCW). Paper used in this book was derived from certified, sustainable forestlands.

TABLE OF CONTENTS

Table of Contents

PREFACE TO THE SIXTH EDITION

To the Student:

This book has been written to help you understand and review high school chemistry. Chemistry is not a particularly easy subject. No book—no matter how well written—can give you *instant* insight. Nevertheless, if you read this book carefully and do all of the problems and review questions, you will have a pretty decent understanding of what chemistry is all about.

I have designed this book to be your "chemistry companion." It is more detailed than most review books, but it is probably less detailed than your hardcover text. The book is divided into 16 chapters. Each chapter begins with an overview (*Key Ideas*) and a summary of what you should learn in the chapter (*Key Objectives*). Each chapter is then divided into two sections. Section I contains basic material that follows the New York State Regents Core syllabus very closely. Section II contains additional material that is *not* in the Regents Core but may well be part of your chemistry course (and is likely to appear on other standardized tests such as the SAT Chemistry Subject Test).

This book provides a large number of problems with detailed solutions. Many of the problems are usually followed by a "Try It Yourself" exercise and its answer. You are encouraged to work out each and every exercise.

Each chapter ends with multiple-choice questions and, usually, one or more constructed-response and free-response questions. These are drawn from both Sections I and II. The answers appear in Appendix 3. Appendix 1 contains the official New York State Regents reference tables that are used throughout the book, and Appendix 2 contains additional (non-Regents) tables that are used for Section II topics. Appendix 4 provides some information on how to answer various types of constructed-response questions.

The Internet is now bursting with websites devoted to high school chemistry. You are strongly encouraged to use search engines, such as Google, to find and use these sites.

If you have any comments about this book or if you find errors (for which I apologize in advance), please e-mail me at:

ast-lrc@optimum.net

I wish you much success.

A. S. Tarendash

To the Teacher:

The sixth edition of this book was written with the same philosophy as the first five editions: A worthwhile review book should provide clear, careful, and detailed explanations that will, in effect, "hold the student's hand" while he or she is studying the subject.

Features of the Sixth Edition

- Most of the 16 chapters are divided into two sections. Section I contains basic material that follows the New York State Regents Core syllabus. Section II contains additional material that goes beyond the New York State Core and may well appear on such standardized tests such as the SAT Subject Test in Chemistry.
- The additional reference tables have been revised to reflect recent discoveries, such as elements 117 and 118, and the naming of elements 113–118.
- Thanks to reader correspondence, errors have been corrected extensively throughout the book.
- The end-of-chapter questions have been revised so that Section I questions contain more recent Regents examination material.

You may find that my ordering of the material is a bit unorthodox. Then again, we all have our own visions as to how a high school chemistry course should be sequenced. In the final analysis, you may not feel that the way I subdivided the material is appropriate for your teaching style or course. If you are more comfortable using your own order or the order shown in your textbook, you should certainly continue with that approach.

Rather than padding each section of the text with multiple-choice questions, I spent considerable time in working out sample problems. The text heavily emphasizes the factor label method (dimensional analysis) because I believe that this one area is where students experience considerable difficulties. In this vein, I have also included an additional appendix (Appendix 4) on how to answer constructed-response and free-response items. This appendix is designed to train your students to compose answers that are logical and orderly and that will not drive you crazy when you grade them!

I reserved the end of each chapter for the multiple-choice and constructed-response questions, including free-response questions. I have *not* included the type of questions found on the SAT subject test. If you are preparing your class to take the SAT subject test, you should obtain a book published by The College Board that contains a variety of *authentic*, released SAT II examinations in all subjects, including chemistry. You will find the chemistry section of that book particularly helpful. You should make every effort to include at least a few SAT II–type questions on your class examinations throughout the year.

A final thought: *Let's Review: Chemistry* was never meant to be one of those thin "Regents prep books." It was always designed to be a review of a *complete* introductory one-year course in high school chemistry as taught in most of the country. To New York teachers: Remember, the New York State Education Department took great pains to indicate that "[the] Core is *not* a syllabus" [*Chemistry Core* p. 3.] but a document to assist in planning a year-long chemistry course. In effect, the Core is merely a blueprint for questions that students might be asked on the Regents Chemistry examination. I firmly believe that no New York State high school chemistry teacher should be using the Core as his or her *sole* curriculum guide. In this vein, many New York teachers have recognized that omitting certain topics makes other basic concepts mysterious to their students. (Can one really teach Lewis electron-dot diagrams in a meaningful way if atomic sublevels and orbitals have been omitted?)

I hope that this book provides some new insights and that it will help you to ensure that you approach your lesson plans and lessons in a spirit of inquiry and problem solving. I also hope that *Let's Review: Chemistry* will be successful for both you and your students.

If you have any comments about this book or if you find errors (for which I apologize in advance), please e-mail me at:

> ast-lrc@optimum.net

Acknowledgments

No book is ever the work of one person. Once again, I wish to acknowledge and thank:

- The editors at Barron's for their guidance, patience, understanding, and gentle prodding;
- My wife, Bea, whose love, support, and help (particularly with the end-of-chapter questions) made the preparation of this edition considerably easier;
- My children, Franci, David, Jeff, and Janet, and my grandchildren, Brittany, Alexa, Danielle, Raina, and Jonathan, just for being near;
- All of my students—past and present—that have inspired me over the last 50 years and have made teaching a happy and satisfying endeavor.

ALBERT S. TARENDASH
Nanuet, New York
January 2017

Chapter One

INTRODUCTION TO CHEMISTRY

KEY IDEAS

This chapter introduces some of the basic concepts in chemistry that serve as a foundation for the chapters that follow. In particular, measurement and problem solving are studied.

KEY OBJECTIVES

At the conclusion of this chapter you will be able to:
- Define the terms *chemistry*, *matter*, *pure substance*, and *mixture*.
- Distinguish among elements, compounds, and mixtures.
- Distinguish between physical and chemical properties.
- Indicate how mixtures are separated.
- Distinguish between physical and chemical changes.
- List the various forms of energy.
- List the common metric units used in chemistry.
- Name the most commonly used metric prefixes and their numerical equivalents.
- Express numbers in scientific notation.
- Perform simple operations on numbers expressed in scientific notation.
- Define the terms *volume* and *density* as they apply to chemistry.
- Describe how measurement readings are taken and estimated.
- Define and apply the terms *accuracy*, *precision*, and *significant digits* (*figures*).
- Apply the rules for adding (subtracting) and multiplying (dividing) measurements.
- Define the term *percent error*, and calculate the percent error of a measurement.
- Understand the factor-label method (FLM), and use it to solve problems.
- Use equations and graphs to solve problems.

Reminder: If you have not already done so, read Preface: To the Student (page xi) at this time.

SECTION I—BASIC (REGENTS-LEVEL) MATERIAL

NYS REGENTS CONCEPTS AND SKILLS

By the time you have finished Section I, you should have mastered the concepts and skills listed below. The Regents chemistry examination will test your knowledge of these items and your ability to apply them.

Concepts are the *basic ideas* that form the body of the Regents chemistry course (what you need to know!).

Skills are the *activities* that demonstrate your mastery of these concepts (how you show that you know them!).

Following each concept or skill is a page reference (given in parentheses) to this chapter.

1.1 Concept:
Matter is classified as a pure substance or as a mixture of substances. (Page 4)

1.2 Concept:
The structure and arrangement of particles determine the phase of a substance.
The three phases of matter (solids, liquids, and gases) have different properties. (Page 4)

Skill:
Use a simple particle model/diagram to differentiate among the properties of a solid, a liquid, and a gas. (Pages 4–5)

1.3 Concept:
A pure substance (element or compound) has a constant composition and constant properties throughout a given sample, and from sample to sample. (Pages 4–5)

Skill:
Use a simple particle model/diagram to differentiate among elements, compounds, and mixtures. (Pages 5–6)

1.4 Concept:
Elements cannot be broken down by chemical change. (Pages 4–5)

1.5 Concept:
Mixtures are composed of two or more different substances that can be separated by physical means. When different substances are mixed together, a homogeneous or heterogeneous mixture is formed. (Pages 5–6)

1.6 Concept:
The proportions of components in a mixture can be varied. Each component in a mixture retains its original properties. (Page 5)

1.7 Concept:
A physical change results in the rearrangement of existing particles in a substance. A chemical change results in the formation of different substances with changed properties. (Page 5)

1.8 Concept:
Energy can exist in different forms, such as chemical, electrical, electromagnetic, thermal, mechanical, and nuclear. (Pages 12–13)

1.9 Concept:
Algebraic and geometric representations can be used to describe and compare data.

Skills:
• Organize, graph, and analyze scientific data. (Pages 27–31)
• Recognize and convert various scales of measurement. (Pages 23–25)

1.10 Concept:
Algebraic and geometric representations can be used to recognize patterns and relationships in mathematics.

Skills:
• Identify direct and inverse relationships. (Pages 28–29)
• Use data trends (such as those shown in graphs) to predict information. (Pages 28–29, 30–31)

1.11 Concept:
Algebraic and geometric concepts can be used to solve problems.

Skills:
• State what assumptions are made when a mathematical equation is used. (Pages 25–27)
• Evaluate whether an answer to a numerical problem is reasonable. (Pages 25–27)

1.12 Concept:
Differences in properties such as density, particle size, molecular polarity, boiling point and freezing point, and solubility permit physical separation of the components of a mixture. (Pages 6–12)

Skill:
Describe the processes and uses of filtration, distillation, and chromatography in the separation of a mixture. (Pages 6–12)

1.1 CHEMISTRY IS . . . ?

Chemistry is the science that focuses on the structure, composition, and properties of matter. In our study, we will also learn about the changes that matter undergoes and the energy that accompanies these changes.

1.2 MATTER AND ENERGY

Our universe is composed of matter and energy. **Matter** is anything that has mass and volume (that is, anything that has *density,* a concept we will develop in Section 1.6).

Matter commonly exists in three phases: *solid*, *liquid*, and *gas*. Generally, solids are composed of particles that are tightly packed and have a regular arrangement. The particles present in gases have no regular arrangement and no appreciable packing. In liquids, the arrangement and packing of the particles are somewhere between those in solids and gases. As a consequence, a solid has a definite shape and volume; a liquid has a definite volume but takes the shape of its container; a gas has neither a definite shape nor a definite volume: its shape and volume are those of its container.

We can use *particle diagrams* to visualize solids, liquids, and gases and to distinguish among them, as shown below.

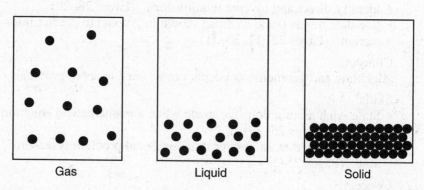

| Gas | Liquid | Solid |

Substances

For convenience, we divide matter into two classes: *pure substances* and *mixtures*.

A pure **substance** is any variety of matter that is homogeneous (uniform) and has a fixed composition by mass. **Elements** are classified as substances because they contain atoms of a single type and cannot be decomposed further. Although **compounds** contain more than one type of element, they are also classified as substances because their compositions are fixed. For example, in every sample of water, the ratio, by mass, of oxygen to hydrogen is eight to one. Because this ratio cannot be changed, we usually say that the elements in a compound are **chemically combined**. The accompanying particle diagrams represent two different elements and a compound.

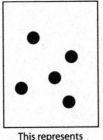

This represents
an **element**.

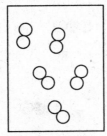

This also represents
an **element**.

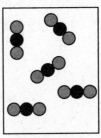

This represents
a **compound**.

Every substance has a unique set of *properties* that allow it to be distinguished from other substances. These properties can be grouped into two categories: physical and chemical.

Physical properties can be measured without changing the identity and composition of a substance, and include color, odor, density, melting point, and boiling point. For example, when the melting point of a substance is measured, its phase changes from solid to liquid. The appearance of the substance changes, but not its composition. Changes that accompany the measurement of physical properties are called **physical changes**. For example, all phase changes are physical changes.

Chemical properties are properties that lead to changes in the identity and composition of a substance. For example, combustibility is a chemical property. When hydrogen gas is burned in air, the hydrogen combines with the oxygen in the air and changes into water, a substance whose composition is entirely different from that of hydrogen. Changes that accompany the measurement of chemical properties are called **chemical changes**; they are also called **chemical reactions**. For example, when a battery provides electrical energy, the changes that occur *inside* it are chemical changes because entirely new substances are produced.

Mixtures

A **mixture** is composed of two or more distinct substances, but the compositions of mixtures may be varied. A solution of sugar and water is a mixture because the ratio of sugar to water can be changed. A solution is an example of a **homogeneous mixture** because each component (sugar and water in this case) is uniformly dispersed throughout the solution. An ice cream soda, on the other hand, is not homogeneous because each component (ice cream, syrup, soda, whipped cream) is not uniformly dispersed. This is an example of a **heterogeneous mixture**. The accompanying particle diagram represents a heterogeneous mixture.

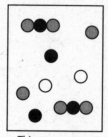

This represents a
heterogeneous mixture.

Separation of Mixtures

One way of distinguishing between mixtures and substances involves the ability to separate mixtures by "physical" methods. For example, it is possible to separate a sand-water mixture by allowing the water to evaporate. When a physical method is used to separate a mixture, the chemical properties of the mixture's components are *not* changed. We discuss these separation techniques below.

Filtration

Suppose we have a sand-salt mixture, and we want to separate the sand from the salt. The technique known as **filtration** will accomplish the task of recovering the sand from the mixture. First, we add water to the mixture, a procedure that dissolves the salt, but not the sand. Second, we fit a funnel with a cone of **filter paper**, and carefully pour the well-stirred mixture into the funnel, as shown in the accompanying diagram.

6

Filtering Diagram

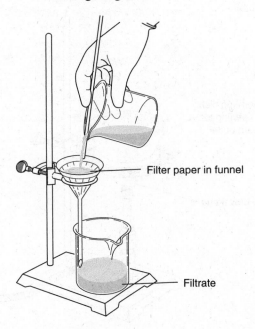

Filter paper in funnel

Filtrate

The salt-water solution that passes through the filter paper into the beaker below is known as the **filtrate**. The sand remains on the filter paper. To remove the last traces of salt from the moist sand, we wash the sand several times with pure water. We then allow the sand to dry, either in air or in a **heating oven**.

Evaporation

The next step is to recover the salt from the salt-water solution. Drying would work but would take a very long time. In situations such as this, we use the technique known as **evaporation**. We pour the solution carefully into an **evaporating dish**, and cover the dish with a **watch glass**. Then we heat the entire assembly *gently over a low flame* until the water boils and is completely evaporated. The watch glass prevents the salt from spattering during the heating process. The accompanying diagram illustrates this technique.

Evaporation-1 Diagram

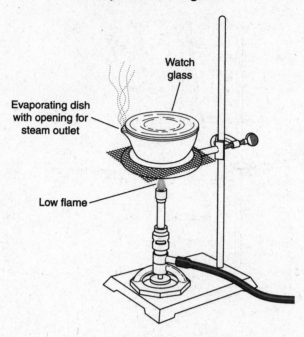

Watch glass

Evaporating dish with opening for steam outlet

Low flame

In the event that *strong heating* is required, a **crucible** is used. Crucibles are frequently made of porcelain and are supported by a **pipestem triangle** while being heated. The accompanying diagram illustrates the use of a crucible.

Evaporation-2 Diagram

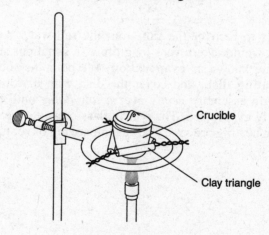

Crucible

Clay triangle

Crystallization

Yet another technique we can use to separate a dissolved solid from its solvent is known as **crystallization**. In this technique, we prepare a concentrated solution at a relatively high temperature and allow the solution to cool slowly. As it cools solid crystals form at the bottom of the vessel containing the solution. At times, the solution becomes *supersaturated* and the crystals do *not* form immediately. In these cases, we can initiate crystallization by adding a single "seed crystal" of the solid to the solution.

Distillation

Suppose we have an aqueous solution containing dissolved solids, but it is the *solvent* that we want to recover. In this instance, we use the technique known as **distillation**. We place the solution in a **boiling flask** and heat it to boiling. As the solvent vapor passes out of the flask, it passes through a **condenser**, a tube that is surrounded by an outer tube containing cold water. The cold water *condenses* the vapor in the tube, and the solvent is collected in a second flask. The collected solvent is known as the **distillate**. The accompanying diagram illustrates one way of distilling a solution.

Distillation Diagram

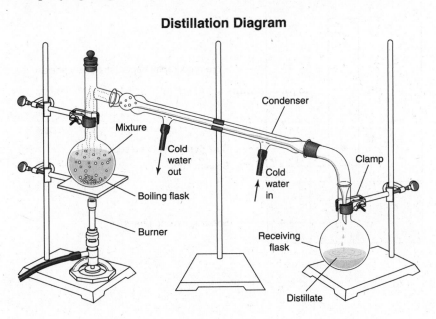

There are a number of other types of distillation. In Chapter 11, we will learn that the process of *fractional distillation* is used to separate petroleum into simpler mixtures of *liquid* hydrocarbons, and that this process is dependent on the *differences in the boiling points* of the hydrocarbon fractions.

Chromatography

Another technique used by chemists to separate mixtures is **chromato-graphy**. The word *chromatography* comes from the Greek words *chroma* (color) and *graphein* (to write). The term was coined by the Russian botanist Mikhail Tsvett, who used the technique in 1906 to separate a mixture of plant pigments. In chromatography, a spot of a mixture is placed on a **stationary phase**, such as paper. The stationary phase is then placed in a sealed container of solvent, which is known as the **moving phase**. As the solvent moves along the stationary phase, it carries the mixture with it. If different components of the mixture have different degrees of attraction for the stationary phase, they will travel at different speeds along it and will be *separated* from one another.

The accompanying diagrams illustrate how one technique, known as *ascending paper chromatography*, is used to separate the components of a mixture.

Distinguishing Types of Matter

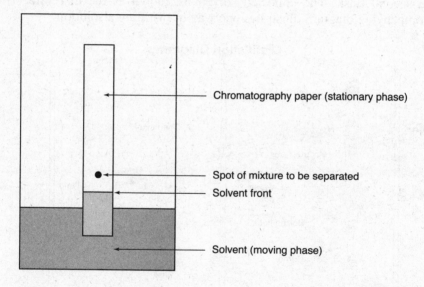

Chromatography paper (stationary phase)

Spot of mixture to be separated

Solvent front

Solvent (moving phase)

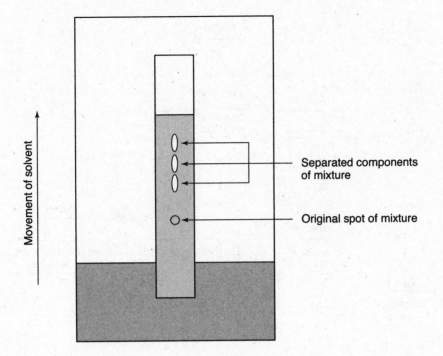

Movement of solvent

Separated components
of mixture

Original spot of mixture

An element *cannot* be separated, and a compound can be separated only by more drastic "chemical" methods. These methods will change the chemical properties of the components that are part of the compound. For example, when the compound water is separated by using an electric current, this familiar liquid is replaced by two gaseous elements, hydrogen and oxygen. The chemical properties of these elements are very different than those of water.

The accompanying diagram presents a scheme for distinguishing among the various types of matter.

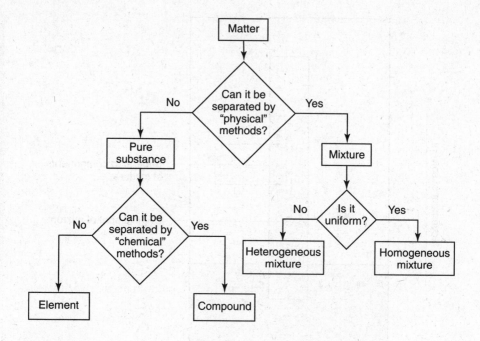

Energy

It is not easy to define **energy** precisely, but we are all familiar with at least some of its forms: thermal (heat), chemical, electrical, electromagnetic radiation (light), nuclear, and mechanical energy. Mechanical energy is divided into **kinetic energy** (the energy associated with motion) and **potential energy** (the energy associated with position). As students of chemistry, we will be particularly interested in heat and in mechanical and electrical energy.

All forms of energy are associated with the concept of mechanical work, which associates a force with a change of position. For example, we use a vacuum cleaner to transform electrical energy into the mechanical work needed to lift and remove dirt from a carpet.

The concept of energy is of great importance to chemists. For example, chemical reactions may absorb energy (*endothermic* reactions) or release it (*exothermic* reactions). Many reactions involve the conversion of one form of energy to another. Yet, as a result of countless experiments, chemists have arrived at the conclusion that energy is never destroyed in a chemical change; rather, it is *conserved*. The role of energy in chemical reactions will be treated more fully in Chapter 5.

Although work and energy may seem very different, they are closely related; in fact, the same unit is used to measure both quantities. In the SI (as shown in the next paragraph) this unit is the joule (J); its multiple is the kilojoule (kJ).

An older method of measuring energy involves calculating how much heat energy is absorbed or released by a substance. This older unit of energy is the calorie (cal); its multiple is the kilocalorie (kcal). Originally, the calorie was defined as the amount of heat energy needed to change the temperature of 1 gram of liquid water by 1 Celsius degree. The calorie is now defined in terms of the joule: 1 calorie = 4.2 joules.

1.3   MEASUREMENT AND THE METRIC SYSTEM

The basis of all science lies in the ability to measure quantities. For example, we can easily measure the length or mass of this book if we have some standardized system of measurement. We use the Système International (SI) and its derivatives because scientists all over the world express measurements in metric units. The SI has established seven fundamental quantities upon which all measurement is based: (1) length, (2) mass, (3) temperature, (4) time, (5) number of particles, (6) electric current, and (7) luminous intensity. For each of these quantities there is a unit of measure based on a standard that can be duplicated easily and does not vary appreciably.

Length

The unit of length is the meter (m), which is approximately 39 inches. In chemistry, we also use various subdivisions of the meter: picometer (pm, 10^{-12} meter), nanometer (nm, 10^{-9} meter), millimeter (mm, 0.001 meter), centimeter (cm, 0.01 meter), and decimeter (dm, 0.1 meter). The prefixes *pico-*, *nano-*, *centi-*, and so on, are known as **metric prefixes**. A list of these prefixes appears in Section 1.4.

Mass

The unit of mass is the kilogram (kg), which has an approximate weight (on Earth) of 2.2 pounds. (We note that mass and weight are not the same quantity. *Mass* is the amount of matter an object contains, while *weight* is the force with which gravity attracts matter.) In chemistry we also use the gram (g; 1000 grams equals 1 kilogram) as a unit of mass.

Temperature

Temperature measures the "hotness" of an object. The SI unit of temperature is the kelvin (K). The fixed points on the Kelvin scale of temperature are the zero point (0 K, known as "absolute zero") and 273.16 K, which is known as the "triple point" of water. (More will be said about the Kelvin scale of temperature in Chapter 6.)

In chemistry, we also use the Celsius temperature scale (°C). Originally the fixed points on this scale were set at the normal freezing and boiling points of water (0°C and 100°C, respectively).

We can convert between the Celsius and Kelvin scales using the simplified equation that appears in Reference Table T in Appendix 1:

$$K = °C + 273$$

Time

The unit of time is the second (s). In chemistry, we also use the familiar minute (min), hour (h), day (d), and year (y) as units of time.

Number of Particles

The unit of number of particles is the mole (mol). The mole is a very large number— 6.02×10^{23}. In chemistry, we frequently need to know the number of particles (such as atoms or molecules) in a sample of matter in order to predict its behavior.

Electric Current

The unit of electric current is the ampere (A). Electric current measures the flow of electric charge. In the SI system of measurement, electric charge is a *derived* quantity that is based on electric current and time. The SI unit of electric charge is the coulomb (C).

Luminous Intensity

The unit of luminous intensity is the candela (cd). As its name implies, luminous intensity measures the brightness of light. We will not need to use this unit in our study of chemistry.

1.4 METRIC PREFIXES

Very often, an SI unit is not convenient for expressing measurements. For example, the meter is not useful for describing the diameter of a red blood cell or the distance between Earth and the Sun. There are two ways to deal with this dilemma: we can use scientific notation (which is introduced in the next section), or we can use metric prefixes to create multiples and subdivisions of any unit of measure. The accompanying table lists some metric prefixes along with their symbols and values.

Factor	Prefix	Symbol	Factor	Prefix	Symbol
10^{15}	peta-	P	0.1	deci-	d
10^{12}	tera-	T	0.01	centi-	c
10^9	giga-	G	0.001	milli-	m
10^6	mega-	M	10^{-6}	micro-	μ
1000	kilo-	k	10^{-9}	nano-	n
100	hecto-	h	10^{-12}	pico-	p
10	deka-	da	10^{-15}	femto-	f

For example, we use the term picogram (pg) to designate 10^{-12} gram (one-trillionth of a gram). This is a very small quantity of matter! An abbreviated list of metric prefixes appears in Reference Table C in Appendix 1.

1.5 SCIENTIFIC NOTATION

Once we have established a system of measurement, we need to be able to express small and large numbers easily. Scientific notation accomplishes this purpose. In scientific notation, a number is expressed as a power of 10 and takes the following form:

$$M \times 10^n$$

M is called the *mantissa*, and n is the *exponent*. The mantissa is a real number that is greater than or equal to 1 and is less than 10 ($1 \leq M < 10$). The exponent is an integer that can be positive, negative, or zero. For example, the number 2300 is written in scientific notation as 2.3×10^3 (*not* as 23×10^2 or 0.23×10^4). The number 0.0000578 is written as 5.78×10^{-5}.

To write a number in scientific notation, we move the decimal place until the mantissa is a number between 1 and 10. If we move the decimal place to the left, the exponent is a positive number; if we move it to the right, the exponent is a negative number.

To *multiply* two numbers expressed in scientific notation, we *multiply the mantissas and add the exponents*. The final result must always be expressed in proper scientific notation. Here are two examples:

$$(2.0 \times 10^5)(3.0 \times 10^{-2}) = 6.0 \times 10^3$$
$$(4.0 \times 10^4)(5.0 \times 10^3) = (20. \times 10^7) = 2.0 \times 10^8$$

Division is accomplished by *dividing the mantissas and subtracting the exponents.*

To *add* (*subtract*) two numbers expressed in scientific notation, *both numbers must have the same exponent. The mantissas are then added* (*subtracted*). The following examples illustrate the application of this rule:

$$(2.0 \times 10^3) + (3.0 \times 10^2) = (2.0 \times 10^3) + (0.30 \times 10^3)$$
$$= 2.3 \times 10^3$$
$$(5.0 \times 10^{-5}) - (2.0 \times 10^{-5}) = 3.0 \times 10^{-5}$$

Every calculator is different. Become familiar with how to input and manipulate scientific notation on your calculator. Doing so will help you work with calculations involving scientific notation.

1.6   VOLUME AND DENSITY

In addition to the fundamental metric quantities, there are many derived quantities that are combinations of the fundamental ones. For example, the speed of an object is the ratio of the distance (length) it travels to a given amount of time. In chemistry, two derived quantities are of immediate interest to us: volume and density.

Volume

The **volume** of an object is the amount of three-dimensional space the object occupies. The SI unit of volume is the cubic meter (m^3), which is too large for practical use in chemistry. We will use the cubic centimeter (cm^3), which is approximately equal to one-thousandth of a quart, and the liter (L), which is approximately equal to 1 quart. One liter is *exactly* equal to 1000 cubic centimeters.

We often designate one-thousandth of a liter as a milliliter (mL). It follows that 1 cubic centimeter equals 1 milliliter ($1\ cm^3 = 1\ mL$).

Density

Density is the ratio of mass to volume:

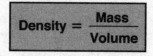

$$\text{Density} = \frac{\text{Mass}}{\text{Volume}}$$

Symbolically, this is written as follows:

$$d = \frac{m}{V}$$

Density measures the "compactness" of a substance. A substance such as lead has a large density because a relatively small volume of lead has a relatively large mass. A substance such as Styrofoam (the material used to make coffee cups) has a small density because a relatively large volume of Styrofoam has a relatively small mass.

Density is a property that depends only on the nature of a substance, not on the size of any particular sample of the substance. For example, a solid gold coin and a solid gold brick have very different masses and volumes, but they have the same density because both are pure gold.

In the SI system, the unit of density is the kilogram per cubic meter (kg/m^3). In chemistry, the densities of solids and liquids are commonly reported in grams per cubic centimeter (g/cm^3), while the densities of gases are given in grams per cubic decimeter (g/dm^3) or, equivalently, in grams per liter (g/L). We will solve some problems involving density in Section 1.8.

1.7 REPORTING MEASURED QUANTITIES

Every scientific discipline, including chemistry, is concerned with making measurements. Since no instrument is perfect, a degree of uncertainty is associated with every measurement. In a well-designed experiment, however, the uncertainty of each measurement is reduced to the smallest possible value.

Accuracy refers to how well a measurement agrees with an accepted value. For example, if the accepted density of a material is 1220 grams per cubic decimeter and a student's measurement is 1235 grams per cubic decimeter, the difference (15 g/dm^3) is an indication of the accuracy of the measurement. The smaller the difference, the more accurate is the measurement.

Precision describes how well a measuring device can reproduce a measurement. The limit of precision depends on the design and construction of the device. No matter how carefully we measure, we can never obtain a result more precise than the limit built into our measuring device. A good general rule is that the limit of precision of a measuring device is equal to plus or minus one-half of its smallest division. For example, in the following diagram:

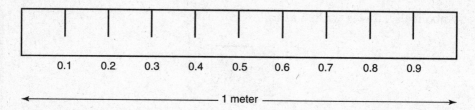

the smallest division of the meter stick is 0.1 meter, and therefore the limit of its precision is ±0.05 meter. When we read any measurement using this meter stick, we must attach this limit to the measurement, for example, 0.27 ± 0.05 meter.

Significant digits (also called **significant figures**) are the digits that are part of any valid measurement. The number of significant digits is a direct result of the number of divisions the measuring device contains. The following diagram shows a meter stick with no divisions. How should the measurement indicated by the arrow be reported?

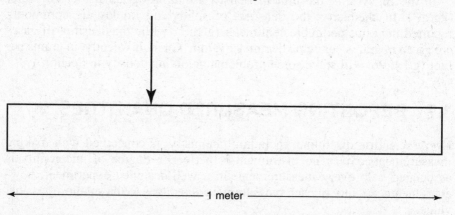

Since there are no divisions, all we know is that the measurement is somewhere between 0 and 1 meter. The best we can do is to make an educated guess based on the position of the arrow. To obtain the best educated guess, we divide the meter stick mentally into ten divisions and report the measurement as 0.3 meter. This meter stick allows us to measure length to one significant digit.

Suppose we now use a meter stick that has been divided into tenths, as shown on the next page, and repeat the measurement.

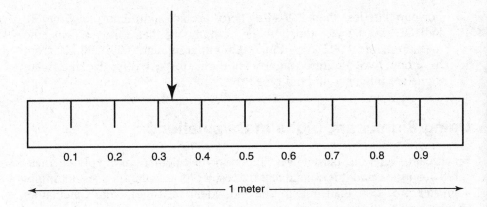

With this device we can report the measurement with less uncertainty because we know that the indicated length lies between 0.3 and 0.4 meter. Allowing ourselves one educated guess, we will report the length as 0.33 meter. This measurement has two significant digits. The more significant digits a measurement has, the more confidence we have in our ability to reproduce the measurement because only the last digit is in doubt.

Measurements that contain zeros can be particularly troublesome. For example, we say that the average distance between Earth and the Moon is 238,000 miles. Do we really know this number to six significant digits? If so, we would have measured the distance to the nearest mile. Actually, this measurement contains only three significant figures. The distance is being reported to the nearest thousand miles. The zeros simply indicate how large the measurement is.

To avoid confusion, a number of rules have been established for determining how many significant digits a measurement has.

Rules for Determining the Number of Significant Digits in a Measurement

1. All nonzero numbers are significant. The measurement 2.735 grams has four significant digits.
2. Zeros located between nonzero numbers are also significant. The measurements 1.0285 liters and 202.03 torr each have five significant digits.
3. For numbers greater than or equal to 1, zeros located at the end of the measurement are significant only if a decimal point is present. The measurement 60 grams has one significant digit. In this case, the zero indicates the size of the number, not its significance. The measurements 60. grams and 60.000 grams, however, have two and five significant digits, respectively.

4. For numbers less than 1, leading zeros are not significant; they merely indicate the size of the number or numbers that follow. Thus, the measurements 0.00**2** kilogram, 0.0**20** kilogram, and 0.000**200** kilogram have one, two, and three significant digits, respectively; the significant digits are indicated in bold type.

Using Significant Digits in Calculations

Significant digits are particularly important in calculations involving measured quantities, and it is crucial that the result of a calculation does not imply a greater precision than any of the individual measurements. Calculators routinely give us answers with ten digits. It is incorrect to believe, however, that the results of most of our calculations have this many significant digits.

Multiplication and Division of Measurements

When two measurements are multiplied (or divided), the answer should contain as many significant digits as the less precise measurement. For example, if the measurement 2.3 centimeters (two significant digits) is multiplied by 7.45 centimeters (three significant digits), the answer must contain only two significant digits:

$$(2.3 \text{ cm}) \cdot (7.45 \text{ cm}) = 17.135 \text{ cm}^2 = 17 \text{ cm}^2$$

(Note that the units are also multiplied.)

TRY IT YOURSELF
Divide the following measurements, and express the result to the correct number of significant figures: 6.443 grams/8.91 liters.

ANSWER
0.723 g/L

Addition and Subtraction of Measurements

When two measurements are added (or subtracted), the answer should contain as many decimal places as the measurement with the smaller number of decimal places. For example, when 8.11 liters (two decimal places) and 2.476 liters (three decimal places) are added, the answer must be taken only to the second decimal place:

$$8.11 \text{ L} + 2.476 \text{ L} = 10.586 \text{ L} = 10.59 \text{ L}$$

TRY IT YOURSELF
Add the following measurements, and express the result to the correct number of significant figures: 246.213 milliliters, 79.91 milliliters, 8786.268 milliliters.

ANSWER
9112.39 mL

Note: If *counted* numbers (such as six atoms) or *defined* numbers (such as 273.16 K) are used in any calculations, they are treated as though they have an infinite number of significant digits or decimal places.

Rounding a Number

Suppose we had a measurement with five significant figures. If we wished to express it to three significant figures, we would need to round this measurement.

We *round* a measurement by looking one digit beyond the precision we need. If the digit is less than 5, we do not change the value of the preceding digit. If it is 5 or greater, we raise the preceding digit by 1.

For example, if our measurement was 127.36 grams and we wanted to round it to three significant figures, we would first examine the fourth digit. Since 3 is less than 5, we would leave the third digit, 7, unchanged, and the rounded measurement would be 127 grams expressed to three significant figures.

If, however, we wished to round the measurement 127.36 grams to four significant figures, we would need to examine the fifth digit. Since 6 is greater than 5, we would raise the preceding digit, 3, by 1. Expressed to four significant figures, this measurement would be 127.4 grams.

TRY IT YOURSELF
Round each of the following measurements to three significant digits:
(a) 903.04 L   (b) 298.86 K   (c) 0.002259 mol

ANSWERS
(a) 903 L   (b) 299 K   (c) 0.00226 mol

Order of Magnitude

There are times when we are interested in the relative size of a measurement rather than its actual value. The *order of magnitude* of a measurement is the power of 10 closest to its value. For example, the order of magnitude of 1284 grams (1.284×10^3) is 10^3, while the order of magnitude of 8756 grams (8.756×10^3) is 10^4. Orders of magnitude are very useful for comparing quantities, such as mass or distance, and for estimating the answers to problems involving complex calculations.

TRY IT YOURSELF
What is the order of magnitude of Avogadro's number (6.02×10^{23})?

ANSWER
10^{24}

Percent Error of a Measurement

The **percent error** of a measurement indicates how closely the measurement agrees with an accepted value of the same quantity. Its definition is as follows:

$$\text{Percent Error} = \frac{\text{Measured Value} - \text{Accepted Value}}{\text{Accepted Value}} \times 100$$

The *measured value* is the value that is experimentally determined, and the *accepted value* is the value that is generally accepted as the most probable value of the measurement. Generally, we are interested only in the *magnitude* of the percent error, not in its algebraic sign.

PROBLEM
A student measures the density of an object to be 5600 grams per liter. The accepted density of the object is 6400 grams per cubic liter. What is the percent error of this measurement?

SOLUTION

$$\begin{aligned}
\text{Percent Error} &= \frac{\text{Measured Value} - \text{Accepted Value}}{\text{Accepted Value}} \times 100 \\
&= \frac{5600 \text{ g/L} - 6400 \text{ g/L}}{6400 \text{ g/L}} \times 100 \\
&= \frac{-800 \text{ g/L}}{6400 \text{ g/L}} \times 100 \\
&= 12\%
\end{aligned}$$

Note that the answer is reported as a positive value even though the numerator of the fraction in this example is negative.

TRY IT YOURSELF
The accepted value for the volume of a gas is 22.4 liters. A student reports her measurement as 23.8 liters. What is the percent error of her measurement?

ANSWER
6.25%

1.8 SOLVING PROBLEMS

Using the Factor-Label Method (FLM) in Chemistry

We now introduce a general technique for solving problems known as the *factor–label method* (FLM). Suppose we want to solve this problem: How many inches equal 40. feet? We know that, by definition, 12 inches equals 1 foot. FLM allows us to use this relation to form two fractions called *conversion factors:*

$$\frac{12 \text{ inches}}{1 \text{ foot}} \quad \text{and} \quad \frac{1 \text{ foot}}{12 \text{ inches}}$$

These fractions are what we will use to solve the problem. First we write a tentative solution as follows:

$$x \text{ inches} = 40. \text{ feet}$$

Because we need to convert feet to inches, we must cancel the unit "feet" on the right and substitute "inches" in its place. This can be done by using the first conversion factor as follows:

$$x \text{ inches} = 40. \ \cancel{\text{feet}} \cdot \frac{12 \text{ inches}}{1 \ \cancel{\text{foot}}} = 480 \text{ inches}$$

Note that units can be multiplied and divided just as numbers are. In the example above, the "feet" units cancel, leaving "inches" in the numerator—which is exactly what we needed to do. All that remains is the arithmetic $(40. \cdot 12)$, and our answer is 480 inches.

Now let us solve a slightly more complicated problem.

PROBLEM
How many hours are there in 5 years?

SOLUTION
First, we need a plan of attack—a "road map" that gives us direction:

$$\text{YEARS} \rightarrow \text{???} \rightarrow \text{HOURS}$$

The symbol "???" must represent some unit that connects years to hours and is easily related to both these units. The unit "days" will do very nicely since there are 365 days in 1 year and 24 hours in 1 day. We set up our problem for solution:

$$x \text{ hours} = 5 \text{ years} \cdot \frac{365 \text{ days}}{1 \text{ year}} \cdot \frac{24 \text{ hours}}{1 \text{ day}} = 43{,}800 \text{ hours}$$

(40,000 hours to 1 significant figure)

Our next problem involves units with exponents.

PROBLEM
How many cubic centimeters are there in 1 cubic decimeter?

SOLUTION
If we examine the metric prefix table in Section 1.4, we can conclude that 1 meter equals 100 centimeters and also equals 10 decimeters. Let us set up the solution to the problem as though the exponent 3 were not present. Our "road map" for this problem is:

$$\text{DECIMETERS} \rightarrow \text{METERS} \rightarrow \text{CENTIMETERS}$$

Using FLM, and standard abbreviations for the units, we can write:

$$x \text{ cm} = 1 \text{ dm} \cdot \left(\frac{1 \text{ m}}{10 \text{ dm}} \right) \cdot \left(\frac{100 \text{ cm}}{1 \text{ m}} \right)$$

Adding the exponent 3 is easy: we simply add it to each term in the equation:

$$x \text{ cm}^3 = 1 \text{ dm}^3 \cdot \left(\frac{1 \text{ m}}{10 \text{ dm}} \right)^3 \cdot \left(\frac{100 \text{ cm}}{1 \text{ m}} \right)^3 = 1000 \text{ cm}^3$$

Our final problem involves units that appear in the numerator and denominator of a fraction.

PROBLEM
A substance called sulfuric acid flows out of a pipe at the rate of 2500 grams per second (2500 g/s). What is the flow rate, in kilograms per hour (kg/h), of the sulfuric acid?

SOLUTION
We will need two "road maps" to solve this problem:

GRAMS → KILOGRAMS

SECONDS → MINUTES → HOURS

We know that there are 60 seconds in 1 minute, 60 minutes in 1 hour, and 1000 grams in 1 kilogram. The FLM solution is:

$$x \ \frac{kg}{h} = 2500 \ \frac{g}{s} \cdot \left(\frac{1 \ kg}{1000 \ g} \right) \cdot \left(\frac{60 \ s}{1 \ min} \right) \cdot \left(\frac{60 \ min}{1 \ h} \right)$$

$$= 9.0 \times 10^3 \frac{kg}{h}$$

TRY IT YOURSELF
The density of water is very nearly 1 gram per cubic centimeter (g/cm^3). What is the density of water in kilograms per cubic meter (kg/m^3)? (*Hint:* Use the techniques of the last two problems solved above.)

ANSWER

$$\text{Density} = 1\frac{g}{cm^3} = 1000 \ \frac{kg}{m^3}$$

As you can see, FLM is a very powerful tool for solving a wide variety of chemistry problems. Whenever you can find simple connections between the units in a problem, you should use FLM to solve that problem. We will use FLM as a problem-solving tool throughout this book.

Using Mathematical Equations in Chemistry

In our study of chemistry, we are frequently required to solve problems in which the variables are related by an equation. We can solve such problems quite easily if we follow five simple steps:

1. Prepare a list of the variables and their values. (Remember to include units!)
2. Write the equation that relates the variables of the problem.
3. Rewrite the equation in order to isolate the unknown variable.
4. Substitute the known values and their units into the rewritten equation.
5. Perform the arithmetic operations indicated by the equation to arrive at a solution.

The following problem shows how this technique is employed.

PROBLEM
The density of a substance is 1200 grams per liter (g/L). Calculate the mass of a sample of this substance if its volume is 0.50 liter (L).

SOLUTION
(The numbers correspond to the steps given on the previous page.)

1. List of values:

$$d = 1200 \text{ g/L}$$
$$V = 0.50 \text{ L}$$
$$m = ???$$

2. Relevant equation:

$$d = \frac{m}{V}$$

3. Rewritten equation:

$$m = d \cdot V$$

4. Substitution of values:

$$m = 1200 \frac{g}{\cancel{L}} \cdot 0.50 \cancel{L}$$

5. Performance of arithmetic operations:

$$m = 6.0 \times 10^2 \text{ g}$$

TRY IT YOURSELF
The density of a substance is 8.0 grams per liter (g/L). What is the volume of a sample of this substance if its mass is 24 grams?

ANSWER
$V = 3.0$ L

PROBLEM
What is the boiling point of mercury (630 K) on the Celsius temperature scale?

SOLUTION

$$K = {}^{\circ}C + 273$$
$${}^{\circ}C = K - 273$$
$$= 630 - 273$$
$$= 357 {}^{\circ}C$$

PROBLEM
What is the normal boiling point of water (100°C) on the Kelvin temperature scale?

SOLUTION

$$K = {}^{\circ}C + 273$$
$$= 100 + 273$$
$$= 373 \text{ K}$$

TRY IT YOURSELF
(a) Convert 155 K to the Celsius temperature scale.
(b) Convert 37°C to the Kelvin temperature scale.

ANSWERS
(a) $-118°C$ (b) 310 K

Using Graphs in Chemistry

Sometimes data are presented as a table of values, and we are asked to interpret the data and draw conclusions from them. In such cases a graph of the data values may provide information about values that do not appear in the original table. In addition, the slope and the x- and y-intercepts of the graph may provide useful information. The following problem illustrates how a graph can be used to obtain additional information.

PROBLEM
A student measures the masses and volumes of several samples of the same substance and obtains the results shown in the table.

Volume/L	Mass/g
6.80	20.0
13.5	40.0
20.9	60.0
28.8	80.0
34.2	100.0

(a) Draw a graph of the data provided in the table.
(b) What is the mass of a sample of the substance if its volume is 25.0 liters?
(c) What is the density of the substance?

SOLUTIONS
(a) We begin by graphing the data points. We plot volume on the x-axis and mass on the y-axis because this arrangement will be useful in answering part (c) of the problem. Note that the axes are "scaled" so that the plotted points fill nearly the entire graph, allowing any trend in the data to be observed.

After we plot the data points, we can see that they form almost a straight line. Instead of playing "connect the dots," we draw a *best-fit* straight line (that is, a line that is most closely associated with its data points) as shown.

27

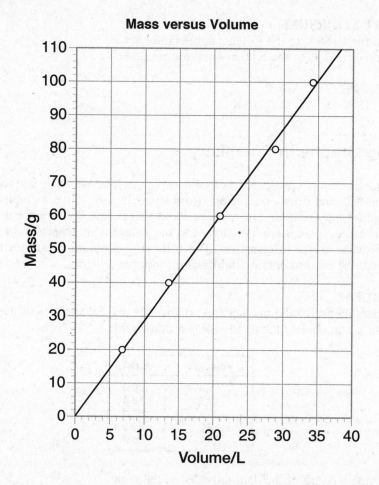

Mass versus Volume

In this example, we draw our best-fit straight line so that it passes through the origin, because a mass of 0.0 g corresponds to a volume of 0.0 L.

(b) By inspecting the graph, we can see that a volume of 25.0 L corresponds to a mass of 71.8 g.

(c) Since every data point does not fit on the line, we cannot use individual data points to calculate the density of the substance. The slope of the line, however, will give us the density because it measures the ratio of mass to volume for the entire set of data. The diagram on the graph shows the quantities Δm and ΔV, and the slope of the line is calculated

from the ratio $\dfrac{\Delta m}{\Delta V}$.

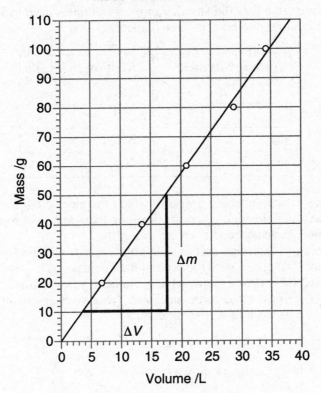

Mass versus Volume

A careful calculation of the slope of the line yields a value of 2.87 g/L for the density of the substance.

TRY IT YOURSELF
Use the graph drawn above to calculate the volume of a sample whose mass is 90.0 grams.

ANSWER
According to the graph, a mass of 90.0 g corresponds to a volume of 31.4 L.

In other situations, we may be presented with data and asked to connect all of the data points, as shown in the following problem.

PROBLEM

The following table lists the atomic numbers (numbers used to identify elements) and melting points of several elements:

Atomic Number	Melting Point /K
3	454
19	337
37	312
55	302

(a) Plot the data points on a graph in which the atomic numbers lie along the x-axis and the melting points lie along the y-axis. Your graph should be scaled so that a trend is clearly observed.

(b) Connect each adjacent pair of data points with a straight line. Describe the trend associated with the plotted data points.

(c) Estimate the melting point of the element whose atomic number is 11. (To see how close to the accepted value your graph came, refer to Reference Table S in Appendix 1.)

SOLUTIONS

(a) The completed graph is shown below:

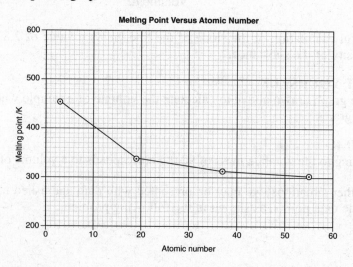

Notice that the y-axis doesn't start at zero since you are plotting melting points that lie between 300 K and 500 K.

(b) The trend is fairly obvious: as the atomic number increases, the melting point decreases. This trend is known as an *inverse relationship*.

(c) According to the graph, the element whose atomic number is 11 has a melting point of approximately 395 K. According to Reference Table S, the accepted value for the melting point is 371 K. (This corresponds to a 6.5% error.)

END-OF-CHAPTER QUESTIONS

1. In which pair are the members classified as substances?
   (1) mixtures and solutions
   (2) compounds and solutions
   (3) elements and mixtures
   (4) compounds and elements

2. Which can *not* be decomposed by a chemical change?
   (1) a compound
   (2) a heterogeneous mixture
   (3) a homogeneous mixture
   (4) an element

3. An example of a heterogeneous mixture is
   (1) an ice cream soda
   (2) a sugar solution
   (3) table salt
   (4) carbon dioxide

4. Which statement describes a characteristic of all compounds?
   (1) Compounds contain one element, only.
   (2) Compounds contain two elements, only.
   (3) Compounds can be decomposed by chemical means.
   (4) Compounds can be decomposed by physical means.

5. Which is a characteristic of all mixtures?
   (1) They are homogeneous.
   (2) They are heterogeneous.
   (3) Their composition is generally fixed.
   (4) Their composition generally can be varied.

6. Which is the equivalent of 750. joules?
   (1) 0.750 kJ   (2) 7.50 kJ   (3) 75.0 kJ   (4) 750. kJ

7. Which temperatures originally represented the fixed points on the Celsius temperature scale?
   (1) 32° and 100°
   (2) 32° and 212°
   (3) 0° and 212°
   (4) 0° and 100°

8. Expressed in proper scientific notation, the number 0.00213 is
   (1) 0.213×10^2
   (2) 2.13×10^{-2}
   (3) 2.13×10^{-3}
   (4) 213×10^0

9. When the numbers 1.2×10^2 and 3.4×10^1 are added, the result, expressed in proper scientific notation, is
   (1) 1.5×10^2 (2) 4.6×10^2 (3) 1.5×10^1 (4) 4.6×10^1

10. When the numbers 4.2×10^2 and 8.4×10^1 are multiplied, the exponent of the result, expressed in proper scientific notation, is
    (1) 1   (2) 2   (3) 3   (4) 4

Base your answers to questions 11 and 12 on the following table, which represents measurements made on four rectangular blocks in a chemistry laboratory.

Block	Mass (g)	Length (cm)	Width (cm)	Height (cm)
A	72.	4.0	3.0	2.0
B	60.	5.0	2.0	1.0
C	30.	10.	1.0	1.0
D	6.0	4.0	2.0	1.5

11. Which block has the smallest density?
    (1) A   (2) B   (3) C   (4) D

12. Which two blocks may be made of the same material?
    (1) A and B   (2) A and C   (3) B and C   (4) B and D

13. A cube has a volume of 8.00 cubic centimeters and a mass of 21.6 grams. The density of the cube is best expressed as
    (1) 2.7 g/cm^3
    (2) 2.70 g/cm^3
    (3) 0.37 g/cm^3
    (4) 0.370 g/cm^3

14. In a laboratory exercise to determine the density of a substance, a student found the mass of the substance to be 6.00 grams and the volume to be 2.0 milliliters. Expressed to the correct number of significant figures, the density of the substance is
    (1) 3.000 g/mL   (2) 3.00 g/mL   (3) 3.0 g/mL   (4) 3 g/mL

15. Which measurement contains a total of three significant figures?
(1) 0.012 g   (2) 0.125 g   (3) 1205 g   (4) 12.050 g

16. The diagram below shows a section of a 100-milliliter graduated cylinder:

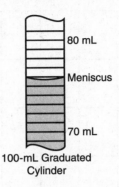

80 mL

Meniscus

70 mL

100-mL Graduated
Cylinder

When the bottom of the meniscus is read to the correct number of significant figures, the volume of water in the cylinder will be recorded as
(1) 75.7 mL   (2) 75.70 mL   (3) 84.3 mL   (4) 84.30 mL

17. A student determined the melting point of a substance to be 328 K. If the accepted value is 323 K, the percent error in her determination was
(1) 0.152%   (2) 1.52%   (3) 1.55%   (4) 5.00%

18. When the rules for significant figures are used, the sum of 0.027 gram and 0.0023 gram should be expressed as
(1) 0.029 g   (2) 0.0293 g   (3) 0.03 g   (4) 0.030 g

19. Which measurement contains three significant figures?
(1) 0.05 g   (2) 0.050 g   (3) 0.056 g   (4) 0.0563 g

20. The graph below represents an experiment in which the decay of a radioactive substance is measured.

Mass versus Time

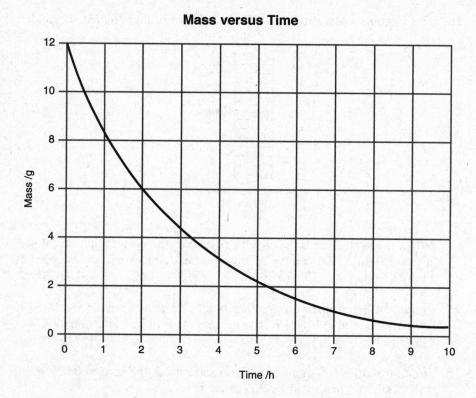

In how many hours will this substance have decayed to one-half its initial value?
(1) 1   (2) 2   (3) 3   (4) 6

21. In an experiment, a student found that the percent of oxygen in a sample was 42.3%. If the accepted value is 39.3%, the magnitude of the experimental percent error is
(1) 42.3/39.3 × 100%   (2) 3.0/42.3 × 100%
(3) 39.3/42.3 × 100%   (4) 3.0/39.3 × 100%

22. According to an accepted chemistry reference, the energy needed to vaporize a substance is 5400 joules per gram. A student determined in the laboratory that this energy was 6200 joules per gram. The student's results had a percent error of
(1) 13   (2) 15   (3) 80.   (4) 87

23. Expressed to the correct number of significant figures, what is the sum of (3.04 g + 4.134 g + 6.1 g)?
   (1) 13 g   (2) 13.3 g   (3) 13.27 g   (4) 13.274 g

24. In the calculation 41.06 centimeters × 10.2 centimeters, how many significant figures should the product of the two values contain?
   (1) 5   (2) 6   (3) 3   (4) 4

25. In an experiment, the density of a gas was determined to be 2.47 grams per liter. Compared to the accepted value of 2.43 grams per liter, the percent error for this determination was
   (1) 0.400   (2) 1.65   (3) 24.7   (4) 98.4

26. The graph below shows the volume and the mass of four different substances at STP.

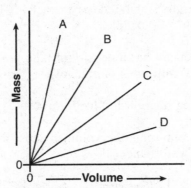

   Which of the four substances has the *lowest* density?
   (1) *A*   (2) *B*   (3) *C*   (4) *D*

27. Five cubes of iron are tested in a laboratory. The tests and results are shown in the table below.

Iron Tests and the Results

Test	Procedure	Result
1	A cube of Fe is hit with a hammer.	The cube is flattened.
2	A cube of Fe is placed into 3 M HCl(aq).	Bubbles of gas form.
3	A cube of Fe is heated to 1811 K.	The cubes melts.
4	A cube of Fe is left in damp air.	The cube rusts.
5	A cube of Fe is placed into water.	The cube sinks.

   Which tests demonstrate chemical properties?
   (1) 1, 3, and 4   (2) 1, 3, and 5   (3) 2 and 4   (4) 2 and 5

28. The ratio of chromium to iron to carbon varies among the different types of stainless steel. Therefore, stainless steel is classified as
(1) a compound
(3) a mixture
(2) an element
(4) substance

Constructed-Response Questions

1. Express the following numbers in scientific notation:
(a) 63,477   (b) 0.000230

2. Express the following numbers in standard decimal form:
(a) 6.54×10^5 (b) 5.55×10^{-3}

3. In a certain foreign country, length is measured in "piddles" and "daddles." One piddle is defined to be exactly 3.33 daddles. How many piddles are there in 7.00 daddles?

4. The compound ethanol has a density of 0.84 gram per milliliter. What is the mass of 25.0 milliliters of ethanol?

5. Solve the following relationship for f: $ab = \dfrac{cd}{ef}$.

6. The table below represents the volumes and corresponding masses of seven samples of a substance.

Volume/cm^3	Mass/g
1.33	8.22
2.41	11.5
3.72	18.7
4.98	25.8
6.81	34.6
8.13	43.8
10.3	54.9

(a) Plot the data points on the grid below and draw the best-fit line through the points.

(b) Determine the density of the substance by measuring the slope ($\Delta m/\Delta V$) of the best-fit line.

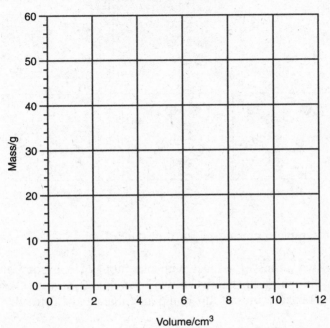

7. How many significant figures are there in each of the following measurements?
   (a) 2.450 mL
   (b) 0.10007 mol
   (c) 0.00023 kg
   (d) 7.01 J
   (e) 50.00 torr

8. Express each of the following operations to the correct number of significant figures:
   (a) 21.2 m + 9.7654 m + 312.22 m
   (b) 0.91765 kJ − 0.9012 kJ
   (c) 6.72 cm × 0.32 cm
   (d) 12.316 g ÷ 2.3 L

Base your answers to questions 9 through 11 on the particle diagrams below, which show atoms and/or molecules in three different samples of matter.

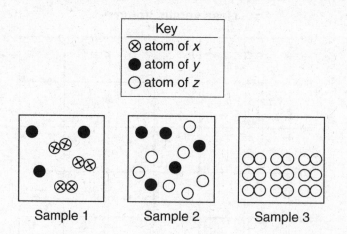

Sample 1   Sample 2   Sample 3

9. Which sample represents a pure substance?

10. When two atoms of y react with one atom of z, a compound forms. Using the number of atoms shown in sample 2, what is the maximum number of molecules of this compound that can be formed?

11. Explain why ⊗⊗ does not represent a compound.

The answers to these questions are found in Appendix 3.

Chapter Two

ATOMS, MOLECULES, AND IONS

KEY IDEAS

This chapter traces the development of the early atomic models of matter from Dalton to Thomson and Rutherford. The nuclear model is then studied with particular emphasis on the placement of protons, neutrons, and electrons within atoms. Molecules, ions, and the Periodic Table of the Elements are also introduced in this chapter.

KEY OBJECTIVES
At the conclusion of this chapter you will be able to:
• Compare the Dalton, Thomson, and Rutherford models of the atom.
• Describe the placement of protons, neutrons, and electrons according to the nuclear model of the atom.
• Define and apply the terms *atomic number*, *mass number*, and *isotope*.
• Calculate the atomic mass of an element, given the mass and abundance of each of its naturally occurring isotopes.
• Define the terms *group* and *period* as they relate to the Periodic Table of the Elements.
• Define and apply the terms *molecule*, *ion*, *monoatomic*, and *polyatomic*.

SECTION I—BASIC (REGENTS-LEVEL) MATERIAL

NYS REGENTS CONCEPTS AND SKILLS

Note: By the time you have finished Section I, you should have mastered the concepts and skills listed below. The Regents chemistry examination will test your knowledge of these items and your ability to apply them.

Concepts are the *basic ideas* that form the body of the Regents chemistry course (what you need to know!).

Skills are the *activities* that demonstrate your mastery of these concepts (how you show that you know them!).

Following each concept or skill is a page reference (given in parentheses) to this chapter.

2.1 Concept:
The modern model of the atom has evolved over a long period of time through the work of many scientists. (Pages 42–44)

Skill:
Relate experimental evidence to models of the atom. (Pages 42–44)

2.2 Concept:
Each atom has a nucleus, with an overall positive charge, surrounded by one or more negatively charged electrons. (Page 44)

Skill:
Use models to describe atoms. (Page 47)

2.3 Concept:
Subatomic particles contained in the nucleus include protons and neutrons. (Pages 44–45)

2.4 Concept:
The proton is positively charged, and the neutron has no charge. The electron is negatively charged. (Page 45)

2.5 Concept:
Protons and electrons have equal but opposite charges. The number of protons equals the number of electrons in an atom. (Page 45)

Skill:
Determine the number of protons or electrons in an atom or ion. (Pages 46–47)

2.6 Concept:
The mass of a proton is nearly equal to the mass of a neutron. An electron is much less massive than a proton or a neutron. (Page 46)

2.7 Concept:
Atoms of an element that contain the same number of protons but different numbers of neutrons are called isotopes of that element. (Page 46)

2.8 Concept:
The number of protons in an atom (atomic number) identifies the element. The sum of the protons and neutrons in an atom (the mass number) identifies an isotope. Common notations that represent isotopes include ^6C, C, carbon-14, C-14. (Page 46)

Skills:
- Interpret and write isotopic notation. (Page 46)
- Given any two of the following values: number of protons, number of neutrons, mass number, calculate the third value. (Pages 46–47)

2.9 Concept:
When an atom gains one or more electrons, it becomes a negative ion. When an atom loses one or more electrons, it becomes a positive ion. (Page 49)

2.1 INTRODUCTION TO THE ATOMIC MODEL OF MATTER

If we wish to explain the *properties* of matter, for example, how substances combine with one another, we will need a detailed description of how matter is constructed. Because of very strong experimental evidence, we accept the *atomic model* as a reliable basis on which to explain the properties of matter.

The term *model* needs further explanation. Scientific models are very similar to road maps, because they answer questions and make predictions. For example, if we wanted to travel from New York to Chicago, a map of the United States would indicate the appropriate interstate highways we ought to take. In addition, we would be able to predict through which major towns and cities we would pass on our journey. But our map would *not* tell us the condition of the highways or provide a street-by-street depiction of Chicago. For this information, we would need more detailed maps. However, no matter how detailed our maps became, they could never provide answers to *all* of our questions because, as models, they only *represent* the landscape; they are not the landscape itself.

A scientific model, such as the atomic model, also gives us only an approximate picture of reality. Nevertheless, we use such models because they provide some answers for us and they allow us to make predictions as to how a part of the universe will behave in a given situation. As we gather more information, we refine and revise our models so that they become even more useful to us. The atomic model of matter has a long history, and with each revision it has provided more scientific information.

2.2 DEVELOPMENT OF THE EARLY MODELS OF THE ATOM

The Dalton Model

In the period 1803–1807, an English school teacher named John Dalton proposed a model of atomic structure based on experimental work. This model can be summarized as follows:

- Each element is composed of indivisible particles called *atoms*.
- In an *element*, all of the atoms are identical; atoms of different elements have different properties, including mass.
- In a chemical reaction, atoms are not created, destroyed, or changed into other types of atoms; the atoms are simply rearranged.
- *Compounds* are formed when atoms of more than one element combine.
- Samples of a given compound *always* have the same relative numbers of atoms. (For example, all samples of water contain two hydrogen atoms for every oxygen atom present.)

According to Dalton's model, atoms are the basic building blocks of matter. The model explains why elements and compounds have fixed compositions by mass. One serious shortcoming of this model, however, is its failure to provide clues as to the *internal structure* of atoms.

Cathode Rays and Electrons

By 1850, experimental data suggested that atoms are composed of smaller **subatomic particles**. In one such experiment, high-voltage electricity was passed across an evacuated glass tube containing metal electrodes, as shown in the accompanying diagram. The electricity produced a stream of radiation that had a negative electric charge and flowed from the negative electrode (the *cathode*) to the positive electrode (the *anode*). This radiation became known as *cathode rays*, and the glass tube is called a *cathode-ray tube*.

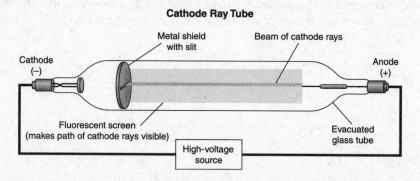

Cathode Ray Tube

Metal shield with slit

Beam of cathode rays

Cathode (−)

Anode (+)

Fluorescent screen (makes path of cathode rays visible)

Evacuated glass tube

High-voltage source

Although cathode rays could not be seen, they were made visible by beaming them at certain *fluorescent* materials. In 1897, the British physicist J. J. Thomson examined cathode rays produced from many different sources and concluded that the rays consisted of streams of identical, negatively charged *particles*, which he called *electrons*. Thomson's experiments enabled him to measure the ratio of the electron's charge to its mass without knowing the value of either the charge or the mass. (This is analogous to our knowing that there are four quarters in every dollar without knowing the number of quarters or dollars we have in our pockets.)

In 1909, the American physicist Robert Millikan determined the charge on an electron by measuring a series of charges on irradiated oil droplets. He discovered that all of the measured charges were multiples of a single number. That number, which is the charge on an electron, is also known as an *elementary charge*. In SI units, the currently accepted value for an elementary charge is 1.6022×10^{-19} coulomb. As a result of Thomson's and Millikan's work, it was possible to calculate the mass of the electron. In SI units, the currently accepted value is 9.1094×10^{-31} kilogram.

The Thomson Model

Thomson's work with cathode rays enabled him to propose a model of the atom that had internal structure. In the Thomson model, an atom consists of a (positively charged) jellylike mass with (negative) electrons scattered throughout it—much as raisins are spread throughout a plum pudding. (You don't know what a plum pudding is? Ask your teacher or any other older person!) This model did not contradict the Dalton model; it merely extended it by providing the atom with some structure.

The Rutherford Model

In 1910, the New Zealand physicist Ernest Rutherford, working at the University of Manchester in England, and two of his students (Hans Geiger and Ernest Marsden) bombarded thin gold foils with the nuclei of helium atoms and observed how these particles were scattered by the foils. (These positively charged particles were named *alpha particles;* they consist of two protons and two neutrons each.) Rutherford and his coworkers observed that most of the alpha particles passed through the foil with little or no deflection. However, a small number of particles were deflected at large angles. This is shown in the accompanying diagram.

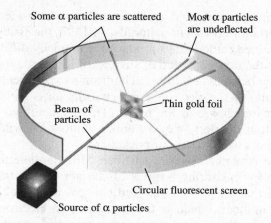

Some α particles are scattered

Most α particles are undeflected

Beam of particles

Thin gold foil

Circular fluorescent screen

Source of α particles

This work showed that the Thomson model of the atom was not consistent with the experimental evidence. Consequently, in 1911, Rutherford proposed a new model with the following characteristics:

• Most of the (volume of the) atom is empty space.
• Most of the mass of the atom is concentrated in a dense, positively charged nucleus.
• Electrons are present in the space surrounding the nucleus.

Rutherford developed the *nuclear model* of the atom, and we credit him with the discovery of the atomic nucleus. The nucleus is so important that we will devote all of Chapter 7 to it.

2.3 THE CURRENT VIEW OF ATOMIC STRUCTURE

We present a simplified view of the atom that is based on the experimental work carried out during the first three decades of the twentieth century. The structure of the atom is based on its principal subatomic particles. An atom consists of a *nucleus* that is positively charged and contains most of the mass of the atom. In the nucleus, we find the *nucleons*: the positively charged *protons* and the *neutrons*, which have no charge. Outside the nucleus are the negatively charged *electrons*. The properties of these particles are summarized in the accompanying table.

Particle	Location	CHARGE		MASS		
		Charge/C*	Relative Charge†	Mass/kg	Mass/u‡	Relative Mass§
Proton	Nucleus	$+1.6022 \times 10^{-19}$	1+	1.6726×10^{-27}	1.0073	1
Neutron	Nucleus	0	0	1.6749×10^{-27}	1.0087	≈ 1
Electron	Outside nucleus	-1.6022×10^{-19}	1−	9.1094×10^{-31}	0.00054858	≈ 0.0005

*The symbol C stands for coulomb, the SI unit of electric charge.
†A **relative charge** is one whose magnitude is compared with the magnitude of an elementary charge, which is assigned a value of 1. In this book, we will adopt the convention that the magnitude of a relative charge is written before its sign.
‡The symbol u stands for atomic mass unit. It will be defined in Chapter 4.
§A **relative mass** is based on a comparison with the mass of a proton, which is assigned a value of 1.

2.4 IDENTIFYING ELEMENTS: NAMES, SYMBOLS, AND ATOMIC NUMBERS

Every element (as well as its atoms) is associated with three unique identifiers: its *name*, *symbol*, and *atomic number*. Reference Table S in Appendix 1 lists most of the elements by atomic number, symbol, and name.

The **name** of an element may be based on a person (e.g., *einsteinium*), a place (e.g., *francium*), or a characteristic of the element itself (e.g., *chlorine* from the Greek word *chloros*, meaning green-yellow).

Each element with a permanent name has a unique **symbol** that consists of one or two letters. The first letter is *always* capitalized; the second letter (if there is one) is *never* capitalized. Newly discovered elements are assigned temporary names and three-letter symbols that are Latin translations of their atomic numbers. (We will learn more about this in Chapter 9.)

The **atomic number** of an element is the number of protons that is contained in the nucleus of each of its atoms. For example, sodium (Na) has the atomic number 11. Each atom of sodium contains 11 protons in its nucleus. Since an atom is electrically neutral, *the number of protons in its nucleus must be equal to the number of electrons surrounding its nucleus*. For example, an atom of sodium also contains 11 electrons surrounding its nucleus.

The **Periodic Table of the Elements** in Appendix 1 lists the elements by atomic numbers and symbols. The Periodic Table is arranged in such a way that the chemical and physical properties of elements display a repeating pattern. The similarity of these properties is most pronounced within a *vertical column* called a **group**. The groups are numbered left to right, from 1 to 18. Elements in Groups 1, 2, and 13–18 are known as the *representative* elements.

45

Some groups have been given special names, as shown in the accompanying table.

Group Number	Group Name	Elements in Group
1	Alkali metals	Li, Na, K, Rb, Cs, Fr
2	Alkaline earth metals	Be, Mg, Ca, Sr, Ba, Ra
17	Halogens	F, Cl, Br, I, At, Ts
18	Noble gases	He, Ne, Ar, Kr, Xe, Rn, Og

Each *horizontal row* is called a **period**. The periods are numbered from top to bottom, from 1 to 7. As one proceeds from left to right across a period, the properties of the elements change from metallic to nonmetallic. We shall examine periodic properties more fully in Chapter 9.

2.5 NEUTRONS, ISOTOPES, AND MASS NUMBERS

Neutrons add mass to an atom but do not change the atom's identity as an element. For example, an atom containing 6 protons, 6 electrons, and 7 neutrons is classified as a carbon atom—as is an atom with 6 protons, 6 electrons, and 8 neutrons. These two atoms are **isotopes** of the element carbon; isotopes contain the same number of protons but have different numbers of neutrons. We may refer to an isotope by its name and its **mass number**, that is, the sum of the number of protons and neutrons in the nucleus of the atom. The first isotope described above has a mass number of 13, and we may refer to it as carbon-13 or C-13. The second isotope, with a mass number of 14, is referred to as carbon-14 or C-14. There are two other ways of designating a particular isotope. In the first method, the mass number is placed as a *superscript* to the *left* of the symbol for the element, for example, ^{13}C or ^{14}C. The second method gives also the atomic number, which is added as a *subscript* to the *left* of the symbol, for example, $^{13}_{6}C$ or $^{14}_{6}C$.

PROBLEM
How many protons, neutrons, and electrons are contained in an atom of uranium-238? The atomic number of uranium (U) is 92.

SOLUTION
From its atomic number, we see that a neutral atom of uranium-238 has 92 protons and 92 electrons. We also know that it has 238 protons and neutrons. If we subtract the atomic number from the mass number, the answer will be the number of neutrons:

$$238 \text{ [protons + neutrons]} - 92 \text{ protons} = 146 \text{ neutrons}$$

A neutral atom of uranium-238 has 92 protons, 146 neutrons, and 92 electrons.

The preceding problem establishes an important relationship:

Mass Number — Atomic Number = Number of Neutrons

We can use a simple model to represent an atom, that is, the contents of the nucleus and the electrons that surround it. We will use the symbols p^+, e^-, and n for protons, electrons, and neutrons, respectively. The model for the sample problem is shown below:

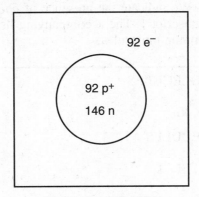

Model of U-238 ATOM

TRY IT YOURSELF
How many protons, neutrons, and electrons are in a neutral atom of $^{63}_{29}Cu$? Draw a model of this atom.

ANSWER
29 protons, 34 neutrons, and 29 electrons

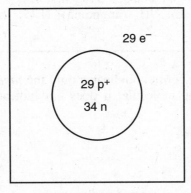

Model of Cu-63 ATOM

2.6 MOLECULES

A **molecule** is the smallest identifiable sample of a substance. In nature, the *noble gases* (the elements of group 18) exist as isolated atoms and are called **monatomic molecules**.

Some substances exist as molecules in which two or more atoms are linked together. Molecules containing two or more atoms are called **polyatomic molecules.** Polyatomic molecules containing exactly two atoms are called **diatomic molecules**.

Every molecule can be represented by a **molecular formula**, in which the symbol of the element is succeeded by a *subscript* that indicates the number of atoms present. We indicate the presence of *one atom* by writing the symbol *without* the subscript 1. The accompanying list gives the molecular formulas of some common molecules.

MONATOMIC MOLECULES
Elements:
He, Ne, Ar, Kr, Xe, Rn

DIATOMIC MOLECULES
Elements:
H_2, O_2, N_2, F_2, Cl_2, Br_2, I_2

Compounds:
CO (carbon monoxide: 1 carbon atom, 1 oxygen atom)
HCl (hydrogen chloride)

POLYATOMIC MOLECULES
Elements:
O_3, P_4, S_8, C_{60}

Compounds:
H_2O (water: 2 hydrogen atoms, 1 oxygen atom)
CO_2 (carbon dioxide), NH_3 (ammonia), H_2O_2 (hydrogen peroxide), CH_4 (methane)

Note that a molecular formula indicates only the *number* of atoms of each element present in the molecule; it does *not* indicate *how* the atoms are linked together.

2.7 IONS

Simple Ions

When atoms combine chemically, the *nuclei* of the atoms cannot change: otherwise, the elements themselves would be changed. In many cases, however, atoms can lose or gain *electrons* when they combine. The loss or gain of electrons by an atom results in the formation of an electrically charged particle called an **ion**. The ion has a charge because the numbers of protons and electrons are no longer equal. If there are more protons than electrons, the ion is *positively* charged; if there are fewer protons than electrons, the ion is *negatively* charged. The following diagrams show how positive and negative ions are formed from atoms:

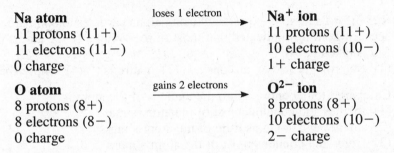

Na atom
11 protons (11+)
11 electrons (11−)
0 charge

loses 1 electron →

Na$^+$ ion
11 protons (11+)
10 electrons (10−)
1+ charge

O atom
8 protons (8+)
8 electrons (8−)
0 charge

gains 2 electrons →

O^{2-} ion
8 protons (8+)
10 electrons (10−)
2− charge

Note that the charge of an ion is written as a *superscript* to the right of the symbol, and because it is a relative charge, the magnitude precedes the sign (as in O^{2-}). We indicate charges of 1+ and 1− by writing only the *sign* of the charge, and omit the magnitude (as in Na$^+$). If no charge or sign appears to the right of an atomic symbol, it is understood that we are dealing with an *atom* (e.g., Na).

The Periodic Table of the Elements in Appendix 1 can be used to find the charges of simple ions. The table below relates the group numbers of the representative elements to their ionic charges:

Group	1	2	13	14	15	16	17
Charge	1+	2+	3+	4±	3−	2−	1−

In Chapter 3 we will learn how to use the Periodic Table to find the charges of other ions.

Polyatomic Ions

A **polyatomic ion** is an electrically charged particle that consists of two or more atoms linked together in much the same way as in a neutral molecule. This linkage serves to make the polyatomic ion behave as a unit rather than as separate atoms or simple ions. The charge on a polyatomic ion does not belong to any individual atom; it belongs to the ion as a whole. For example, the polyatomic *hydroxide ion* (OH^-) has a $1-$ charge that belongs to the entire OH unit, rather than to the oxygen or the hydrogen atom. Reference Table E in Appendix 1 includes the names and charges of selected polyatomic ions.

END-OF-CHAPTER QUESTIONS

1. Rutherford's famous experiment using alpha particles to bombard a thin sheet of gold foil indicated that most of the volume of the atoms in the foil is taken up by
   (1) electrons   (2) protons   (3) neutrons   (4) empty space

2. Compared to the entire atom, the nucleus of the atom is
   (1) smaller and contains most of the atom's mass
   (2) smaller and contains little of the atom's mass
   (3) larger and contains most of the atom's mass
   (4) larger and contains little of the atom's mass

3. Which of the following particles has the *least* mass?
   (1) an electron   (2) a proton   (3) a deuteron   (4) a neutron

4. What is the total number of protons and neutrons in an atom of ^{79}Se?
   (1) 34   (2) 45   (3) 79   (4) 113

5. What particles are found in the nucleus of an atom?
   (1) protons and electrons   (2) protons and neutrons
   (3) neutrons and electrons   (4) neutrons and positrons

6. The atoms in a sample of an element must contain nuclei with the same number of
   (1) electrons   (2) protons   (3) neutrons   (4) nucleons

7. The atomic number of an atom is always equal to the total number of
   (1) neutrons in the nucleus
   (2) protons in the nucleus
   (3) neutrons plus protons in the atom
   (4) protons plus electrons in the atom

8. What is the total number of electrons in an atom with an atomic number of 13 and a mass number of 27?
   (1) 13 (2) 14 (3) 27 (4) 40

9. All atoms of potassium must have the same
   (1) atomic mass (2) atomic weight
   (3) mass number (4) atomic number

10. Compared to an atom of calcium-40, an atom of potassium-39 contains fewer
    (1) protons (2) neutrons
    (3) occupied sublevels (4) occupied principal energy levels

11. The number of protons in the nucleus of ^{32}P is
    (1) 15 (2) 17 (3) 32 (4) 47

12. The number of protons in an atom of ^{36}Cl is
    (1) 17 (2) 18 (3) 35 (4) 36

13. The nucleus of an atom of $^{127}_{53}I$ contains
    (1) 53 neutrons and 127 protons
    (2) 53 protons and 127 neutrons
    (3) 53 protons and 74 neutrons
    (4) 53 protons and 74 electrons

14. When an atom becomes a 2^+ ion, it
    (1) gains 2 electrons (2) gains 2 protons
    (3) loses 2 electrons (4) loses 2 protons

15. How many protons, neutrons, and electrons are contained in a $^{34}S^{2-}$ ion?
    (1) 17 protons, 17 neutrons, 16 electrons
    (2) 18 protons, 16 neutrons, 20 electrons
    (3) 16 protons, 18 neutrons, 18 electrons
    (4) 16 protons, 34 neutrons, 16 electrons

16. Which particle has a negative charge and a mass that is approximately 1/1836 the mass of a proton?
    (1) a neutron (2) an alpha particle
    (3) an electron (4) a positron

17. The element found in Group 13 and in Period 2 of the Periodic Table is
    (1) Be (2) Mg (3) B (4) Al

18. Which element is in Group 2 and Period 7?
    (1) Mg (2) Mn (3) Ra (4) Rn

19. In which group of the Periodic Table is oxygen located?
    (1) 1 (2) 2 (3) 16 (4) 17

20. Which of the following elements is classified as an alkali metal?
(1) magnesium   (2) potassium   (3) argon   (4) phosphorus

21. Iodine belongs to the family of elements known as
(1) halogens   (2) alkaline earth elements
(3) noble gases   (4) transition elements

22. All of the following gases exist as diatomic molecules *except*
(1) hydrogen   (2) fluorine   (3) nitrogen   (4) argon

23. When a magnesium atom loses 2 electrons, it forms an ion whose charge is
(1) 1−   (2) 2−   (3) 1+   (4) 2+

24. The name of the polyatomic ion whose formula is HCO_3^- is
(1) hydrogen sulfate   (2) sulfate
(3) hydrogen carbonate   (4) carbonate

25. Which statement best describes electrons?
(1) They are positive subatomic particles and are found in the nucleus.
(2) They are positive subatomic particles and are found surrounding the nucleus.
(3) They are negative subatomic particles and are found in the nucleus.
(4) They are negative subatomic particles and are found surrounding the nucleus.

26. An atom of carbon-12 and an atom of carbon-14 differ in
(1) atomic number
(2) mass number
(3) nuclear charge
(4) number of electrons

27. Hydrogen has three isotopes with mass numbers of 1, 2, and 3 and has an average atomic mass of 1.00794 amu. This information indicates that
(1) equal numbers of each isotope are present
(2) more isotopes have an atomic mass of 2 or 3 than of 1
(3) more isotopes have an atomic mass of 1 than of only 2 or 3
(4) isotopes have an atomic mass of only 1

28. The atomic number of an atom is always equal to the number of its
(1) protons, only
(2) neutrons, only
(3) protons plus neutrons
(4) protons plus electrons

29. Which subatomic particle has no charge?
(1) alpha particle
(2) beta particle
(3) neutron
(4) electron

30. When the electrons of an excited atom return to a lower energy state, the energy emitted can result in the production of
(1) alpha particles
(2) isotopes
(3) protons
(4) spectra

31. Which phrase describes the charge and mass of the neutron?
(1) a charge of +1 and no mass
(2) a charge of +1 and an approximate mass of 1 u
(3) no charge and no mass
(4) no charge and an approximate mass of 1 u

32. What is the number of electrons in a potassium atom?
(1) 18   (2) 19   (3) 20   (4) 39

33. Which substance *cannot* be broken down by a chemical change?
(1) ethane   (2) propanone   (3) silicon   (4) water

34. What is the number of electrons in an Al^{3+} ion?
(1) 10   (2) 13   (3) 3   (4) 16

Constructed-Response Questions

1. Fill in the gaps in the following table:

Symbol	^{39}K				$^{31}P^{3-}$			
Protons		23				82		28
Neutrons		28		136		124	45	31
Electrons						80	36	
Atomic no.			79	86				
Mass no.			197					
Net charge	0	0	0	0			2−	2+

2. Draw a model for $^{31}P^{3-}$.

| Chapter
Three | # FORMULAS, EQUATIONS, AND CHEMICAL REACTIONS |

KEY IDEAS

This chapter introduces the language of chemistry: formulas and equations. A chemical formula provides information about the atomic composition of a substance, while an equation allows us to describe a chemical reaction using the names or formulas of the reactants and products involved in the process.

KEY OBJECTIVES
At the conclusion of this chapter you will be able to:
- Distinguish between ionic compounds and molecular compounds.
- Define the following terms: *ionic formula*, *molecular formula*, *empirical formula*, *structural formula*, and *binary compound*.
- Write the formulas for ionic compounds containing simple and/or polyatomic ions and name these compounds using the IUPAC system.
- Define the term *oxidation number* (*state*) and assign oxidation numbers to elements, simple ions, and the elements contained in compounds and polyatomic ions.
- Write the formulas for molecular compounds and name these compounds using the IUPAC system.
- Write chemical equations using names and chemical formulas.
- Balance chemical equations.
- Classify a chemical reaction as *synthesis*, *decomposition*, *single replacement*, or *double replacement*.

SECTION I—BASIC (REGENTS-LEVEL) MATERIAL

NYS REGENTS CONCEPTS AND SKILLS

Note: By the time you have finished this chapter, you should have mastered the concepts and skills listed below. The Regents chemistry examination will test your knowledge of these items and your ability to apply them.

Concepts are the *basic ideas* that form the body of the Regents chemistry course (what you need to know!).

Skills are the *activities* that demonstrate your mastery of these concepts (how you show that you know them!).

Following each concept or skill is a page reference (given in parentheses) to this chapter.

3.1 Concept:
A chemical compound can be represented by a specific chemical formula and assigned a name based on the IUPAC system.
(Pages 57, 59–64)

3.2 Concept:
Types of chemical formulas include empirical, molecular, and structural. (Pages 57–58)

3.3 Concept:
In all chemical reactions there is a conservation of mass, energy, and charge. (Pages 65–67)

Skills:
• Interpret balanced chemical equations in terms of conservation of matter, charge, and energy. (Pages 65–67)
• Balance a chemical equation, given the formulas of reactants and products. (Pages 65–67)
• Balance a chemical equation using smallest-whole-number coefficients. (Pages 65–67)
• Use a particle diagram to represent a balanced chemical equation. (Pages 67–68)

3.4 Concept:
Types of chemical reactions include synthesis, decomposition, single replacement, and double replacement. (Pages 68–72)

Skill:
Identify types of chemical reactions. (Pages 68–72)

3.1 CHEMICAL FORMULAS

A **chemical formula** is an important part of the language of chemistry because it tells us something about the composition of an element or a compound. A formula can be as simple as He (helium) or as complicated as $C_{27}H_{46}O$ (cholesterol).

Elemental Formulas

The formula for an atom of an element is its symbol, such as Ca, Fe, or Kr. In cases where the element exists as a diatomic or polyatomic molecule, we use molecular formulas such as F_2, P_4, and S_8.

Ionic Formulas

Ionic compounds such as NaCl and $Al(NO_3)_3$ consist of a combination of positive and negative ions. The positive ions consist of simple *metallic* ions (such as Na^+) or *polyatomic* ions (such as NH_4^+). The negative ions consist of simple *nonmetallic* ions (such as S^{2-}) or polyatomic ions (such as ClO_3^-). The compound CaF_2 contains *exactly two elements* (calcium and fluorine) and is known as a **binary compound**. The ions in ionic compounds do *not* exist as individual molecules.

An **ionic formula** is the chemical formula for an ionic compound. The formula tells us the relative numbers of ions present in the compound. In NaCl, the ratio of Na^+ ions to Cl^- ions is 1 to 1; in $Al(NO_3)_3$, the ratio of Al^{3+} ions to NO_3^- ions is 1 to 3.

Molecular Formulas

Molecular compounds such as NH_3 and H_2O usually consist of combinations of *nonmetallic* elements in which the bonds between the atoms are formed by the *sharing* of electrons. As the name implies, molecular compounds exist as individual molecules.

The formula NH_3 is a **molecular formula** because NH_3 is a molecular compound, that is, NH_3 exists in the form of individual molecules. A molecular formula tells us how many atoms of each element are present in one molecule of the compound. For example, one molecule of C_2H_6 (ethane) contains two carbon and six hydrogen atoms. Note that the compounds NH_3 and C_2H_6 are also binary compounds (why?).

Empirical Formulas

In an **empirical formula** the elements appear in smallest whole-number ratios. For example, the molecular formula H_2O is also an empirical formula because the ratio of hydrogen atoms to oxygen atoms (2 to 1) cannot be reduced further. In contrast, the molecular formula N_2O_4 has NO_2 as its empirical formula. Note that an empirical formula need not be related in any special way to its parent formula: it may represent another compound, or it may not even exist as a compound.

PROBLEM

What is the empirical formula of $C_6H_{12}O_6$?

SOLUTION

Divide the formula by 6 to obtain CH_2O.

TRY IT YOURSELF

What are the empirical formulas of NH_3 and C_6H_6?

ANSWER

NH_3 and CH

Structural Formulas

A **structural formula** shows how the atoms in a molecule are bonded to one another. (It is usual to represent chemical bonds by lines and unshared pairs of electrons by pairs of dots.) A structural formula may also show the shape of the molecule. For example, the water molecule may be represented as shown in the accompanying diagram:

$$\overset{\cdot\cdot}{\underset{H\qquad H}{O}}$$

The diagram shows that each hydrogen atom is bonded to the oxygen atom, that the molecule is "V-shaped," and that the oxygen atom contains two unshared pairs of electrons.

Particle diagrams can also be used to represent structural formulas, as shown in the accompanying particle diagram of CO_2:

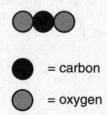

= carbon

= oxygen

The particle diagram shows that each oxygen atom is bonded to the carbon atom, and that the molecule is linear.

3.2 WRITING AND NAMING CHEMICAL FORMULAS

Writing Formulas for Ionic Compounds

Since all compounds are electrically neutral, the total electric charge on an ionic compound must be 0. When we write the formula of an ionic compound, we must be certain that the *total* charge on the positive ions equals the *total* charge on the negative ions.

In the compound that contains K^+ ions and F^- ions, this is easily accomplished if the ion ratio is 1 to 1: the formula is KF. In the compound containing Li^+ ions and O^{2-} ions, the ion ratio must be 2 to 1: the formula is Li_2O.

PROBLEM
Write the formula of the compound that contains Al^{3+} ions and S^{2-} ions.

SOLUTION
We need a common total charge for both positive and negative ions. If we multiply 3 and 2 (the ionic charges), we obtain 6, a number that is evenly divisible by 3 and 2. Two Al^{3+} ions will have a total charge of 6+, and three S^{2-} ions will have a total charge of 6–.

The formula for the compound is Al_2S_3.

TRY IT YOURSELF
The element lead forms *two* ions: Pb^{2+} and Pb^{4+}. Write the formulas for the compounds that contain these ions combined with the O^{2-} ion.

ANSWER
PbO and PbO_2

Note that, in general, when we write a formula the positive ion appears first, followed by the negative ion. (There are a small number of exceptions to this rule, but we will deal with them as necessary.)

Formulas for ionic compounds that contain polyatomic ions follow the same rules as those for simple ions. When more than one polyatomic ion is needed in a formula, it is enclosed in parentheses, as in $Fe(NO_3)_2$. This notation tell us that the ratio of Fe^{2+} ions to NO_3^- ions is 1 to 2.

PROBLEM
Write the formula for the compound that contains Zn^{2+} ions and PO_4^{3-} ions.

SOLUTION
The ratio of Zn^{2+} ions to PO_4^{3-} ions must be 3 to 2 in order for the compound to have a total charge of 0. The formula is $Zn_3(PO_4)_2$.

TRY IT YOURSELF
Write the formula for the compound that contains NH_4^+ ions and SO_4^{2-} ions.

ANSWER
$(NH_4)_2SO_4$

We can also work backwards: given the formula, we can determine the charges on the ions present in the compound. In Chapter 2 we related the charge on an ion of a representative element to its group number (page 49). Moreover, the charges on selected polyatomic ions are given in Reference Table E in Appendix 1.

PROBLEM
Given the ionic formula CuI_2, what are the charges on the ions?

SOLUTION
Refer to the Periodic Table of the Elements in Appendix 1. The representative element I is located in Group 17, its ion has a 1– charge. Since there are two I^- ions in the formula, the total negative charge is 2–. To balance this charge, a total positive charge of 2+ must be present on the single Cu ion, that is, Cu^{2+}.

TRY IT YOURSELF
Determine the ionic charges in the compound Ag_2S.

ANSWER
Ag^+ and S^{2-}

Naming Binary Ionic Compounds

We name a binary ionic compound by combining the names of the two elements that make up the compound. The ion that has a positive charge is placed first; the ion with the negative charge is second; and the suffix *-ide* is added at the end of the name. Thus, NaCl is named sodium chloride, CaO is calcium oxide, and AlF_3 is aluminum fluoride.

A list of some elements with negative oxidation numbers and their name endings is given below:

H^-	hydride	C^{4-}	carbide	N^{3-}	nitride	O^{2-}	oxide
F^-	fluoride	P^{3-}	phosphide	S^{2-}	sulfide	Cl^-	chloride
Br^-	bromide	I^-	iodide				

Many metallic elements can have more than one positive charge. (Refer to the Periodic Table of the Elements in Appendix 1, and examine "Selected Oxidation States" in the upper right corner of each element.)

A metallic element such as iron (Fe) can form more than one compound. For example, iron can exist as the Fe^{2+} or Fe^{3+} ion and, therefore, can form two oxides: FeO and Fe_2O_3. How should we name these compounds? Clearly, both are entitled to be called iron oxide! Since it is iron whose charge varies, we provide Fe with a *Roman numeral in parentheses*, which corresponds to its positive charge. Therefore, FeO is given the name iron(II) oxide, and Fe_2O_3 is called iron(III) oxide. If a metallic ion can have only *one charge* (such as Na^+ or Mg^{2+}), no Roman numeral is used in the name. This method of naming compounds is known as the *IUPAC system,* where IUPAC stands for the International Union of Pure and Applied Chemistry.

TRY IT YOURSELF
Write the formulas for the following compounds:
(a) copper(II) sulfide   (b) iron(III) bromide
(c) tin(IV) oxide   (d) aluminum nitride

ANSWERS
(a) CuS   (b) $FeBr_3$ (c) SnO_2 (d) AlN

TRY IT YOURSELF
Name the following compounds:
(a) MgF_2 (b) KH   (c) NiI_3 (d) V_2O_5

(Use the Periodic Table and Reference Table S in Appendix 1.)

ANSWERS
(a) magnesium fluoride   (b) potassium hydride
(c) nickel(III) iodide   (d) vanadium(V) oxide

Naming Compounds that Contain Polyatomic Ions

Ionic compounds that contain polyatomic ions are named in the same fashion as binary ionic compounds. The trick is to recognize that a polyatomic ion is actually present in the formula. At this point, you should review the names and formulas of the polyatomic ions contained in Reference Table E in Appendix 1.

PROBLEM
Name the following compounds:
(a) K_2SO_4 (b) NH_4NO_3 (c) $Fe(C_2H_3O_2)_3$

SOLUTIONS
(a) Note that two K^+ ions are needed to balance the SO_4^{2-} ion for charge. The name of this compound is potassium sulfate.

(b) This compound contains two polyatomic ions: NH^+ and NO_3^-.
The name of this compound is ammonium nitrate.
(c) This compound contains the Fe^{3+} ion and the $C_2H_3O_2^-$ ion.
The name of this compound is iron(III) acetate.

TRY IT YOURSELF
Name the following compounds:
(a) Na_3PO_4
(b) $Hg(OH)_2$
(c) $(NH_4)_2SO_3$

ANSWERS
(a) sodium phosphate
(b) mercury(II) hydroxide
(c) ammonium sulfite

Writing Formulas for Molecular Compounds

Atoms in molecular compounds do not have *real* charges (as ions do in ionic compounds). Nevertheless, atoms are electrical in nature, and they can behave as though they possessed charges, even in molecular compounds. The type of charge found in molecular compounds is known as an *apparent charge*. To be able to work with both real and apparent charges, the concept of **oxidation numbers** (also known as **oxidation states**) was developed.

An oxidation number can be positive, negative, or zero. When we write oxidation numbers, we place the sign of the number *before* the number itself, for example, +2. (This distinguishes oxidation numbers from actual charges, in which the sign is placed *after* the number, as in 2+.) The following list gives six simplified rules for assigning oxidation numbers to elements.

1. The oxidation number of every atom in a *free element* is 0. By "free element" we mean an element that is not combined with another element. Therefore, Na, He, O_2, and P_4 are free elements, and each of the atoms that comprise them has an oxidation number of 0.
2. In a molecular or ionic compound, the *algebraic sum* of the oxidation numbers must be 0 since all compounds are electrically neutral.
3. The oxidation number of a simple ion is its charge. For example, the oxidation number of Cl^- is -1, and the oxidation number of Al^{3+} is +3.
4. In a polyatomic ion, the *algebraic sum* of the individual oxidation numbers must equal the *charge* on the ion.
5. Hydrogen in combination usually has an oxidation number of +1. An exception is the group of compounds known as *hydrides* (such as sodium hydride, NaH), in which the hydride ion has a charge of 1^-.

6. Oxygen in combination usually has an oxidation number of –2. Exceptions include *peroxides* (such as hydrogen peroxide, H_2O_2), in which the oxidation number of oxygen is –1, and oxygen–fluorine compounds, in which the oxidation number of oxygen has a *positive* value.

The Periodic Table of the Elements in Appendix 1 shows the common oxidation states of the elements. These values appear in the upper right side of each element's box.

We consider *binary* molecular compounds first. The rules are simple: (1) Elements with positive oxidation numbers generally precede elements with negative oxidation numbers, and (2) the algebraic sum of the numbers must be 0.

For example, sulfur can combine with oxygen to form *two* compounds. Since O must have an oxidation number of –2, S must have a *positive* oxidation number. Referring to the Periodic Table, we see that the positive oxidation numbers of S are +4 and +6. Therefore, the two possible formulas are SO_2 and SO_3.

TRY IT YOURSELF
Write the formulas for the binary compounds corresponding to the following combinations of elements:
(a) N and O   (b) P and Cl   (c) H and S   (d) H and N

ANSWERS
(a) There are *five* possibilities: N_2O, NO, N_2O_3, NO_2, N_2O_5.
(b) There are *two* possibilities: PCl_3, PCl_5.
(c) H_2S
(d) NH_3 (This is relatively straightforward, but we write the formula placing nitrogen as the first element. The name of this compound is *ammonia*.)

We can also work backwards: Given the formula of a molecular compound we can determine the oxidation number of each element in the compound. We can also do this for a polyatomic ion, since the atoms within the ion behave pretty much as those in a molecular compound: they *share* electrons.

PROBLEM
In the compound Al_2O_3, aluminum has an oxidation number of +3. Show that the sum of the oxidation numbers is 0.

SOLUTION
Since there are two Al atoms, the oxidation number total for Al is $2 \cdot (+3) = +6$. The oxidation number total for O is $3 \cdot (-2) = -6$. The sum of these two oxidation number totals is 0: $(+6) + (-6) = 0$.

Chapter Three FORMULAS, EQUATIONS, REACTIONS

63

PROBLEM

Find the oxidation number of each of the following elements:

(a) S in H_2SO_3 (b) Cr in $Na_2Cr_2O_7$

(c) Fe and Cl in $FeCl_3$ (d) P in $PO_4{}^{3-}$

(e) O in OF_2

SOLUTIONS

(a) The oxidation number total for H is $2 \cdot (+1) = +2$. The oxidation number total for O is $3 \cdot (-2) = -6$. These numbers add to -4.

   If the compound is to be neutral, S must have an oxidation number of $+4$.

(b) The oxidation number total for Na is $2 \cdot (+1) = +2$. The oxidation number total for O is $7 \cdot (-2) = -14$. These numbers add to -12.

   If the compound is to be neutral, the two Cr atoms must have an oxidation number total of $+12$. Therefore, each Cr atom has an oxidation number of $+6$.

(c) We must rely on the fact that Cl will be negative in this compound. According to the Periodic Table, the negative oxidation number of Cl is -1. In this compound, then, the oxidation number total for Cl is $3 \cdot (-1) = -3$.

   For the compound to be neutral, the oxidation number of Fe must be $+3$.

(d) The oxidation number total for O is $4 \cdot (-2) = -8$. In this polyatomic ion the sum of the oxidation number totals must equal the charge on the ion, that is, -3.

   Therefore, P must have an oxidation number of $+5$.

(e) According to the Periodic Table, F has an oxidation number of -1. The oxidation number total for F is $2 \cdot (-1) = -2$.

   If the compound is to be neutral, O must have an oxidation number of $+2$. (Note that this oxidation number is *not* shown on the Periodic Table.)

TRY IT YOURSELF

Find the oxidation number of each of the following elements:

(a) Cl in $ClO_4{}^-$ (b) N in N_2O_5 (c) C in $C_6H_{12}O_6$

ANSWERS

(a) $+7$.   (b) $+5$.

(c) If you followed all of the rules, you calculated an oxidation number for C equal to 0! This is entirely possible: according to our rules, some *combined* elements may also have oxidation numbers of 0.

Naming Molecular Compounds

The IUPAC system for naming molecular compounds uses *prefixes* that indicate how many atoms of an element are present. The following prefixes are used:

Prefix	Number of Atoms
mono	1
di	2
tri	3
tetra	4
penta	5
hexa	6

The prefix "mono" is never used for the first element. For example, SO_2, CO, and N_2O_3, are named, respectively, sulfur dioxide, carbon monoxide, and dinitrogen trioxide.

TRY IT YOURSELF
Name the following molecular compounds:
(a) PCl_5
(b) SO_3
(c) SF_6
(d) H_2O

ANSWERS
(a) phosphorus pentachloride
(b) sulfur trioxide
(c) sulfur hexafluoride
(d) dihydrogen monoxide—but we will continue to use its common name: *water*!

3.3 CHEMICAL EQUATIONS

An *equation* provides a "recipe" for carrying out a chemical reaction. Consider the following description of a chemical reaction:

Methane (CH_4) gas reacts with oxygen gas to produce
carbon dioxide gas, liquid water, and heat.

This is quite a mouthful! Let us develop a chemical equation for this reaction, step by step.

Developing an Equation

First, we write a *word equation:*

methane(g) + oxygen(g) → carbon dioxide(g) + water(ℓ) + heat

Note that an arrow separates what we react (the *reactants*) from what we produce (the *products*), and that the *phases* of the substances are included in parentheses: "s" for solid, "ℓ" for liquid, "g" for gas, and "aq" for a substance dissolved in water.

Next, we replace the name of each substance by its formula:

$$CH_4(g) + O_2(g) \rightarrow CO_2(g) + H_2O(\ell) + \text{heat}$$

And that's about it. We have changed our original, wordy statement into a shortened form that tells what happens when methane gas reacts with oxygen. This equation is still incomplete. We will "balance" it in Section 3.4.

TRY IT YOURSELF

Write word and formula equations for the following process:

Solid zinc reacts with a water solution of hydrogen sulfate to produce hydrogen gas, a water solution of zinc sulfate, and heat.

ANSWER

zinc(s) + hydrogen sulfate(aq) → hydrogen(g) + zinc sulfate(aq) + heat

$Zn(s) + H_2SO_4(aq) \rightarrow H_2(g) + ZnSO_4(aq) + \text{heat}$

3.4   BALANCING A CHEMICAL EQUATION

All chemical reactions must obey three *conservation laws:* conservation of matter (mass), conservation of energy, and conservation of electric charge. This means that the total mass, energy content, and electric charge of the reactants must equal those of the products.

If we inspect the equation written in Section 3.3 for the reaction between CH_4 and O_2 carefully, we find something seriously wrong with it. On the left side of the equation we have four hydrogen atoms and two oxygen atoms. On the right side we have two hydrogen atoms and three oxygen atoms. This is intolerable—our equation violates the law of conservation of matter!

If, however, we inspect the equation written in answer to "Try It Yourself," we find that the numbers of zinc, hydrogen, sulfur, and oxygen atoms on each side of the equation are equal. This equation is said to be *balanced,* while the other one is unbalanced. To correct this serious defect, we must *balance* the equation for the reaction between CH_4 and O_2.

To balance an equation, we use a *coefficient:* a number that is placed before a substance to indicate how many units of the substance we need. For example, the notation $5NH_3$ means that five molecules of NH_3 are required.

We now proceed to balance the faulty equation by inspection. (First, we will drop the heat term since it will not play a direct part in what we do.) We have four hydrogen atoms on the left, so we must produce four on the right. Since hydrogen appears only in the H_2O, it follows that we need $2H_2O$ to account for all of the hydrogen:

$$CH_4(g) + O_2(g) \rightarrow CO_2(g) + 2H_2O(\ell)$$
(The equation is balanced for hydrogen.)

If we count the oxygen atoms on the *right* in our new equation, we see that we have four—two from the CO_2 and two from the $2H_2O$. (Remember: the coefficient applies to the *entire* H_2O, not only the hydrogen.) We can easily produce four oxygen atoms on the *left* side by using $2O_2$:

$$CH_4(g) + 2O_2(g) \rightarrow CO_2(g) + 2H_2O(\ell)$$
(The equation is completely balanced.)

Now inspect the three equations given below. Are they balanced?

$$2CH_4(g) + 4O_2(g) \rightarrow 2CO_2(g) + 4H_2O(\ell)$$
$$6CH_4(g) + 12O_2(g) \rightarrow 6CO_2(g) + 12H_2O(\ell)$$
$$\frac{1}{2}CH_4(g) + O_2(g) \rightarrow \frac{1}{2}CO_2(g) + H_2O(\ell)$$

Clearly, the first two equations are balanced because we count the same number of atoms of each kind on the left and right sides. Note that the coefficients are *proportional* to the coefficients in the equation we just balanced.

Although the last equation looks strange (it has *fractional* coefficients), it is balanced because its coefficients are also in proportion to those in our original equation. (In Chapter 5 we will see how such equations are used.)

Our original (balanced) equation:

$$CH_4(g) + 2O_2(g) \rightarrow CO_2(g) + 2H_2O(\ell)$$

is called an equation balanced with *smallest whole-number coefficients*. For the present, we will use this form.

TRY IT YOURSELF
Write an equation for each of the following reactions, and balance it using smallest whole-number coefficients:
(a) nitrogen(g) + hydrogen(g) → ammonia(g)
(b) hydrogen(g) + oxygen(g) → water(ℓ)
(c) copper(II) oxide(s) + carbon(s) → copper(s) + carbon dioxide(g)
(d) magnesium(s) + hydrogen sulfate(aq) → hydrogen(g) + magnesium sulfate(aq)
(e) ammonia(g) + oxygen(g) → nitrogen monoxide(g) + water(ℓ)

ANSWERS
(a) $N_2(g) + 3H_2(g) \rightarrow 2NH_3(g)$
(b) $2H_2(g) + O_2(g) \rightarrow 2H_2O(\ell)$
(c) $2CuO(s) + C(s) \rightarrow 2Cu(s) + CO_2(g)$
(d) $Mg(s) + H_2SO_4(aq) \rightarrow H_2(g) + MgSO_4(aq)$
(e) $4NH_3(g) + 5O_2(g) \rightarrow 4NO(g) + 6H_2O(\ell)$

Particle Diagrams and Balanced Chemical Equations

We have used particle diagrams to represent atoms and molecules. We can also use particle diagrams to represent balanced equations. For example, consider the balanced equation $2H_2 + O_2 \rightarrow 2H_2O$. The following diagram specifies how each molecule in the equation is represented:

$$\text{⬤⬤} = H_2$$

$$\text{◯◯} = O_2$$

$$\text{⬤◯⬤} = H_2O$$

Now let us use these diagrams to form the balanced equation:

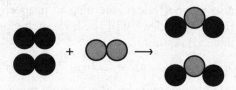

We can see that the equation is balanced because we can count the number of each type of atoms on each side of the equation.

TRY IT YOURSELF

(a) First, balance the chemical equation $C + O_2 \rightarrow CO$ using smallest whole-number coefficients. (b) Then use the following particle diagrams to represent the balanced equation:

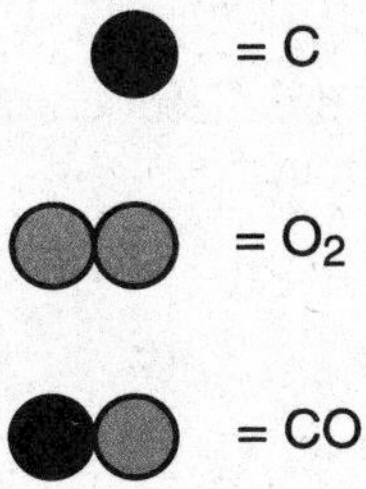

= C

= O_2

= CO

ANSWERS

(a) $2C + O_2 \rightarrow 2CO$ (b)

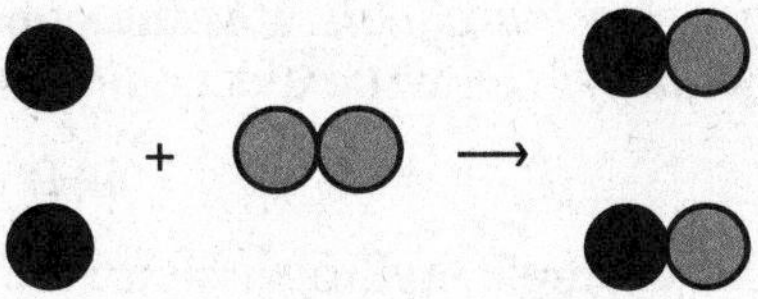

+ $\longrightarrow$

3.5 CLASSIFYING CHEMICAL REACTIONS

As noted in Section 3.4, an equation describes a specific chemical reaction. Since there are so many chemical reactions, chemists find it useful to place them in various categories based on their similarities. Throughout this book, we will be examining these categories. We begin, in this section, with the four simplest types of chemical reactions.

Direct Combination (Synthesis) Reactions

A *direct combination* (or *synthesis*) reaction involves the combination of *two or more reactants* to produce *one product*. The reactants can be elements or compounds. In general, we can write a direct combination reaction in the form

$$X + Y \rightarrow XY$$

As an example, the production of iron(II) chloride from iron and chlorine is a direct combination reaction:

$$Fe(s) + Cl_2(g) \rightarrow FeCl_2(s)$$

Another example involves the production of carbon dioxide from the combination of carbon monoxide and oxygen:

$$2CO(g) + O_2(g) \rightarrow 2CO_2(g)$$

TRY IT YOURSELF
Write a balanced equation for the direct combination of sodium(s) and bromine(ℓ) to produce a solid product.

ANSWER
$2Na(s) + Br_2(\ell) \rightarrow 2NaBr(s)$

Decomposition Reactions

A *decomposition* reaction involves the breakdown of a *single reactant* into *two or more products*. It is the opposite of a direct combination reaction. In general, we can write a decomposition reaction in the form

$$AB \rightarrow A + B$$

As an example, hydrogen peroxide decomposes into water and oxygen:

$$2H_2O_2\ (\ell) \rightarrow 2H_2O(\ell) + O_2(g)$$

TRY IT YOURSELF
Write a balanced equation for the decomposition of ammonia gas (NH_3) into its elements.

ANSWER
$2NH_3(g) \rightarrow N_2(g) + 3H_2(g)$

Single-Replacement Reactions

In a *single-replacement* reaction, an *uncombined element* replaces another element that is part of a compound. As a result, the replaced element becomes uncombined. In general, we can write a single-replacement reaction in the form

$$E + FG \rightarrow EG + F$$

As an example, the element zinc replaces the element copper from an aqueous solution of the compound copper(II) sulfate:

$$Zn(s) + CuSO_4(aq) \rightarrow ZnSO_4(aq) + Cu(s)$$

Single-replacement reactions occur only under certain conditions. For example, the replacement of zinc by copper (the reverse of the reaction shown above) will *not* occur. Since zinc replaces copper from solution, zinc is said to be a *more active metal* than copper. In other words, zinc will form a *positive* ion (in solution) more easily than will copper.

We can use Reference Table J in Appendix 1 to predict which single-replacement reactions of metals will occur. (Note that, even though H_2 is not a metallic element, it is included in the table because it forms H^+ ions; in fact, the table is based on H_2 as a standard.) The *higher* a metallic element appears in the table, the more active it is; and a metal will always replace one that is *lower* in the table.

PROBLEM
(a) Will magnesium (Mg) replace lead (Pb) [as $Pb(NO_3)_2$] in aqueous solution? If so, write a balanced equation for the reaction.
(b) Will silver (Ag) replace sodium (Na) [as NaCl] in aqueous solution? If so, write a balanced equation for the reaction.

SOLUTIONS
(a) Yes, since Mg is higher than Pb in Reference Table J in Appendix 1. The equation is

$$Mg + Pb(NO_3)_2 \rightarrow Pb + Mg(NO_3)_2$$

(b) No, since Ag is lower in the table than Na.

We can also compare the activities of *nonmetals* (i.e., their abilities to form *negative* ions in solution) by using Reference Table J. Consider the reaction between liquid bromine (Br_2) and aqueous potassium iodide (KI):

$$Br_2(\ell) + 2KI(aq) \rightarrow I_2(s) + 2KBr$$

Since Br_2 is higher than I_2 in Reference Table J, Br_2 is a more active non-metal than I_2; that is, Br_2 replaces I_2 because Br_2 forms a negative ion more easily than does I_2.

PROBLEM
Write the single-replacement reaction that occurs between gaseous Cl_2 and NaBr in aqueous solution.

SOLUTION
Since Cl_2 is higher than Br_2 on reference Table J, a single-replacement reaction will occur:

$$Cl_2(g) + 2NaBr(aq) \rightarrow Br_2(\ell) + 2NaCl$$

Double-Replacement Reactions

In a *double-replacement* reaction, *two elements in different compounds* replace each other. In general, we can write a double-replacement reaction in the form

$$PQ + RS \rightarrow PS + RQ$$

As an example, when aqueous solutions of barium nitrate and sodium sulfate are mixed, barium and sodium replace each other:

$$Ba(NO_3)_2(aq) + Na_2SO_4(aq) \rightarrow BaSO_4(s) + 2NaNO_3(aq)$$

Double-replacement reactions also occur only under certain conditions: at least one of the products must form a solid (known as a **precipitate**), a gas, or a molecular compound. In the example above, the reaction occurs because a precipitate of $BaSO_4$ forms. We can use Reference Table F in Appendix 1 to predict when a precipitate will form. All of the ionic compounds that are classified as **insoluble** will form precipitates; **soluble** compounds will remain in solution.

PROBLEM
A solution of $AgNO_3$ is mixed with a solution of NaCl. Does a double-replacement reaction occur?

SOLUTION
First, we write the equation as though the reaction does occur:

$$AgNO_3\ (\ ) + NaCl(\ ) \rightarrow AgCl(\ ) + NaNO_3(\ )$$

Next, we refer to Table F. According to the table:

- All compounds containing Na^+ or NO_3^- ions are *soluble*: they remain in solution.
- Compounds containing Ag^+ and Cl^- or Br^- or I^- are *insoluble*: they form precipitates.

Since AgCl is an insoluble product, the reaction will occur.
    Finally, we complete the equation:

$$AgNO_3\ (aq) + NaCl(aq) \rightarrow AgCl(s) + NaNO_3(aq)$$

TRY IT YOURSELF
(a) A solution of sodium sulfide is mixed with a solution of zinc nitrate. Does a reaction occur? Use Reference Table F in Appendix 1.
(b) A solution of potassium chloride is mixed with a solution of sodium iodide. Does a reaction occur? Use Reference Table F in Appendix 1.

ANSWERS

(a) Yes. The balanced equation is:

$$Na_2S(aq) + Zn(NO_3)_2(aq) \rightarrow ZnS(s) + 2NaNO_3(aq)$$

(b) No. The products (potassium iodide and sodium chloride) are soluble, as are the reactants (potassium chloride and sodium iodide).

SECTION II—ADDITIONAL MATERIAL

3.1A   OTHER WAYS OF NAMING IONIC COMPOUNDS

There are other, older systems for naming ionic compounds. For metallic elements that can form exactly *two* positive ions, the suffixes *-ous* and *-ic* are added to distinguish between the lower and higher charge. The first suffix (*-ous*) is used with the element's *smaller* charge, and the second (*-ic*) with its *larger* charge. These suffixes are usually (but not always) attached to the *Latin* names of the elements. Some examples are shown below:

$FeCl_2$ ferr*ous* chloride $FeCl_3$ ferr*ic* chloride
Cu_2O cupr*ous* oxide CuO cupr*ic* oxide
Hg_2Br_2 mercur*ous* bromide $HgBr_2$ mercur*ic* bromide

We will *not* employ this system in naming compounds; rather, we will use the IUPAC system throughout this book.

END-OF-CHAPTER QUESTIONS

1. Which represents a substance dissolved in water?
    (1) HCl(aq)   (2) HCl(ℓ)   (3) HCl(g)   (4) HCl(s)

2. Which may represent a crystalline material?
    (1) K_2SO_4(s)   (2) $Br_2(\ell)$ (3) NaCl(aq)   (4) CO_2(g)

3. Which represents a compound?
    (1) Ca   (2) Cr   (3) CO   (4) Co

4. Which is the formula of a binary compound?
    (1) KOH   (2) $NaClO_3$ (3) Al_2S_3 (4) $Bi(NO_3)_3$

5. In a sample of solid $Ba(NO_3)_2$, the ratio of barium ions to nitrate ions is
   (1) 1:1 (2) 1:2 (3) 1:3 (4) 1:6

6. A chemical formula is an expression used to represent
   (1) mixtures, only
   (2) elements, only
   (3) compounds, only
   (4) compounds and elements

7. In which compound does chlorine have the highest oxidation number?
   (1) $KClO$ (2) $KClO_2$ (3) $KClO_3$ (4) $KClO_4$

8. What is the oxidation number of sulfur in H_2SO_4?
   (1) 0 (2) -2 (3) $+6$ (4) $+4$

9. If element X forms the oxides XO and X_2O_3, the oxidation numbers of element X are
   (1) $+1$ and $+2$
   (3) $+1$ and $+3$
   (2) $+2$ and $+3$
   (4) $+2$ and $+4$

10. For the compound $(NH_4)_2SO_4$, the oxidation numbers of all the atoms must add to
    (1) 1 (2) 0 (3) 3 (4) 11

11. What is the oxidation number of sulfur in $Na_2S_2O_7$?
    (1) -2 (2) $+2$ (3) $+6$ (4) $+4$

12. Sulfur exhibits a negative oxidation state in
    (1) H_2S (2) H_2SO_3 (3) S_8 (4) SO_3

13. In which compound does hydrogen have an oxidation number of -1?
    (1) NH_3 (2) KH (3) HCl (4) H_2O

14. The formula for iron(III) oxide is
    (1) FeO_3 (2) Fe_2O_3 (3) Fe_3O (4) Fe_3O_2

15. Which is the formula for potassium hydride?
    (1) KH (2) KH_2 (3) KOH (4) $K(OH)_2$

16. Which is the formula for dinitrogen monoxide?
    (1) NO (2) N_2O (3) NO_2 (4) N_2O_4

17. Which formula represents mercury(I) chloride?
    (1) Hg_2Cl (2) $HgCl_2$ (3) Hg_2Cl_2 (4) Hg_2Cl_4

18. Which is the formula for titanium(III) oxide?
    (1) Ti_2O_3 (2) TiO (3) Ti_3O_2 (4) Ti_2O_4

19. Which is the formula for sulfur dioxide?
    (1) SO (2) SO_2 (3) SO_3 (4) SO_4

20. Which is the correct formula for carbon monoxide?
    (1) CO (2) CO_2 (3) C_2O (4) C_2O_3

21. What is the correct name for the compound with the formula $CrPO_4$?
    (1) chromium(II) phosphate   (2) chromium(III) phosphate
    (3) chromium(II) phosphide   (4) chromium(III) phosphide

22. In an equation, which symbol indicates a mixture?
    (1) $NH_3(s)$ (2) $NH_3(\ell)$ (3) $NH_3(aq)$ (4) $NH_3(g)$

23. Given this unbalanced equation:

    $$Ca(OH)_2 + (NH_4)_2SO_4 \rightarrow CaSO_4 + NH_3 + H_2O$$

    What is the *sum* of the coefficients when the equation is completely balanced using smallest whole-number coefficients?
    (1) 5   (2) 7   (3) 9   (4) 11

24. When the equation

    $$\_Na(s) + \_H_2O(\ell) \rightarrow \_NaOH(aq) + \_H_2(g)$$

    is correctly balanced using smallest whole-number coefficients, the coefficient of the water is
    (1) 1   (2) 2   (3) 3   (4) 4

25. Given this unbalanced equation:

    $$(NH_4)_3PO_4 + Ba(NO_3)_2 \rightarrow Ba_3(PO_4)_2 + NH_4NO_3$$

    What is the coefficient of NH_4NO_3 when the equation is correctly balanced using smallest whole-number coefficients?
    (1) 6   (2) 2   (3) 3   (4) 4

26. When the equation

    $$\_NH_3 + \_O_2 \rightarrow \_HNO_3 + \_H_2O$$

    is correctly balanced using smallest whole-number coefficients, the coefficient of O_2 is
    (1) 1   (2) 2   (3) 3   (4) 4

27. When the equation

$$\_SiO_2 + \_C \rightarrow \_SiC + \_CO$$

is correctly balanced using smallest whole-number coefficients, the sum of all the coefficients is

(1) 5   (2) 7   (3) 8   (4) 9

28. When the equation

$$\_Al_2(SO_4)_3 + \_ZnCl_2 \rightarrow \_AlCl_3 + \_ZnSO_4$$

is correctly balanced using smallest whole-number coefficients, the sum of the coefficients is

(1) 9   (2) 8   (3) 5   (4) 4

29. When the equation

$$\_H_2 + \_Fe_3O_4 \rightarrow \_Fe + \_H_2O$$

is correctly balanced using smallest whole-number coefficients, the coefficient of H_2 is

(1) 1   (2) 2   (3) 3   (4) 4

30. When the equation

$$\_C_2H_4 + \_O_2 \rightarrow \_CO_2 + \_H_2O$$

is correctly balanced using smallest whole-number coefficients, what is the coefficient of O_2?

(1) 1   (2) 2   (3) 3   (4) 4

31. Given this unbalanced equation:

$$\_AlPO_4 + \_Ca(OH)_2 \rightarrow \_Al(OH)_3 + \_Ca_3(PO_4)_2$$

When the equation is correctly balanced using smallest whole-number coefficients, the sum of the coefficients is

(1) 7   (2) 8   (3) 3   (4) 4

32. Which equation illustrates the conservation of mass?

(1) $H_2 + Cl_2 \rightarrow HCl$ (2) $H_2 + Cl_2 \rightarrow 2HCl$

(3) $H_2 + O_2 \rightarrow H_2O$ (4) $H_2 + O_2 \rightarrow 2H_2O$

33. When the equation

$$C_2H_6 + O_2 \rightarrow CO_2 + H_2O$$

is correctly balanced, the coefficient of O_2 is

(1) 7   (2) 10   (3) 3   (4) 4

34. When the equation

$$Cu + H_2SO_4 \rightarrow CuSO_4 + H_2O + SO_2$$

is correctly balanced, what is the coefficient of $CuSO_4$?
(1) 1   (2) 2   (3) 3   (4) 4

35. Given this balanced equation:

$$2Na + 2H_2O \rightarrow 2X + H_2$$

What is the correct formula for the product represented by the letter X?
(1) NaO   (2) Na_2O (3) NaOH   (4) Na_2OH

36. The products, X and Y, of this balanced equation:

$$CS_2 + 3O_2 \rightarrow X + 2Y$$

could be
(1) CO_2 and SO_3 (2) CO and SO
(3) CO and SO_2 (4) CO_2 and SO_2

Base your answers to questions 37–40 on the four types of reactions given below. Match each reaction with its appropriate reaction type, (1), (2), (3), or (4). A choice may be used more than once or not at all.

(1) synthesis
(2) decomposition
(3) single replacement
(4) double replacement

37. $CaCO_3 \rightarrow CaO + O_2$

38. $Cl_2 + 2KI \rightarrow I_2 + 2KCl$

39. $MgSO_4 + Ba(NO_3)_2 \rightarrow BaSO_4 + Mg(NO_3)_2$

40. $Zn + 2AgNO_3 \rightarrow 2Ag + Zn(NO_3)_2$

41. Given the balanced equation representing a reaction:

$$K_2CO_3(aq) + BaCl_2(aq) \rightarrow 2KCl(aq) + BaCO_3(s)$$

Which type of reaction is represented by this equation?
(1) synthesis
(2) decomposition
(3) single replacement
(4) double replacement

Constructed-Response Questions

1. Below is a grid of positive and negative ions. In the space where a positive and a negative ion intersect, write the correct formula for the compound formed by those two ions. A number of correct formulas have already been inserted to assist you.

		bromide Br^-	carbonate CO_3^{2-}	chlorate ClO_3^-	chloride Cl^-	chromate CrO_4^{2-}	nitrate NO_3^-	phosphate PO_4^{3-}	sulfate SO_4^{2-}	sulfide SO_4^{2-}
aluminum	Al^{3+}					$Al_2(CrO_4)_3$				
ammonium	NH_4^+					$(NH_4)_2CrO_4$				
barium	Ba^{2+}					$BaCrO_4$				
calcium	Ca^{2+}					$CaCrO_4$				
copper (II)	Cu^{2+}					$CuCrO_4$				
iron (II)	Fe^{2+}					$FeCrO_4$				
iron (III)	Fe^{3+}					$Fe_2(CrO_4)_3$				
lead (II)	Pb^{2+}	$PbBr_2$	$PbCO_3$	$Pb(ClO_3)_2$	$PbCl_2$	$PbCrO_4$	$Pb(NO_3)_2$	$Pb_3(PO_4)_2$	$PbSO_4$	PbS
lead (IV)	Pb^{4+}					$Pb(CrO_4)_2$				
magnesium	Mg^{2+}					$MgCrO_4$				
mercury (I)	Hg_2^{2+}					Hg_2CrO_4				
mercury (II)	Hg^{2+}					$HgCrO_4$				
potassium	K^+					K_2CrO_4				
silver	Ag^+					Ag_2CrO_4				
sodium	Na^+					Na_2CrO_4				
zinc	Zn^{2+}					$ZnCrO_4$				

2. Determine the oxidation numbers of the elements in SO_3^{2-}.

3. Determine the oxidation numbers of the elements in $KMnO_4$.

4. Balance each of the following equations using smallest whole-number coefficients.
    (a) $Zn(OH)_2 + H_3PO_4 \rightarrow H_2O + Zn_3(PO_4)_2$
    (b) $CO_2 + H_2O \rightarrow C_6H_{12}O_6 + O_2$

5. Name each of the following compounds using the IUPAC system:
    (a) NaI   (b) CS_2 (c) $FeCl_3$ (d) P_4O_{10} (e) Cl_2O_7

6. Write a balanced equation for a synthesis reaction with magnesium and Br_2 as reactants.

7. Write a balanced equation for the decomposition of FeO.

8. Write a balanced equation for the single-replacement reaction that occurs between zinc and aluminum. Assume that NO_3^- is the negative ion.

9. Write a balanced equation for the double-replacement reaction between Na_3PO_4 and $Pb(NO_3)_2$. Use Reference Table F in Appendix 1 to determine which compound is the precipitate.

Base your answers to questions 10 through 12 on the balanced chemical equation below.

$$2H_2O \rightarrow 2H_2 + O_2$$

10. What type of reaction does this equation represent?

11. How does the balanced chemical equation show the law of conservation of mass?

12. What is the total number of molecules of O_2 produced when 8 molecules of H_2O are completely consumed?

The answers to these questions are found in Appendix 3.

Chapter
Four

CHEMICAL CALCULATIONS

KEY IDEAS

This chapter focuses on problem solving in chemistry. The concept of the mole is stressed throughout the chapter. A variety of problems involving single substances and equations is explored.

KEY OBJECTIVES

At the conclusion of this chapter you will be able to:

- Define the terms *atomic mass unit*, *isotopic mass*, and *average atomic mass*.
- Calculate the atomic mass of an element, given the masses of its naturally occurring isotopes and the abundances of these isotopes.
- Calculate the formula mass of a substance.
- Define the term *mole* in relation to number of particles and the mass of a substance.
- Define the term *molar mass* (also called *gram-formula mass*).
- Calculate the molar masses of various types of substances.
- Solve mole-mass problems.
- Solve percent composition problems.
- Calculate the molecular formula of a substance, given its empirical formula and molar mass.
- Calculate the empirical formula of a substance from its percent composition by mass.
- Solve mole and mass problems involving chemical equations.
- Solve problems involving percent yields and limiting reactants.

SECTION I—BASIC
(REGENTS-LEVEL) MATERIAL

NYS REGENTS CONCEPTS AND SKILLS

Note: By the time you have finished Section I, this chapter, you should have mastered the concepts and skills listed below. The Regents chemistry examination will test your knowledge of these items and your ability to apply them.

Concepts are the *basic ideas* that form the body of the Regents chemistry course (what you need to know!).

Skills are the *activities* that demonstrate your mastery of these concepts (how you show that you know them!).

Following each concept or skill is a page reference (given in parentheses) to this chapter.

4.1 Concept:
The mass of each proton and of each neutron is approximately equal to one atomic mass unit. An electron is much less massive than a proton or a neutron. The mass of an atom is very nearly equal to its mass number. (Pages 83–84)

4.2 Concept:
The average atomic mass of an element is the weighted average of the masses of its naturally occurring isotopes. (Pages 83–84)

Skills:
• Given the atomic mass of an element, determine the mass number of the element's most abundant isotope. (Pages 83–84)
• Calculate the atomic mass of an element, given the masses and ratios of the element's naturally occurring isotopes. (Pages 83–85)

4.3 Concept:
The empirical formula of a compound is the simplest whole number ratio of atoms of the elements in a compound. It may be different from the molecular formula, which is the actual ratio of atoms in a molecule of that compound. (Pages 85–86)

Skills:
• Determine the molecular formula of a compound, given the empirical formula and the molecular mass. (Pages 86–87)
• Determine the empirical formula from the molecular formula. (Pages 86–87)

4.4 Concept:
The formula mass of a substance is the sum of the atomic masses of its atoms. The molar mass (gram formula mass) of a substance is the mass of 1 mole of that substance. (Pages 87–88)

Skills:
- Calculate the formula mass and the molar mass (gram-formula mass) of a substance. (Pages 83–84)
- Determine the number of moles of a substance, given its mass. (Pages 85–86)
- Determine the mass of a given number of moles of a substance. (Pages 85–86)

4.5 Concept:
The percent composition by mass of each element in a compound can be calculated mathematically. (Pages 87–88)

Skill:
Determine the percent composition(s) of one or more elements in a compound. (Pages 87–88)

4.6 Concept:
A balanced chemical equation represents conservation of atoms. The coefficients in a balanced chemical equation can be used to determine mole ratios in the reaction. (Pages 90–91)

Skill:
Solve simple mole-mole stoichiometry problems, given a balanced equation. (Pages 90–91)

4.1 INTRODUCTION

As its name implies, this chapter is devoted to solving various numerical problems in chemistry. The ability to solve numerical problems is essential to chemists because it provides the means to translate theory into practice. We will solve many of the problems in this chapter by using the factor-label method (FLM) introduced in Chapter 1. You may wish to review the method at this time.

4.2 AVERAGE ATOMIC MASS

How do scientists measure the masses of atoms and their components? They use an instrument called a *mass spectrometer*. First, though, a standard mass must be established and it, must be given a value. Chemists have agreed that a neutral atom of the isotope carbon-12 is the appropriate standard. Its mass has been set at exactly 12 atomic mass units (12 u or 12 amu). It follows that 1 atomic mass unit is 1/12 of the mass of a neutral carbon-12 atom. All other atomic masses are measured relative to this value. With this system, the masses of a proton and a neutron are nearly equal to 1 atomic mass unit, while the electron has a mass close to 0.0005 atomic mass unit.

We are now able to report the mass of any atom or subatomic particle. But how should we report the masses of the atoms in a sample of an element? There are problems here because elements may contain atoms with different mass numbers (i.e., isotopes). The solution is to measure the mass of each individual isotope in the element and then calculate a *weighted average* based on the *abundance* of each isotope in the sample of the element.

This weighted average is known as the **average atomic mass** of the element. For example, the average atomic mass of the element carbon is 12.0111 atomic mass units, indicating that a sample of carbon contains isotopes whose mass numbers are other than 12. We can think of the quantity 12.0111 atomic mass units as the mass of an *average* atom the element carbon.

If we examine The Periodic Table of the Elements in Appendix 1, we note that the average atomic masses of most elements are nearly whole numbers. If an average atomic mass is rounded to a whole number, it usually represents the mass number of the most abundant isotope of the element.

PROBLEM
What is the most abundant isotope of the element oxygen (average atomic mass = 15.9994 u)?

SOLUTION
The average atomic mass is very nearly equal to 16 atomic mass units, so we conclude that oxygen-16 is the most abundant isotope of the element oxygen.

TRY IT YOURSELF
What is the most abundant isotope of the element gold (average atomic mass = 196.9665 u)?

ANSWER
Gold-197

PROBLEM
An element consists of two isotopes. Isotope *A* has an abundance of 75.00 percent, and its mass is 14.000 atomic mass units. Isotope *B* has an abundance of 25.00 percent, and its mass is 15.000 atomic mass units. What is the average atomic mass of the element?

SOLUTION

The term *percentage abundance* indicates how many atoms of an isotope are contained in 100 atoms of the element. In this example, 100 atoms of the element would contain 75 atoms of isotope A and 25 atoms of isotope B.

To calculate the average atomic mass of the element, we multiply the mass of each isotope by the *decimal equivalent* of its percentage abundance and then add the results:

$$\text{Average atomic mass} = (14.000 \text{ u} \cdot 0.7500) + (15.000 \text{ u} \cdot 0.2500)$$
$$= 14.25 \text{ u}$$

TRY IT YOURSELF

The element boron occurs in nature as two isotopes. Boron-10 has a mass of 10.0130 atomic mass units, and its abundance is 19.90 percent; boron-11 has a mass of 11.0093 atomic mass units, and its abundance is 80.10 percent. From these data, calculate the average atomic mass of the element boron.

ANSWER

10.81 u

From this point forward, we will replace the term *average atomic mass* with the simpler term **atomic mass**.

4.3 THE FORMULA MASS OF A SUBSTANCE

The **formula mass** is the sum of the masses of all of the atoms in a given formula. We recall that atomic masses are measured in atomic mass units (u). The formula mass of the molecule O_2 equals 32 atomic mass units since there are two oxygen atoms in the formula and each oxygen atom has a mass of 16.00 atomic mass units.

Suppose we want to find the formula mass of the compound $CuSO_4$. We need to do a little planning: first, we obtain the atomic masses from the Periodic Table of Elements in Appendix 1 of this book, and then we set up a table to keep things organized:

Element	Atomic Mass	×	Number of Atoms in Formula	=	Mass of Element
Cu	63.55 u	×	1	=	63.55 u
S	32.07 u	×	1	=	32.07 u
O	16.00 u	×	4	=	64.00 u
			Formula mass (total)	=	159.62 u

TRY IT YOURSELF
Calculate the formula mass of NH_3.

ANSWER
17.03 u

4.4 THE MOLE CONCEPT AND MOLAR MASS

The *mole* is one of the most useful and important concepts in chemistry. Mole (abbreviated as *mol*) is a word with two faces: it represents a number and a mass. In Chapter 6, we will see that the mole has a third face: the volume of a gas under certain specified conditions.

One mole refers to *Avogadro's number* ($N_A = 6.02 \times 10^{23}$) of particles of anything. For example, 1 mole of silver refers to 6.02×10^{23} *atoms* of Ag, while 1 mole of CO_2 refers to 6.02×10^{23} *molecules* of CO_2. And 1 mole of slices of pizza? That's correct, 6.02×10^{23} slices!

One mole of particles has a special mass associated with it: *the formula mass of the substance expressed in grams*. For example, 1 mole of H_2O has a mass of 18.00 grams (because its formula mass is 18.00 u). In this context, we refer to 1 mole as the **molar mass** (also known as the **gram-formula mass**), whose units are grams per mole (g/mol). In this book, we will use these two terms interchangeably and we will use the symbol $\mathcal{M}$ to represent them. If the formula happens to be that of an element, such as calcium, the molar mass will refer to the atomic mass of the element; if the formula represents a molecule or an ionic compound, the molar mass will refer to the formula mass of the substance.

PROBLEM
Calculate the molar mass of each of the following:
(a) Ne   (b) Cl_2 (c) SO_3 (d) KBr

SOLUTIONS
(a) We use the atomic mass of Ne: $\mathcal{M} = 20.18$ g/mol.
(b) We calculate the molar mass of Cl_2: $\mathcal{M} = 70.91$ g/mol.
(c) We calculate the molar mass of SO_3: $\mathcal{M} = 80.07$ g/mol.
(d) We calculate the molar mass of KBr: $\mathcal{M} = 118.0$ g/mol.

TRY IT YOURSELF
Calculate the molar mass of $CaCO_3$.

ANSWER
$\mathcal{M} = 100.1$ g/mol

4.5 PROBLEMS INVOLVING A SINGLE SUBSTANCE

We can solve a variety of problems involving a single substance because any substance can be represented by a chemical formula, and any formula can be associated with the mole concept. For example, the formula H_2SO_4 can be interpreted as follows: 1 mole of H_2SO_4 contains 2 moles of H atoms, 1 mole of S atoms, and 4 moles of O atoms. Similarly, 1 mole of $CaCl_2$ contains 1 mole of Ca^{2+} ions and 2 moles of Cl^- ions.

Mole-Mass Conversions

Reference Table T in Appendix 1 gives us a formula for calculating among mass, molar mass, and number of moles:

$$\text{number of moles} = \frac{\text{given mass (g)}}{\text{molar mass (g/mol)}}$$

In symbols, we can represent this formula as:

$$n = \frac{m}{\mathcal{M}}$$

where n is the number of moles, m is the mass of the substance in grams, and $\mathcal{M}$ is the molar mass in grams per mole.

PROBLEM
(a) Calculate the number of moles in 250. grams of $CaCO_3$ ($\mathcal{M} = 100.$ g/mol).
(b) Calculate the mass of 6.50 moles of KBr ($\mathcal{M} = 118$ g/mol).

SOLUTIONS

(a) $n = \dfrac{m}{\mathcal{M}} = \dfrac{250.\ \text{g}}{100.\ \text{g/mol}} = 2.50\ \text{mol}$

(b) $n = \dfrac{m}{\mathcal{M}}$
$m = n \cdot \mathcal{M} = (6.50\ \text{mol}) \cdot (118\ \text{g/mol}) = 767\ \text{g}$

TRY IT YOURSELF
Calculate the molar mass of a substance if 0.25 mole of the substance has a mass of 45 grams.

ANSWER

$\mathcal{M} = 180$ g/mol

Mole-mass conversion problems can also be solved by using the factor-label method (FLM). When FLM is used, the molar mass becomes the conversion factor. For example, the molar mass of NH_3 is 17.03 grams per mole.

Therefore, we can form two possible fractions:

$$\frac{17.03 \text{ g NH}_3}{1 \text{ mol NH}_3} \quad \text{or} \quad \frac{1 \text{ mol NH}_3}{17.03 \text{ g NH}_3}$$

Either of these two fractions can be used in solving a mole-mass problem.

PROBLEM

How many moles of molecules are contained in 67.25 grams of NH_3 ($\mathcal{M} = 17.03$ grams per mole)?

SOLUTION

The appropriate conversion factor is $\dfrac{1 \text{ mol NH}_3}{17.03 \text{ g NH}_3}$. Our problem is then set up for solution:

$$x \text{ mol NH}_3 = 67.25 \text{g NH}_3 \cdot \frac{1 \text{ mol NH}_3}{17.03 \text{ g NH}_3} = 3.949 \text{ mol NH}_3$$

TRY IT YOURSELF

Calculate the mass of 3.00 moles of nitrogen gas (N_2) using the FLM.

ANSWER

84.04 g N_2

Empirical Formulas, Molecular Formulas, and Molar Mass

In Chapter 3, we learned how to calculate the empirical formula of a substance, given its molecular formula. Now, we learn how to calculate the molecular formula of a substance, given its empirical formula and its molar mass. The following steps show how this is accomplished:

1. Calculate the molar mass of the *empirical formula*.
2. *Divide* the molar mass of the substance by the molar mass of the empirical formula.
3. *Multiply* the empirical formula by the number obtained in Step 2.

PROBLEM

The empirical formula of a molecular substance is CH_2, and the molar mass of the substance is 70.05 grams per mole. Determine the molecular formula of the substance.

SOLUTION

1. The molar mass of the empirical formula CH_2 is 14.01 g/mol.

2. $\dfrac{70.05 \text{ g/mol}}{14.01 \text{ g/mol}} = 5.000 = 5$

3. $5 \cdot (CH_2) = C_5H_{10}$

TRY IT YOURSELF

The empirical formula of a substance is NO_2, and the molar mass of the substance is 92.01 g/mol. Determine the molecular formula of the substance.

ANSWER

N_2O_4

Percent Composition by Mass

The term *percent* is always associated with a number meaning "parts per hundred." We calculate all percentages in a very simple way: we divide the quantity under consideration by the total quantity involved, and multiply the result by 100. For example, if 7 persons in 20 wear eyeglasses, then the percentage of persons wearing eyeglasses is calculated from this relationship: $(7/20) \cdot 100 = 35$ percent. Reference Table T in Appendix 1 provides a general formula for percent composition.

$$\% \text{ composition by mass} = \frac{\text{mass of part}}{\text{mass of whole}} \times 100$$

Similarly, we can calculate the percent composition of an element in a compound by dividing the mass of the element by the molar mass of the substance and multiplying the result by 100.

PROBLEM

Calculate the percentage of oxygen by mass in $CuSO_4$.

SOLUTION

We construct the accompanying table to calculate the molar mass of $CuSO_4$.

Element	Atomic Mass	×	Number of Atoms in Formula	=	Mass of Element
Cu	63.55 g/mol	×	1	=	63.55 g/mol
S	32.07 g/mol	×	1	=	32.07 g/mol
O	16.00 g/mol	×	4	=	64.00 g/mol
			Molar mass (total)	=	159.62 g/mol

We note that the mass of oxygen in the formula is 64.00 g/mol, while the molar mass is 159.62 g/mol. The percent composition of oxygen in $CuSO_4$ is found from this relationship:

$$\frac{64.00 \text{ g/mol}}{159.62 \text{ g/mol}} \cdot 100 = 40.10\%$$

TRY IT YOURSELF

Calculate the percentage of carbon by mass in the sugar glucose ($C_6H_{12}O_6$).

ANSWER

40.00%

4.6 PROBLEMS INVOLVING CHEMICAL EQUATIONS

Let us consider this balanced chemical equation:

$$N_2 + 3H_2 \rightarrow 2NH_3$$

Previously, we interpreted the coefficients (1, 3, and 2) as numbers of molecules. Since moles and numbers of molecules are closely related, these coefficients can also be read as numbers of *moles* of molecules; then the equation can be read as follows:

1 *mole* of N_2 molecules and 3 *moles* of H_2 molecules react to yield 2 *moles* of NH_3 molecules.

Since we can convert among moles, mass, and numbers of particles, it is possible for us to solve a wide variety of problems by applying the factor-label method to chemical equations.

Mole-Mole Problems

Any two substances in a balanced chemical equation (reactants, products, or both) can be connected by means of their respective coefficients, since the coefficients stand for the numbers of moles involved in the chemical reaction. For example, let us consider the following problem:

PROBLEM
In the equation $N_2 + 3H_2 \rightarrow 2NH_3$, how many moles of N_2 are needed to produce 5.0 moles of NH_3?

SOLUTION
From the equation we know that 1 mol of N_2 will produce 2 mol of NH_3. We can now solve the problem by using FLM. Our solution map is as follows:

$$5.0 \text{ mol } NH_3 \xrightarrow{\substack{\text{equation}\\\text{coefficients}}} \text{mol } N_2$$

We need to use the appropriate coefficients of the equation because a mole-mole conversion is involved. The solution is as follows:

$$\text{mol } N_2 = 5.0 \text{ mol } NH_3 \cdot \frac{1 \text{ mol } N_2}{2 \text{ mol } NH_3} = 2.5 \text{ mol } N_2$$

Note that the equation coefficients form the conversion factor in the solution to the problem.

TRY IT YOURSELF
In the equation $2NO + O_2 \rightarrow 2NO_2$, how many moles of O_2 are needed to produce 3.50 moles of NO_2?

ANSWER
1.75 mol

SECTION II—ADDITIONAL MATERIAL

4.1A CONVERTING BETWEEN MOLES AND NUMBERS OF PARTICLES

Since 1 mole is equal to Avogadro's number of particles ($N_A = 6.02 \times 10^{23}$), we can use this number as the conversion factor.

PROBLEM
How many particles are in 2.00 moles of SO_2?

SOLUTION

$$N \text{ molecules SO}_2 = 2.00 \text{ mol SO}_2 \cdot \frac{6.02 \times 10^{23} \text{ molecules SO}_2}{1 \text{ mol SO}_2} = 1.20 \times 10^{24} \text{ molecules SO}_2$$

TRY IT YOURSELF

How many moles of molecules are equivalent to 9.03×10^{23} molecules of H_2O?

ANSWER

1.50 mol H_2O

We can also solve more complicated problems involving moles, mass, and number of particles by the factor-label method. In this case, *both* the molar mass and Avogadro's number are used as conversion factors.

PROBLEM

What is the mass of 2.40×10^{24} molecules of O_2?

SOLUTION

We need to convert the number of molecules to moles using Avogadro's number, and then convert the number of moles to mass using the molar mass of oxygen. The factor-label method allows us to solve this problem in *one step*:

$$(2.40 \times 10^{24} \text{ molecules O}_2) \cdot \left(\frac{1 \text{ mol O}_2}{6.02 \times 10^{23} \text{ molecules O}_2} \right) \cdot \left(\frac{32.0 \text{ g O}_2}{1 \text{ mol O}_2} \right) = 128 \text{ g O}_2$$

TRY IT YOURSELF

What is the number of atoms contained in 12.01 grams of helium (He)?

ANSWER

1.81×10^{24} atoms of He

4.2A EMPIRICAL FORMULA FROM PERCENT COMPOSITION

Recall that the **empirical formula** of a compound is a formula reduced to simplest numbers by division. For example, the empirical formula for N_2O_4 is NO_2, and the empirical formula for C_6H_6 is CH. The empirical formula for H_2O, however, remains H_2O because it cannot be reduced further.

Since we calculate percent compositions from chemical formulas, we are able to reverse the process and determine formulas from percent composition. However, when we use this process, we are able to determine only *empirical* formulas. The technique we use relies on our connection of a formula with the mole concept.

PROBLEM
Calculate the empirical formula of a substance that contains 50.00 percent sulfur and 50.00 percent oxygen by mass.

SOLUTION
If we had 100.0 g of this compound, we would know from the percent composition given in the problem that the compound contains 50.00 g of S and 50.00 g of O, but formulas indicate numbers of *moles*, not masses in grams. If, however, we divide the given masses by the atomic masses of the two elements, we transform mass into moles as follows:

$$S_{\dfrac{50.00\,\cancel{g}}{32.0\,\dfrac{\cancel{g}}{mol}}}\; O_{\dfrac{50.00\,\cancel{g}}{16.0\,\dfrac{\cancel{g}}{mol}}} = S_{1.56\,mol}\,O_{3.125\,mol}$$

As strange as this formula looks, it is accurate because it shows the *relative number of moles* of each element. Now, our task is to make the formula *look* like a formula, that is, to convert it into an expression with whole numbers. Usually we accomplish this by dividing each element by the *smallest* number of moles, as illustrated below:

$$S_{\dfrac{1.56}{1.56}}\,O_{\dfrac{3.125}{1.56}} = S_{1.00}O_{2.00} = SO_2$$

TRY IT YOURSELF
A hydrocarbon consists of 80.00 percent carbon and 20.00 percent hydrogen by mass.

(a) Express the formula of the compound in terms of the relative numbers of moles of carbon and hydrogen.

(b) Determine the empirical formula of the hydrocarbon.

ANSWERS

(a) $C_{6.661}H_{19.84} = C_{1.000}H_{2.979}$ (b) CH_3

4.3A   MOLE-MASS PROBLEMS

PROBLEM
In the equation $4Al + 3O_2 \rightarrow 2Al_2O_3$, how many grams of aluminum will combine with 1.50 moles of oxygen?

SOLUTION

From the equation we know that 4 mol of Al will combine with 3 mol of O_2. We can solve the problem by constructing a "solution map" and using FLM. Here is our solution map:

$$1.50 \text{ mol } O_2 \xrightarrow{\overset{\text{equation}}{\text{coefficients}}} \text{mol Al} \longrightarrow \text{g Al}$$

The solution is as follows:

$$\text{g Al} = 1.50 \, \cancel{\text{mol } O_2} \left(\frac{4 \, \cancel{\text{mol Al}}}{3 \, \cancel{\text{mol } O_2}} \right) \cdot \left(\frac{27.0 \text{ g Al}}{1 \, \cancel{\text{mol Al}}} \right) = 54.0 \text{ g Al}$$

Notice that the equation coefficients and the molar mass of Al form the conversion factors in the solution to the problem.

TRY IT YOURSELF

In the equation given in the problem above, how many moles of oxygen are needed to produce 51.0 grams of Al_2O_3?

ANSWER

0.750 mol

4.4A MASS-MASS PROBLEMS

PROBLEM

In the equation $CH_4 + 2O_2 \rightarrow CO_2 + 2H_2O$, how many grams of CO_2 are formed when 8.0 grams of CH_4 reacts with an excess of O_2?

SOLUTION

Again, the problem is solved by using FLM. Since equations are given in terms of *moles*, not grams, we must convert to and from moles to arrive at an answer. Here is our solution map:

$$8.0 \text{ g } CH_4 \longrightarrow \text{mol } CH_4 \xrightarrow{\overset{\text{equation}}{\text{coefficients}}} \text{mol } CO_2 \longrightarrow \text{g } CO_2$$

The problem is solved as follows:

$$\text{g } CO_2 = 8.0 \, \cancel{\text{g } CH_4} \cdot \frac{1 \, \cancel{\text{mol } CH_4}}{16.0 \, \cancel{\text{g } CH_4}} \cdot \frac{1 \, \cancel{\text{mol } CO_2}}{1 \, \cancel{\text{mol } CH_4}} \cdot \frac{44.0 \text{ g } CO_2}{1 \, \cancel{\text{mol } CO_2}} = 22 \text{ g } CO_2$$

The term *excess* in the problem statement means that we have more than enough O_2 to react with the CH_4. Therefore, the mass of CO_2 that will be produced depends solely on the quantity of CH_4 present (i.e., 8.0 g). For this reason, CH_4 is called a *limiting reactant*. At the end of the reaction, all of the CH_4 will have been consumed and some of the O_2 will remain unreacted. We will explore limiting reactant problems in more detail later in this section (see pages 96–98).

TRY IT YOURSELF
In the equation $2H_2O_2 \rightarrow 2H_2O + O_2$, how many grams of O_2 will be formed from the decomposition of 17.0 grams of H_2O_2?

ANSWER
8.00 g

4.5A PERCENT YIELD

In the real world, reactions do not always go as planned and the yield produced by a reaction may be *less* than the yield predicted by a pencil-and-paper solution (e.g., a mass-mass problem). Usually, the result of a real-world reaction is reported in terms of *percent yield,* given by the following expression:

$$\text{percent yield} = \frac{\text{actual yield}}{\text{predicted yield}} \times 100$$

PROBLEM
Now, suppose that something goes wrong in the reaction in the "Try It Yourself" immediately preceding, and we produce only 6.00 grams of O_2. What is the percent yield of this reaction?

SOLUTION

$$\% \text{ yield} = \frac{6.00 \text{ g}}{8.00 \text{ g}} \times 100 = 75.0\%$$

TRY IT YOURSELF

In the equation $CH_4 + 2O_2 \rightarrow CO_2 + 2H_2O$:

(a) How many grams of H_2O are expected when 8.00 grams of CH_4 reacts with an excess of O_2?

(b) If only 17.0 grams of H_2O are produced, what is the percent yield of this reaction?

ANSWERS

(a) 18.0 g   (b) 94.4%

4.6A   LIMITING REACTANTS

When we discussed mass-mass problems earlier in this chapter, we introduced the concept of a *limiting reactant*, that is, a reactant that is *not* present in excess and is entirely consumed in a reaction. We now focus on a technique that will allow us to determine *which reactant* is present in a limiting quantity.

Let us consider this equation:

$$N_2 + 3H_2 \rightarrow 2NH_3$$

Suppose 2.0 moles of N_2 and 2.0 moles of H_2 are brought together to react. Which of the reactants (N_2 or H_2) is present in a limiting quantity?

We begin by calculating the *mole ratio* of the reactants (N_2 to H_2) *as they appear in the problem:*

$$\text{mole ratio (problem)} \ \frac{2.0 \text{ mol } N_2}{2.0 \text{ mol } H_2} = 1.0 \ \frac{\text{mol } N_2}{\text{mol } H_2}$$

Now we calculate the *mole ratio* of the reactants *as they appear in the equation*. To do this, we use the equation coefficients:

$$\text{mole ratio (equation)} \ \frac{1 \text{ mol } N_2}{3 \text{ mol } H_2} = 0.33 \ \frac{\text{mol } N_2}{\text{mol } H_2}$$

We now *compare* the two mole ratios. If the mole ratio (problem) is *larger* than the mole ratio (equation), then the *reactant in the numerator is present in excess*; if the mole ratio (problem) is *smaller* than the mole ratio (equation), then the *reactant in the denominator is present in excess*. The reactant that is *not* present in excess is the *limiting reactant*.

In the example shown above, the mole ratio given in the *problem* is larger than the mole ratio given in the *equation*, so the reactant in the *numerator* (N_2) is present in excess. *Therefore, H_2 is the limiting reactant* and all of it will be consumed in the reaction.

The next question we need to ask is: What quantity of N_2 will actually react? We solve this by the factor-label method, using the limiting reactant (H_2) and the equation coefficients:

$$\text{mol } N_2 \text{ reacting} = 2.0 \; \cancel{\text{mol } H_2} \cdot \frac{1 \text{ mol } N_2}{3 \; \cancel{\text{mol } H_2}} = 0.67 \text{ mol } N_2$$

Therefore, we conclude that all 2.0 moles of H_2 but only 0.67 mole of N_2 will react according to the equation given above.

PROBLEM

Consider the reaction $C_3H_8 + 5O_2 \rightarrow 3CO_2 + 4H_2O$, in which 1.5 moles of C_3H_8 and 15 moles of O_2 are brought together.

(a) Calculate the mole ratio C_3H_8/O_2 according to the problem and the equation.

(b) Determine which reactant is present in excess and which is present in limiting quantity.

(c) Calculate the quantity of the excess reactant that reacts with the limiting reactant.

SOLUTIONS

(a) mole ratio (problem) = $\dfrac{1.5 \text{ mol } C_3H_8}{15 \text{ mol } O_2} = 0.10 \; \dfrac{\text{mol } C_3H_8}{\text{mol } O_2}$

  mole ratio (equation) = $\dfrac{1 \text{ mol } C_3H_8}{5 \text{ mol } O_2} = 0.20 \; \dfrac{\text{mol } C_3H_8}{\text{mol } O_2}$

(b) Since the mole ratio of C_3H_8 to O_2 is smaller in the problem than in the equation, it follows that O_2 is present in excess and that C_3H_8 is the limiting reactant.

(c) mol O_2 reacting = $1.5 \; \cancel{\text{mol } C_3H_8} \cdot \dfrac{5 \text{ mol } O_2}{1 \; \cancel{\text{mol } C_3H_8}} = 7.5 \text{ mol } O_2$

As a final example, let us tackle a mass-mass problem that involves limiting reactants.

PROBLEM

How many grams of water can be made according to this equation:

$$2H_2 + O_2 \rightarrow 2H_2O$$

if 100.0 grams of H_2 and 160.0 grams of O_2 are brought together to react?

SOLUTION
Warning! This is not a simple mass-mass problem! We need first to determine the limiting reactant by converting the masses to moles (using the molar masses of H_2 and O_2) and then to calculate the H_2/O_2 mole ratios:

$$\text{moles } H_2 \text{ in problem} = 100. \text{ g } H_2 \cdot \frac{1 \text{ mol } H_2}{2.00 \text{ g } H_2} = 50.0 \text{ mol } H_2$$

$$\text{moles } O_2 \text{ in problem} = 160. \text{ g } O_2 \cdot \frac{1 \text{ mol } O_2}{32.0 \text{ g } O_2} = 5.00 \text{ mol } O_2$$

$$\text{mole ratio (problem)} = \frac{50.0 \text{ mol } H_2}{5.00 \text{ mol } O_2} = 10.0 \frac{\text{mol } H_2}{\text{mol } O_2}$$

$$\text{mole ratio (equation)} = \frac{2 \text{ mol } H_2}{1 \text{ mol } O_2} = 2.00 \frac{\text{mol } H_2}{\text{mol } O_2}$$

Since the mole ratio in the problem is *larger* than the mole ratio in the equation, we conclude that H_2 is present in excess and that O_2 is the limiting reactant. Then, since O_2 is the limiting reactant, we know that all of it will be consumed, and this fact will determine how much water is produced.

We now proceed as we did in the problems just given, using the solution map:

$$\text{mol } O_2 \rightarrow \text{mol } H_2O \rightarrow \text{g } H_2O$$

the equation coefficients, and the molar mass of water:

$$5.00 \text{ mol } O_2 \cdot \frac{2 \text{ mol } H_2O}{1 \text{ mol } O_2} \cdot \frac{18.0 \text{ g } H_2O}{1 \text{ mol } H_2O} = 180. \text{ g } H_2O$$

END-OF-CHAPTER QUESTIONS

Some questions have the symbol "§2" in front of the question number. This symbol means that the question is based on Section II material.

1. The standard of atomic mass is the isotope
   (1) ^{16}S, which is equal to 16.000 u
   (2) ^{32}S, which is equal to 32.000 u
   (3) ^{12}C, which is equal to 12.000 u
   (4) ^{14}N, which is equal to 14.000 u

2. Element X has three naturally occurring isotopes. The table below lists the mass numbers and percent abundances of these isotopes.

Mass Number	Percent Abundance
10	10.0
11	20.0
12	70.0

The average atomic mass of element X is closest to
(1) 11.0 u   (2) 11.6 u   (3) 12.0 u   (4) 12.4 u

3. What is the total mass of oxygen in 1.00 mole of $Al_2(CrO_4)_3$?

(1) 192 g   (2) 112 g   (3) 64.0 g   (4) 48.0 g

4. How many moles of $N_2(g)$ molecules will contain exactly 4.0 moles of nitrogen atoms?
(1) 1.0   (2) 2.0   (3) 3.0   (4) 4.0

5. The gram formula mass of NH_4Cl is closest to
(1) 22.4 g/mol   (2) 28.0 g/mol   (3) 53.5 g/mol   (4) 95.5 g/mol

6. Which is an empirical formula?
(1) P_2O_5 (2) P_4O_6 (3) C_2H_4 (4) C_3H_6

7. Which molecular formula is also an empirical formula?
(1) H_2O_2 (2) H_2O (3) C_2H_6 (4) C_6H_6

8. What mass contains 6.0×10^{23} atoms?
(1) 6.0 g of C   (2) 16 g of S
(3) 3.0 g of He   (4) 28 g of Si

9. How many moles of *hydrogen atoms* are in 1 mole of $C_6H_{12}O_6$ molecules?
(1) $24(6.0 \times 10^{23})$ (2) $12(6.0 \times 10^{23})$
(3) 24   (4) 12

10. What is the molar mass of $CuSO_4 \cdot 5H_2O$?
(1) 160. g/mol   (2) 178 g/mol   (3) 186 g/mol   (4) 250. g/mol

11. What is the molecular formula of a compound whose empirical formula is CH_4 and whose molar mass is 16 grams per mole?

    (1) CH_4 (2) C_2H_4 (3) C_4H_8 (4) C_8H_{18}

12. The percent by mass of hydrogen in NH_3 is closest to

    (1) $\frac{17}{1} \times 100$ (2) $\frac{17}{3} \times 100$ (3) $\frac{1}{17} \times 100$ (4) $\frac{3}{17} \times 100$

13. What is the approximate percent by mass of sulfur in H_2SO_4? [formula mass = 98]

    (1) 16   (2) 33   (3) 65   (4) 98

14. A hydrated salt is a solid that includes water molecules within its crystal structure. A student heated a 9.10-gram sample of a hydrated salt to a constant mass of 5.41 grams. What percent by mass of water did the salt contain?

    (1) 3.69 %   (2) 16.8%   (3) 40.5%   (4) 59.5%

§2 15. A compound consists of 85 percent silver and 15 percent fluorine by mass. What is its empirical formula?

    (1) AgF (2) AgF_2 (3) Ag_2F (4) Ag_6F

16. The percent by mass of carbon in CO_2 is equal to

    (1) $44/12 \times 100$ (2) $12/44 \times 100$
    (3) $28/12 \times 100$ (4) $12/28 \times 100$

§2 17. A sample of a compound contains 24 grams of carbon and 64 grams of oxygen. What is the empirical formula of this compound?

    (1) CO (2) CO_2 (3) C_2O_2 (4) C_2O_4

18. The percent by mass of oxygen in $Ca(OH)_2$ (molar mass = 74 g/mol) is closest to

    (1) 16   (2) 22   (3) 43   (4) 74

19. What is the molecular formula of a compound that has a molar mass of 92 grams per mole and an empirical formula of NO_2?

    (1) NO_2 (2) N_2O_4 (3) N_3O (4) N_4O_8

§2 20. At STP, 32 grams of O_2 will occupy the same volume as

    (1) 64 g of H_2 (2) 32 g of SO
    (3) 8.0 g of CH_4 (4) 4.0 g of He

21. The percent by mass of oxygen in MgO (molar mass = 40 g/mol) is closest to
(1) 16   (2) 24   (3) 40   (4) 60

§2 **22.** What is the empirical formula of an ion whose composition by mass is 57.14. percent sulfur and 42.86. percent oxygen?
(1) SO_2^{2-} (2) SO_3^{2-} (3) $S_2O_3^{2-}$ (4) SO_4^{2-}

23. A compound has the empirical formula NO_2. Its molecular formula could be
(1) NO_2 (2) N_2O (3) N_4O_2 (4) N_4O_4

24. Which represents the greatest mass of chlorine?
(1) 1 mole of Cl   (2) 1 atom of Cl
(3) 1 gram of Cl   (4) 1 molecule of Cl

25. What is the percent by mass of hydrogen in CH_3COOH (molar mass = 60 g/mol)?
(1) 1.7   (2) 5.0   (3) 6.7   (4) 7.1

26. An example of an empirical formula is
(1) C_2H_2 (2) H_2O_2 (3) C_2Cl_2 (4) $CaCl_2$

27. A 254-gram sample of I_2 contains approximately the same number of molecules as
(1) 14 g of N_2 (2) 2.0 g of H_2
(3) 36 g of H_2O (4) 40 g of Ne

28. Given this reaction

$$CH_4(g) + 2O_2(g) \rightarrow CO_2(g) + 2H_2O(g)$$

How many moles of oxygen are needed for the complete combustion of 3.0 moles of $CH_4(g)$?
(1) 6.0   (2) 2.0   (3) 3.0   (4) 4.0

29. Given this equation:

$$6CO_2 + 6H_2O \rightarrow C_6H_{12}O_6 + 6O_2$$

What is the total number of moles of water needed to make 2.5 moles of $C_6H_{12}O_6$?
(1) 2.5   (2) 6.0   (3) 12   (4) 15

30. Given this equation:

$$2C_2H_2(g) + 5O_2(g) \rightarrow 4CO_2(g) + 2H_2O(g)$$

How many moles of oxygen are required to react with 1.0 mole of C_2H_2?

(1) 2.5   (2) 2.0   (3) 5.0   (4) 10.

§2 **31.** Given this reaction:

$$S + O_2 \rightarrow SO_2$$

What is the total number of grams of oxygen needed to react completely with 2.0 moles of sulfur?

(1) 20   (2) 32   (3) 64   (4) 128

32. In the reaction:

$$Zn + 2HCl \rightarrow ZnCl_2 + H_2$$

how many moles of hydrogen will be formed when 4 moles of HCl is consumed?

(1) 6   (2) 2   (3) 8   (4) 4

§2 **33.** Given this reaction:

$$2H_2 + O_2 \rightarrow 2H_2O$$

The total number of grams of O_2 needed to produce 54 grams of water is

(1) 36   (2) 48   (3) 61   (4) 75

34. Given this reaction:

$$2Al + 3H_2SO_4 \rightarrow 3H_2 + Al_2(SO_4)_3$$

The total number of moles of H_2SO_4 needed to react completely with 5.0 moles of aluminum is

(1) 2.5   (2) 5.0   (3) 7.5   (4) 9.0

§2 **35.** According to the reaction $H_2 + Cl_2 \rightarrow 2HCl$, the production of 2.0 moles of HCl would require 70. grams of Cl_2 and

(1) 1.0 g of H_2 (2) 2.0 g of H_2
(3) 3.0 g of H_2 (4) 4.0 g of H_2

§2 **36.** Given this reaction:

$$Cu + 4HNO_3 \rightarrow Cu(NO_3)_2 + 2H_2O + 2NO_2$$

What is the total mass of H_2O produced when 32 grams of copper is completely consumed?
(1) 9.0 g   (2) 18 g   (3) 36 g   (4) 72 g

§2 **37.** Given this reaction:

$$2C_2H_6 + 7O_2 \rightarrow 4CO_2 + 6H_2O$$

What is the total number of CO_2 molecules produced when 1 mole of C_2H_6 is consumed?
(1) 6.02×10^{23} (2) $2(6.02 \times 10^{23})$
(3) $3(6.02 \times 10^{23})$ (4) $4(6.02 \times 10^{23})$

§2 **38.** Given this reaction:

$$4Al + 3O_2 \rightarrow 2Al_2O_3$$

How many moles of Al_2O_3 will be formed when 27 grams of aluminum reacts completely with O_2?
(1) 1.0   (2) 2.0   (3) 0.50   (4) 4.0

39. The percent, by mass, of water in $BaCl_2 \cdot 2H_2O$ (molar mass = 243 g/mol) is equal to
(1) $18/243 \times 100$ (2) $36/243 \times 100$
(3) $243/18 \times 100$ (4) $243/36 \times 100$

40. What species contains the greatest percent by mass of hydrogen?
(1) OH   (2) H_2O (3) H_3O^+ (4) H_2O_2

41. In which list are the elements arranged in order of increasing atomic mass?
(1) Cl, K, Ar   (2) Fe, Co, Ni
(3) Te, I, Xe   (4) Ne, F, Na

42. What is the percent by mass of oxygen in H_2SO_4? [formula mass = 98]
(1) 16%   (2) 33%   (3) 65%   (4) 98%

43. A hydrate is a compound that includes water molecules within its crystal structure. During an experiment to determine the percent by mass of water in a hydrated crystal, a student found the mass of the hydrated crystal to be 4.10 grams. After heating to constant mass, the mass was 3.70 grams. What is the percent by mass of water in this crystal?
(1) 90%   (2) 11%   (3) 9.8%   (4) 0.40%

44. Given the equation:
$$2C_2H_2(g) + 5O_2(g) \rightarrow 4CO_2(g) + 2H_2O(g)$$
How many moles of oxygen are required to react completely with 1.0 mole of C_2H_2?
(1) 2.5   (2) 2.0   (3) 5.0   (4) 10

45. The atomic mass of an element is calculated using the
(1) atomic number and the ratios of its naturally occurring isotopes
(2) atomic number and the half-lives of each of its isotopes
(3) masses and the ratios of its naturally occurring isotopes
(4) masses and the half-lives of each of its isotopes

46. Given the reaction:
$$PbCl_2(aq) + Na_2CrO_4(aq) \rightarrow PbCrO_4(s) + 2NaCl(aq)$$
What is the total number of moles of NaCl formed when 2 moles of Na_2CrO_4 react completely?
(1) 1 mole   (2) 2 moles
(3) 3 moles   (4) 4 moles

47. In which compound is the percent by mass of oxygen greatest?
(1) BeO   (2) MgO
(3) CaO   (4) SrO

48. The formula mass of a compound is the
(1) sum of the atomic masses of its atoms
(2) sum of the atomic numbers of its atoms
(3) product of the atomic masses of its atoms
(4) product of the atomic numbers of its atoms

49. The atomic mass of an element is the weighted average of the atomic masses of
(1) the least abundant isotopes of the element
(2) the naturally occurring isotopes of the element
(3) the artificially produced isotopes of the element
(4) the natural and artificial isotopes of the element

Constructed-Response Questions

1. The grid below is based on the mole concept developed in this chapter and is partially filled in. Complete the grid by using the data given in the grid.

Substance	Molar Mass ($\mathcal{M}$) (g/mol)	Mass of Substance (m) (g)	Moles of Substance (n) (mol)	Number of Particles of Substance (N)
O_3		24		
NH_3		170		
F_2		38		
CO_2			0.10	
NO_2			0.20	
Ne				1.5×10^{23}
N_2O				1.2×10^{24}
***		8.5		3.0×10^{23}

*** One of the seven substances listed above.

2. Calculate the molar mass of each of the following:
   (a) H_2SO_3 (b) $C_{21}H_{30}O_2$ (the active ingredient in marijuana)

3. In a 64-gram sample of SO_2:
   (a) How many moles are present?
   (b) How many molecules are present?

4. Calculate the mass of 3.33 moles of C_2H_5OH ($\mathcal{M} = 46.1$ g/mol).

§2 5. Calculate the empirical formula of a compound containing 22.7 percent sodium, 21.6 percent boron, and 55.7 percent oxygen by mass.

§2 6. What is the molecular formula of a binary compound of sulfur and chlorine if it contains 47.5 percent sulfur by mass and its molar mass is 135.1 grams per mole?

§2 7. The net process of photosynthesis may be represented by this balanced equation:

$$6CO_2 + 6H_2O \rightarrow C_6H_{12}O_6 + 6O_2$$

   (a) How many grams of $C_6H_{12}O_6$ are produced when 0.131 mole of CO_2 is consumed?
   (b) How many grams of O_2 can be obtained when 65.0 grams of CO_2 is consumed?

§2 **8.** Given the equation

$$2H_2 + O_2 \rightarrow 2H_2O$$

how many grams of water can be produced from the reaction of 77.7 grams of oxygen with 15.0 grams of hydrogen?

§2 **9.** Given this equation:

$$N_2 + 3H_2 \rightarrow 2NH_3$$

if 27.7 grams of ammonia is produced by the reaction of 18.4 grams of hydrogen with an excess of nitrogen, what is the percent yield of the reaction?

10. (a) Calculate the molar mass of $Mg(OH)_2$
    (b) How many moles of $Mg(OH)_2$ are present in an 8.40-gram sample?

11. The table below gives information about two isotopes of element X.

Isotope	Mass	Relative Abundance
X-10	10.01	19.91%
X-11	11.01	80.09%

Calculate the average atomic mass of element X. (Show a correct numerical setup, and express your answer to the correct number of significant figures.)

Base your answers to questions 12 and 13 on the information below. Gypsum is a mineral that is used in the construction industry to make drywall (sheet-rock). The chemical formula for this hydrated compound is $CaSO_4 \cdot 2H_2O$. A hydrated compound contains water molecules within its crystalline structure. Gypsum contains 2 moles of water for each mole of calcium sulfate.

12. Calculate the gram formula mass of $CaSO_4 \cdot 2H_2O$.

13. Calculate the percent composition by mass of water in this compound.

Base your answers to questions 14–16 on the information below and on your knowledge of chemistry. The two naturally occurring isotopes of antimony are Sb-121 and Sb-123. The table below shows the atomic mass and the percent natural abundance for these isotopes.

Naturally Occurring Isotopes of Antimony

Isotope	Atomic Mass (u)	Natural Abundance (%)
Sb-121	120.90	57
Sb-123	122.90	43

Antimony and sulfur are both found in the mineral stibnite, Sb_2S_3. To obtain antimony, stibnite is roasted (heated in air), producing oxides of antimony and sulfur. The following unbalanced equation represents one of the reactions that occurs during the roasting:

$$Sb_2S_3(s) + O_2(g) \rightarrow Sb_2O_3(s) + SO_2(g)$$

14. Determine the percent composition by mass of antimony in stibnite (molar mass = 340. g/mol).

15. Show a correct numerical setup for calculating the atomic mass of antimony.

16. Balance the equation given above using smallest whole-number coefficients.

The answers to these questions are found in Appendix 3.

<table>
<tr><td>Chapter
Five</td><td></td></tr>
</table>

Chapter Five

ENERGY AND CHEMICAL REACTIONS

KEY IDEAS

This chapter focuses on the role energy plays in chemical reactions and the factors that determine whether a chemical process will occur under a given set of conditions.

KEY OBJECTIVES
At the conclusion of this chapter you will be able to:
- Define the terms *system* and *surroundings* as they relate to chemical processes.
- Define the terms *internal energy* and *heat*.
- Distinguish between *heat* and *temperature*.
- Distinguish between exothermic and endothermic reactions.
- Define the term *specific heat*, and use specific heats to solve calorimetry problems.
- Relate the first law of thermodynamics to the law of conservation of energy.
- Define the term *heat of reaction*, and solve problems involving heats of reaction.
- Define the terms *standard heat of formation* and *formation reaction*, and use the appropriate reference tables to solve problems related to the standard heat of formation.
- Interpret a potential energy diagram.
- Define the term *activation energy*.
- Define the term *spontaneous reaction*, and name and describe the factors that drive spontaneous reactions.
- Define the term *entropy*, and predict whether a given reaction leads to an increase or a decrease in entropy.

SECTION I—BASIC (REGENTS-LEVEL) MATERIAL

NYS REGENTS CONCEPTS AND SKILLS

Note: By the time you have finished Section I, you should have mastered the concepts and skills listed below. The Regents chemistry examination will test your knowledge of these items and your ability to apply them.

Concepts are the *basic ideas* that form the body of the Regents chemistry course (what you need to know!).

Skills are the *activities* that demonstrate your mastery of these concepts (how you show that you know them!).

Following each concept or skill is a page reference (given in parentheses) to this chapter.

5.1 Concepts:
  • Heat is a transfer of energy (usually thermal energy) from a body of higher temperature to a body of lower temperature. (Page 112)
  • Thermal energy is the energy associated with the random motion of atoms and molecules. (Page 112)

  Skills:
  • Distinguish between heat energy and temperature. (Page 112)
  • Calculate the heat involved in a temperature change. (Pages 112–113)

5.2 Concepts:
  • Chemical and physical changes can be exothermic or endothermic. (Page 112)
  • Energy released or absorbed during a chemical reaction (the heat of reaction) is equal to the difference between the potential energy of the products and the potential energy of the reactants. (Page 113)

  Skill:
  Distinguish between endothermic and exothermic reactions in terms of:
  (a) a reaction equation (Page 114);
  (b) ΔH (Pages 113–114);
  (c) experimental data (Page 114).

5.3 Concept:
  Energy released or absorbed by a chemical reaction can be represented by a potential energy diagram. (Pages 114–116)

  Skill:
  Read and interpret potential energy diagrams in terms of:
  (a) potential energies of reactants and products (Page 114);
  (b) activation energy (Page 114);
  (c) heat of reaction (ΔH) (Page 114).

5.4 Concept:
  Entropy is a measure of the randomness or disorder of a system. A system with greater disorder has greater entropy. (Pages 117–118)

Skill:
Compare the entropies of solids, liquids, and gases.
(Pages 117–118)

5.5 Concept:
Systems in nature tend to undergo changes toward lower energy
and higher entropy. (Pages 117–118)

5.1 ENERGY AND ITS MEASUREMENT

In Chapter 1, we related the concept of *energy* to the concept of *mechanical work*. In this chapter, we will study more closely the concept of energy and its role in chemistry. First, we need to define some basic terms that will be used throughout the chapter.

The part of the universe that a chemist chooses to study is called a **system**. The rest of the universe is known as the **surroundings**. For example, suppose you decide to study the melting of ice in a beaker, as shown in the accompanying diagram.

Surroundings

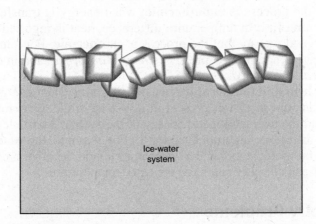

Ice-water
system

The ice and water constitute the system; the beaker, the air, and everything else constitute the surroundings.

In chemical processes, a system can *interact* with its surroundings in a number of ways, as shown in the following examples:

• A beaker of liquid absorbs heat from a burner; in this example, *energy* is transferred between the system (the beaker of liquid) and its surroundings.
• A cylinder gas is compressed by a piston; in this example, *work* is done *on* the system (the gas) by its surroundings.

111

• Water is poured into a solution; in this example, *mass* has been transferred to the system (the solution) from its surroundings.

As we see from the examples given above, energy transfers involve transfers of heat and/or work. These transfers affect the *total* energy within a system, known as the **internal energy** (E), which has two main components: **thermal energy**, the energy associated with random molecular motions, and **chemical energy**, the energy associated with chemical bonds and attractions between the particles of a system.

Since energy and work are so closely related, the same unit is used to measure both quantities. In the SI system, this unit is the **joule** (J); its multiple is the **kilojoule** (kJ). In this book, we will use the joule and the kilojoule *exclusively*.

Heat

Heat is the energy transferred between a system and its surroundings as a result of a *temperature difference*. In the first example given above, heat passes from the burner to the beaker of liquid because the temperature of the burner is higher than the temperature of the beaker and liquid. Heat is represented by the symbol q. When energy is transferred *into* a system as a result of a temperature difference, heat is reported as a *positive* number, and we say that the process is **endothermic**; when energy is transferred *out of* a system as a result of a temperature difference, heat is reported as a *negative* number, and we say that the process is **exothermic**. For simplicity, we will state that in endothermic processes *heat is absorbed*, and in exothermic processes *heat is released*.

Note that **heat and temperature are not the same**. Heat energy is related to the total amount of matter present in a system, while temperature is not: it is related to the average kinetic energy of the system's particles. (We will discuss this difference again in Chapter 6.) For example, the water in a lake can transfer far more heat than the water in a cup; on the other hand, both the lake and the cup of water can have the same temperature.

Calorimetry Problems

To measure the heat absorbed or released by a substance that is not undergoing a phase change, we need to know three things about the substance: its identity (what it is!), its mass, and the change in its temperature. These three quantities can be combined into a simple equation that will allow us to solve many problems involving heat. In mathematical form, the equation is as follows:

$$q = c_p \cdot m \cdot \Delta T_C$$

This equation (in a slightly different form) is found on Reference Table T in Appendix 1 of this book. In this equation, where q represents the amount of heat, in joules; m is the mass of the substance, in grams; and ΔT_C is the temperature change of the substance, in Celsius degrees (C°). (*Note*: The unit "C°" stands for temperature *changes*, while the unit "°C" represents temperature *readings*.) The symbol c_p stands for the **specific heat capacity** (or more simply, the **specific heat**) of the substance. The substance is identified by the value of its specific heat. The smaller the specific heat of a substance, the more readily the substance will change temperature as it absorbs or releases heat.

Reference Table B in Appendix 1 of this book lists the specific heat of water as 4.2 joules per gram · Kelvin. *Note that a Kelvin (K) and a Celsius degree (C°) are completely equivalent.*

PROBLEM
How much heat is needed to raise the temperature of 20.0 grams of liquid water from 5.0°C to 20.0°C?

SOLUTION
$q = c_p \cdot m \cdot \Delta T_C$
$q = \text{???}$ (This is the quantity we are asked to determine.)
$c_p = 4.2 \text{ J/g} \cdot \text{K}$ (See Reference Table B in Appendix 1.)
$m = 20.0 \text{ g}$
$\Delta T_C = T_{final} - T_{initial} = 20.0°C - 5.0°C = 15.0C°$

We now solve the problem by direct substitution of the values into the equation:

$$q = c_p \cdot m \cdot \Delta T_C$$

$$= \frac{4.2 \text{ J}}{\text{g} \cdot K} \cdot 20.0 \text{ g} \cdot 15.0 C°$$

$$= 1260 \text{ J } (1300 \text{ J to 2 significant figures.})$$

5.2 HEAT OF REACTION

When applied to a chemical reaction, ΔH is known as the **heat of reaction**. It is the difference in enthalpy (potential energy) between the products of the reaction and the reactants:

$$\Delta H = H_{products} - H_{reactants}$$

In an *endothermic* reaction, the enthalpy of the products is *larger* than the enthalpy of the reactants and the sign of ΔH is *positive*. In an *exothermic* reaction, the enthalpy of the products is *smaller* than the enthalpy of the reactants and the sign of ΔH is *negative*.

We indicate ΔH for an exothermic reaction in either of two ways:

$$2CO(g) + O_2(g) \rightarrow 2CO_2(g) \qquad \Delta H = -566.0 \text{ kJ}$$

or

$$2CO(g) + O_2(g) \rightarrow 2CO_2(g) + 566.0 \text{ kJ}$$

The two forms mean exactly the same thing: 2 moles of CO gas combine with 1 mole of O_2 gas to produce 2 moles of CO_2 gas and release 566.0 kilojoules of heat.

To illustrate an endothermic reaction, we need only write the reactions given above in reverse:

$$2CO_2(g) \rightarrow 2CO(g) + O_2(g) \qquad \Delta H = +566.0 \text{ kJ}$$

or

$$566.0 \text{ kJ} + 2CO_2(g) \rightarrow 2CO(g) + O_2(g)$$

This example also illustrates the important principle that a reaction that is exothermic in one direction is endothermic in the opposite direction. Reference Table I in Appendix 1 lists selected heats of reaction.

5.3 POTENTIAL ENERGY DIAGRAMS

A **potential energy diagram** is used to illustrate the progress of a chemical reaction and to provide qualitative information about energy changes within a reaction. We need to define one more term: The **activation energy** (or **energy of activation**), represented by the symbol E_a, is the minimum energy needed by the reactants in order for a reaction to occur.

Let us consider the *exothermic* reaction

$$\text{Reactants} \rightarrow \text{Products} + \text{Heat}$$

and the accompanying potential energy diagram for this reaction. We note that the potential energy of the system is plotted along the y-axis, and a quantity called the *reaction coordinate* (which monitors the progress of the reaction) along the x-axis.

Reaction Profile of an Exothermic Reaction

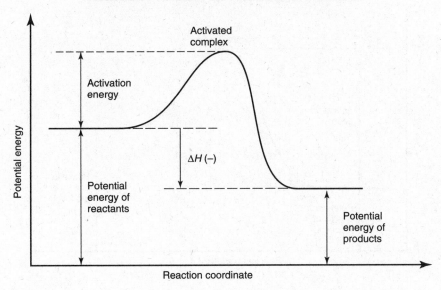

Let us trace along the curve in the diagram and describe the events that occur during the course of this reaction:

- At the beginning, the reactants contain a specific amount of energy (potential energy of reactants).
- Additional energy is *absorbed* by the reactants and serves to break certain bonds and initiate the reaction (activation energy).
- As the reactants absorb energy, they are transformed into an intermediate known as the *activated complex*.
- As the activated complex is converted into products, new bonds are formed and energy is released.
- The products also contain a specific amount of energy (potential energy of products).
- Since *more* energy is released than absorbed, the products are at a *lower* energy state than the reactants and the overall reaction is exothermic.
- The *difference* between the energy of the products and the energy of the reactants is the heat of reaction, ΔH.

Many exothermic reactions are *self-sustaining*; that is, they continue to occur because the heat they liberate provides the activation energy for the reactants. A burning match is an example of a self-sustaining reaction.

Now let us consider the *endothermic* reaction

$$\text{Heat} + \text{Reactants} \rightarrow \text{Products}$$

Reaction Profile of an Endothermic Reaction

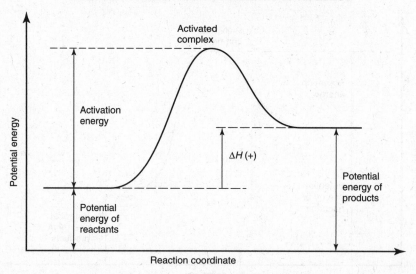

and the accompanying potential energy diagram for this reaction. The sequence of events in an endothermic reaction is *exactly the same* as the sequence of events in an exothermic reaction. At the end of the reaction, however, *less* energy is released than absorbed and the products are at a *higher* energy state than the reactants.

5.4 SPONTANEOUS REACTIONS

When we think of the word *spontaneous* in everyday language, we tend to think of an event happening by itself, without outside interference. In chemistry, nothing happens by itself!

The term **spontaneous reaction** refers to a reaction that can occur under a given set of conditions without the application of external work. During the course of a spontaneous reaction, heat will be absorbed from, or released to, the surroundings without any outside assistance. For example, at $+25°C$ and 1 atmosphere, ice is able to melt but water is *not* able to freeze.

$H_2O(s) \rightarrow H_2O(\ell)$ is a spontaneous reaction at $+25°C$ and 1 atmosphere.
$H_2O(\ell) \rightarrow H_2O(s)$ is *not* a spontaneous reaction at $+25°C$ and 1 atmosphere.

At $-25°C$ and 1 atmosphere, the opposite is true:

$H_2O(s) \rightarrow H_2O(\ell)$ is *not* a spontaneous reaction at $-25°C$ and 1 atmosphere.
$H_2O(\ell) \rightarrow H_2O(s)$ is a spontaneous reaction at $-25°C$ and 1 atmosphere.

We note that, when a reaction is spontaneous in one direction, it is not spontaneous in the reverse direction. The fact that a reaction is not spontaneous does *not* mean that it can never occur. If external work is applied to a system, the reaction can be *forced* to take place. For example, water will not decompose into hydrogen and oxygen spontaneously. The application of a direct electric current, however, will force this decomposition.

All spontaneous reactions are able to contribute a portion of their energy to perform *useful work*. For example, certain reactions can be used to produce electric energy that will drive a mechanical device, such as a CD player. The portion of energy that is available for useful work is known as the *Gibbs free energy* and is symbolized by the letter *G*. As a spontaneous reaction proceeds, its capacity to do further work *decreases* (i.e., its Gibbs free energy is lowered). After a time, the system is no longer able to convert any energy to work and the reaction ceases to be spontaneous. At this point, we say that the system is in a state of *equilibrium*. A practical example is the operation of a battery: as it functions, its capacity decreases until it "dies" and can function no more.

What factors enable a reaction to occur spontaneously under a given set of conditions? In nature, two fundamental tendencies—the *energy factor* and the *disorder factor*—govern every system.

The Energy Factor

We know from experience that systems tend to change from higher to lower energy states. Water drops from famous Niagara Falls from a higher to a lower potential energy state under the influence of gravity. Many chemical reactions release energy, with the result that the products have less energy than the reactants.

From a chemical point of view, there is a strong tendency for a reaction to occur spontaneously when it is *exothermic*: that is, when ΔH is *negative*.

The Disorder Factor: Entropy

We also know from experience that systems tend to reach a state of higher disorder: smoke spontaneously diffuses through the air, sugar dissolves uniformly in water, and your bedroom becomes chaotic after only a few days of neglect!

Disorder is measured by a quantity called **entropy**, and it is represented by the symbol *S*. When a system achieves greater disorder, its entropy increases.

Events Leading to Higher Disorder

Chemically, a system can reach a state of higher disorder by means of a number of events or circumstances:

- The temperature of the system increases, leading to an increase in the random motion of the particles present.
- There is a phase change: gases have the most disorder; solids, the least.
- The products of a chemical reaction are simpler in structure than the reactants.
- There are more products than reactants in a chemical reaction.
- A substance is placed in solution.

As an example, let us consider the reaction

$$2KClO_3(s) \rightarrow 2KCl(s) + 3O_2(g)$$

This reaction leads to an increase in disorder because the products are simpler in structure, there are more of them, and one product is a gas.

From a chemical point of view, there is a strong tendency for a reaction to occur spontaneously when its entropy *increases*: that is, when ΔS is *positive*.

PROBLEM
Predict whether each of the following changes is accompanied by an increase or a decrease in disorder:
(a) $NH_4HS(s) \rightarrow NH_3(g) + H_2S(g)$
(b) $2H_2(g) + O_2(g) \rightarrow 2H_2O(g)$
(c) $C_6H_{12}O_6(s) \rightarrow C_6H_{12}O_6(aq)$
(d) $He(g) [300 \, K] \rightarrow He(g) [200 \, K]$

SOLUTIONS
(a) Increase: The solid is converted into two gases.
(b) Decrease: The number of particles is decreased from 3 mol to 2 mol.
(c) Increase: The solid is dissolved in water.
(d) Decrease: The temperature is lowered from 300 K to 200 K.

SECTION II—ADDITIONAL MATERIAL

5.1A   ADDITIONAL CALORIMETRY PROBLEMS

PROBLEM

How much heat is released by 200.0 grams of solid aluminum as it cools from 200.0°C to 150.0°C?

SOLUTION
Use Reference Table W-1 in Appendix 2, which contains a more complete list of specific heats.

$$q = c_p \cdot m \cdot \Delta T_C$$

$q = {}$??? (This is the quantity we are asked to determine.)

$c_p = 0.897 \text{ J/g} \cdot \text{C}°$ (aluminum)

$m = 200.0 \text{ g}$

$\Delta T_C = T_{final} - T_{initial} = 150.0°\text{C} - 200.0°\text{C} = -50.0\text{C}°$

$$q = c_p \cdot m \cdot \Delta T_C$$

$$= \left(0.897 \frac{\text{J}}{\cancel{g} \cdot \cancel{C}°}\right) \cdot (200.0 \ \cancel{g}) \cdot (-50.0 \cancel{C}°) = -8970 \text{ J}$$

That is, 8970 J of heat is released.

TRY IT YOURSELF

How much heat is absorbed by 5.00 grams of copper when it is heated from 0.0°C to 200.0°C?

ANSWER

385 J

5.2A TRANSFER OF ENERGY AND EQUILIBRIUM TEMPERATURE

What happens if two objects with different temperatures are brought into contact? Experience tells us that heat is always transferred from the hotter object to the colder one until they both reach the same final temperature. When this occurs, we say that the objects have reached *thermal equilibrium*, and the final temperature is known as the *equilibrium temperature*. This will always occur; it does not matter what the objects are composed of or how massive they are. (The compositions of the objects and their masses will determine the value of the final temperature, however.)

Problems Involving Transfer of Energy

Problems involving heat transfer also utilize the relationship $q = c_p \cdot m \cdot \Delta T_C$. However, this relationship must be used twice—once for the object that releases ("loses") heat, and once for the object that absorbs ("gains") it. Because heat is assumed to be conserved in these problems, the heat "lost" by the hotter object is set equal to the heat energy "gained" by the colder object. The equation looks like this:

$$|q|_{lost} = |q|_{gained}$$
$$q = c_p \cdot m \cdot |\Delta T_C|_{lost} = c_p \cdot m \cdot |\Delta T_C|_{gained}$$

The vertical bars (| |) mean that we are interested only in the (absolute) value of the quantity, not in whether it is positive or negative. The quantity $|\Delta T_C|$ is evaluated by subtracting the smaller temperature from the larger one.

PROBLEM
A 10.0-gram block of copper at 60.0°C is placed in contact with an identical 10.0-gram block of copper at 20.0°C. What is the equilibrium temperature of the two blocks?

SOLUTION
We could guess at the answer to this (40.0°C—the midpoint temperature between 20.0°C and 60.0°C), and we would be correct! But let us see why this is so. The two blocks have equal masses (10.0 g) and are composed of the same substance (copper). The change in heat energy should therefore affect the temperature of each block equally (but in opposite directions). If both blocks must reach the same final temperature, the midpoint temperature of 40.0°C is the only temperature that satisfies these requirements. Now let us *prove* that this is the case by solving the pair of equations given above:

$$|q|_{lost} = |q|_{gained}$$
$$q = c_p \cdot m \cdot |\Delta T_C|_{lost} = c_p \cdot m \cdot |\Delta T_C|_{gained}$$

Hotter Object (loses heat)
$c_p = 0.385$ J/g · C°
$m = 10.0$ g
$|\Delta T_C|_{lost} = 60.0°C - T_{final}$

Colder Object (gains heat)
$c_p = 0.385$ J/g · C°
$m = 10.0$ g
$|\Delta T_C|_{gained} = T_{final} - 20.0°C$

Substitution into the equations yields:

$$(0.385 \text{ J/g} \cdot C°) \cdot (10.0 \text{ g}) \cdot (60.0°C - T_{final}) = (0.385 \text{ J/g} \cdot C°) \cdot (10.0 \text{ g}) \cdot (T_{final} - 20.0°C)$$

Simplifying yields: $2T_{final} = 80.0°C$
And our answer is $T_{final} = 40.0°C$

TRY IT YOURSELF
Calculate the equilibrium temperature when a 5.00-gram sample of lead at 20.0°C is placed in contact with a 10.0-gram block of silver at 40.0°C. (Use Reference Table W-1.)

ANSWER
35.7°C

Note that the equilibrium temperature is closer to the initial temperature of the silver (40.0°C) than to that of the lead (20.0°C). This is due to the larger mass of the silver block and to the higher specific heat of silver.

5.3A THE ROLE OF ENERGY IN CHEMICAL REACTIONS

If we are to understand anything about how chemical reactions occur, we must first recognize the fundamental role played by energy.

The First Law of Thermodynamics

The role of energy in a chemical reaction is embodied in a fundamental rule that relates *internal energy*, *heat*, and *work*. This rule is known as the **first law of thermodynamics** (or, more briefly, as the *first law*). The first law is represented by this equation:

$$\Delta E = q + w$$

This equation tells us that the *change* in the *internal energy* of a system is equal to the sum total of the *heat absorbed* by the system and the *work done on* the system. In this situation, the heat and work are expressed as *positive* numbers; if heat is *released*, or work is *done by* the system, these two quantities will be *negative* numbers.

The first law is another way of expressing the *law of conservation of energy*: If a *system* gains (loses) a given amount of energy, then the energy lost (gained) by the *surroundings* must be *exactly* the same amount; that is, *the energy of the universe is always constant*. We can express this idea by the equation

$$\Delta E_{system} + \Delta E_{surroundings} = \Delta E_{universe} = 0$$

Throughout history, many attempts have be made to circumvent the first law. Numerous patent applications have been made for machines that promise to deliver more energy than the amount with which they were supplied. As of this date, *not one patent has been granted* for a machine of this type. We also hear of "foolproof" reducing diets that promise weight loss without calorie-counting or exercise. According to the first law, this is utter nonsense! If the quantity of food energy ingested is greater than the quantity of energy expended through exercise, the excess energy is stored as—you guessed it—extra pounds.

How does a chemist *measure* ΔE? When a reaction occurs in a *sealed* container so that the *volume is constant*, the heat transferred under these conditions (q_v) is equal to the internal energy change:

$$\Delta E = q_v$$

If heat is absorbed by the system at constant volume, the internal energy will *increase* ($\Delta E > 0$); if heat is released by the system at constant volume, the internal energy will *decrease* ($\Delta E < 0$).

Under laboratory conditions, however, most reactions take place in *open* containers at nearly *constant pressure*. In this case, the heat transferred (q_p) is equal to a change in a quantity that is closely related to the internal energy and is known as **enthalpy** (H). (Sometimes, the term *heat content* is used as a synonym for enthalpy, but this is not really accurate, since it implies that heat is a *substance*, rather than a *transfer of energy*.) The equation relating an enthalpy change to the heat transferred is

$$\Delta H = q_p$$

If heat is absorbed by the system at constant pressure, the enthalpy will *increase* ($\Delta H > 0$); if heat is released by the system at constant pressure, the enthalpy will *decrease* ($\Delta H < 0$). In this book, we will assume that all chemical reactions occur under conditions of *constant pressure*.

5.4A   ADDITIONAL ASPECTS OF HEATS OF REACTION

When a heat of reaction is measured under *standard thermodynamic conditions* (pressure = 100.00 kPa; all reactants and products are present in their pure states), we use the special symbol $\Delta H°$. Generally a reference temperature of 298.15 K (25°C) is also used. Standard heats of reaction are usually measured in kilojoules.

PROBLEM
How much heat is released when 0.5000 mole of $CH_4(g)$ is reacted with O_2 if $\Delta H° = -890.4$ kilojoule per mole of CH_4 reacted?

SOLUTION
Since 1.000 mol of $CH_4(g)$ *releases* 890.4 kJ of heat, 0.5000 mol of $CH_4(g)$ will release *half* this amount, or 445.2 kJ.

TRY IT YOURSELF
How much heat is absorbed by 0.300 mole of $KNO_3(s)$ when it dissolves in H_2O if $\Delta H° = +34.89$ kilojoules per mole of KNO_3 dissolved?

ANSWER
10.5 kJ

Hess's Law

In the nineteenth century, the Swiss chemist G.H. Hess found that, when two or more chemical reactions are combined ("added"), the heat of reaction of the *combined* reaction is simply the *sum* of the *individual* heats of reaction. The example below illustrates this technique:

$$N_2(g) + O_2(g) \rightarrow 2NO(g) \qquad \Delta H_1 = +180 \text{ kJ}$$

$$\underline{2NO(g) + O_2(g) \rightarrow 2NO_2(g) \qquad \Delta H_2 = -112 \text{ kJ}}$$

$$N_2(g) + 2O_2(g) \rightarrow 2NO_2(g) \qquad \Delta H_{rxn} = \Delta H_1 + \Delta H_2 = +180 \text{ kJ} - 112 \text{ kJ}$$
$$= +68 \text{ kJ}$$

The letters "rxn" form a widely used abbreviation for the word *reaction*. Notice that $NO(g)$ does not appear in the combined reaction because this compound was present *in equal amounts on opposite sides of the individual* reactions. Hess's law is really an application of the first law of thermodynamics, which in itself is a statement of the law of conservation of energy applied to chemical reactions.

PROBLEM
Given the following reactions:

$$2SO_2(g) + O_2(g) \rightarrow 2SO_3(g) \qquad \Delta H°_1 = -196 \text{ kJ}$$
$$2S(s) + 3O_2(g) \rightarrow 2SO_3(g) \qquad \Delta H°_2 = -790 \text{ kJ}$$

Calculate $\Delta H°_{rxn}$ for this reaction: $S(s) + O_2(g) \rightarrow SO_2(g)$.

SOLUTION
The compound $SO_3(g)$ does *not* appear in the combined reaction. Therefore, one of the individual reactions will have to be reversed, along with the *sign* of its $\Delta H°$. We will reverse the *first* reaction because $SO_2(g)$ needs to appear on the *right side*, and $S(s)$ on the *left side,* of the combined reaction. This step gives us:

$$2SO_3(g) \rightarrow 2SO_2(g) + O_2(g) \qquad -\Delta H°_1 = +196 \text{ kJ}$$
$$2S(s) + 3O_2(g) \rightarrow 2SO_3(g) \qquad \Delta H°_2 = -790 \text{ kJ}$$

The combined reaction is:

$$2S(s) + 2O_2(g) \rightarrow 2SO_2(g) \qquad \Delta H°_{rxn} = (+196 \text{ kJ} - 790 \text{ kJ}) = -594 \text{ kJ}$$

If we compare the combined reaction with the reaction asked for in the problem statement, we find that all of the coefficients are *doubled*. Therefore, all we need do is multiply the entire equation—and its $\Delta H°$—by 1/2 and our answer is:

$$S(s) + O_2(g) \rightarrow SO_2(g) \qquad \Delta H°_{rxn} = -297 \text{ kJ}$$

TRY IT YOURSELF
Given the following reactions:

$$H_2S(g) + \frac{3}{2}O_2(g) \rightarrow H_2O(\ell) + SO_2(g) \qquad \Delta H°_1 = -563 \text{ kJ}$$

$$CS_2(\ell) + 3O_2(g) \rightarrow CO_2(g) + 2SO_2(g) \qquad \Delta H°_2 = -1075 \text{ kJ}$$

Calculate $\Delta H°_{rxn}$ for this reaction: $CS_2(\ell) + 2H_2O(\ell) \rightarrow 2H_2S(g) + CO_2(g)$
(*Hint:* Get rid of the fraction first!)

ANSWER
+51 kJ

Standard Heat of Formation of a Compound

The **standard heat of formation** of a compound is a special type of heat of reaction. It is defined as the heat released or absorbed when *1 mole* of a compound is formed from its *elements* under standard conditions.

The symbol for standard heat of formation is $\Delta H°_f$, where the subscript f stands for the word *formation*. The elements that form the compound are in their *stable states* (i.e., oxygen would be a gas; iron, a solid; etc.) at 298.15 K. The standard heat of formation of an *element* is *assigned* a value of 0 kilojoule per mole.

Reference Table W-4 in Appendix 2 lists the standard heats of formation of various compounds. In the table, standard heat of formation is measured in *kilojoules per mole of product formed*. Let us examine what these values mean. According to the table, $\Delta H°_f$ for ammonia [$NH_3(g)$] is –45.9 kilojoule per mole. This means that, when *1 mole* of $NH_3(g)$ is formed from its elements (nitrogen and hydrogen) under standard conditions, 45.9 kilojoules of heat is released.

We can write a **formation reaction** to illustrate this event:

$$\frac{1}{2}N_2(g) + \frac{3}{2}H_2(g) \rightarrow NH_3(g) + 45.9 \text{ kJ}$$

We note that the reaction is written so that *1 mole of product* is formed. Therefore, we require that 0.5 mole of $N_2(g)$ react with 1.5 moles of $H_2(g)$.

PROBLEM
(a) Write the formation reaction for KCl(s).
(b) How much heat is released when 2.00 moles of KCl(s) is formed from its elements under standard conditions?
(c) Suppose we could *decompose* 1.00 mole of KCl(s) into its elements under standard conditions. What would $\Delta H°$ for this reaction be?

SOLUTIONS
(a) Under standard conditions, potassium exists as a solid, and chlorine as diatomic gas molecules. Therefore, the formation reaction is

$$K(s) + \frac{1}{2} Cl_2(g) \rightarrow KCl(s)$$

(b) From Reference Table W-4 we see that $\Delta H°_f$ for KCl(s) is -436.5 kJ/mol, as the formation of 2.00 mol of KCl will release 873 kJ.
(c) Decomposition is the *opposite* of formation. Therefore, the decomposition of 1.00 mol of KCl(s) would require the *absorption* of 436.5 kJ. (In other words, $\Delta H° = +436.5$ kJ/mol KCl(s) decomposed.)

PROBLEM
Can an element have a $\Delta H°_f$ value different from 0?

SOLUTION
Yes, if the element is not in its stable state. For example, note $\Delta H°_f$ for white and gray tin in Reference Table W-4.

Heats of Reaction from Standard Heats of Formation

It is possible to use the standard heats of formation of individual substances to calculate the standard enthalpy change ($\Delta H°$) of a reaction involving these substances. For example, we will consider this reaction:

$$4NH_3(g) + 5O_2(g) \rightarrow 4NO(g) + 6H_2O(g)$$

Reference Table W-4 gives us the standard heats of formation for the three compounds in the equation; oxygen is an element and so has a standard heat of formation of 0. These ΔH_f values are given in the second column of the accompanying table.

Substance	$\Delta H°_f$ / (kJ/mol)	× Equation Coefficient / (mol)	= Product of 2nd and 3rd Columns / (kJ)
$NH_3(g)$	−45.9	4	−183.6
$O_2(g)$	0.0	5	0.0
$NO(g)$	+91.3	4	+365.2
$H_2O(g)$	−241.8	6	−1450.8

In the third column we have listed the coefficients of the four substances as they appear in the equation. The last column shows the *products* of the second and third columns.

We calculate the heat of reaction by *subtracting* the sum of the values for the *reactants* (NH_3 and O_2) of the reaction from the sum of the values for the *products* (NO and H_2O), as given in the last column of the table:

$$\Delta H°_{rxn} = (-1450.8 \text{ kJ} + 365.2 \text{ kJ}) - (-183.6 \text{ kJ} + 0.0 \text{ kJ}) = -902.0 \text{ kJ}$$

Why does this method work? First, we assume that a reaction takes place by breaking down the reactants into their elements and then reassembling those elements as products:

$$\text{Reactants} \rightarrow \text{Elements} \rightarrow \text{Products}$$

The reaction does not *really* take place in this way, but *from an energy point of view* this assumption, which follows from the first law of thermodynamics, is perfectly logical. We make this assumption when we *subtract* the reactant values from the product values. Second, the coefficients of the equation tell us how many moles of each substance are involved in the reaction. Since a standard heat of formation is based on *1 mole* of substance, we need to *multiply* by the coefficient for each substance in the equation.

If you MUST have a mathematical relationship for this technique, here it is:

$$\Delta H°_{rxn} = \left(\Sigma n \cdot \Delta H°_f\right)_{products} - \left(\Sigma m \cdot \Delta H°_f\right)_{reactants}$$

Don't let the fancy symbols throw you! The "Σ" symbol means that you add the individual values of the products and reactants separately *before* you subtract. The symbols m and n represent the coefficients associated with all of the substances in the reaction.

TRY IT YOURSELF
Calculate $\Delta H°$ for this reaction:

$$C_2H_4(g) + H_2(g) \rightarrow C_2H_6(g)$$

using the values given in Reference Table W-4.

ANSWER
−136.4 kJ (Did you forget that hydrogen is an *element* in its stable state?)

5.5A THE SECOND LAW OF THERMODYNAMICS

The role of entropy in spontaneous processes is governed by a rule known as the **second law of thermodynamics** (or, more briefly, as the *second law*). This rule states that in any spontaneous process the total entropy of the system and its surroundings, that is, the entropy of the *universe*, must always *increase*. We can state the second law in the form of an equation:

$$\Delta S_{system} + \Delta S_{surroundings} = \Delta S_{universe} > 0$$

Unlike energy, entropy is *not conserved* in a spontaneous process. One consequence of the second law is the assurance that heat will always flow spontaneously from a hotter object to a colder one, an assumption we made in solving calorimetry problems in Section 5.1.

Another consequence of the second law is that it allows some spontaneous processes to occur, even when a system experiences a *decrease* in entropy. For example, when water freezes to ice at −10°C, the entropy of the *system* decreases as the solid is formed. However, the entropy of the surroundings *increases to an even greater extent*, ensuring that there will be an increase in the entropy of the universe.

During the course of a chemical reaction, (1) energy may be absorbed or released and (2) the entropy of the system may increase or decrease. If we focus on the reaction itself and *ignore* the surroundings, we know that a release of energy and an increase in entropy favor a spontaneous reaction, while an absorption of energy and a decrease in entropy do not. Given these possible changes in energy and entropy, we need to raise the following question: How can we *predict* when a reaction will be spontaneous? During the nineteenth century, J. Willard Gibbs, a professor of physics at Yale University, developed a relationship that allows us to make this prediction.

The Gibbs Free-Energy Change

The *Gibbs free-energy change*, denoted by the symbol ΔG, enables us to predict whether a reaction will be spontaneous under a given set of conditions. The free-energy change is defined as follows:

$$\Delta G = \Delta H - T \Delta S$$

127

Note that ΔG takes both ΔH and ΔS into account. Moreover, the absolute temperature plays an important part and is associated with the disorder factor. As the temperature rises, the disorder factor becomes more important.

When the equation is applied, a *negative* ΔG indicates that the reaction is spontaneous; if ΔG is *positive*, the reaction is *not* spontaneous. Let us see how the equation can be used in practice.

PROBLEM

The reaction $N_2O_4(g) \rightarrow 2NO_2$ (g) is *endothermic*. At low temperatures it is not spontaneous, yet it is spontaneous at higher temperatures. How does the free-energy equation explain this fact?

SOLUTION

This reaction leads to an *increase* in disorder. (Why?) Therefore, the disorder factor favors a spontaneous reaction, but the energy factor does not.

At low temperatures, the disorder factor ($T\Delta S$) is small and the unfavorable energy factor is more important. At higher temperatures, however, $T\Delta S$ becomes the more important factor and the reaction occurs spontaneously.

We can summarize these results in the accompanying table, which relates the *algebraic signs* of the variables to the possibility of a spontaneous reaction. Note that absolute temperature is *not* included in this table because its sign is always *positive*. As noted on page 119, "rxn" is an abbreviation for the word *reaction*.

Rxn	ΔH	ΔS	ΔG	Result	Example
A	(−)	(+)	(−)	Rxn is spontaneous at *all* temperatures.	$2NO_2(g) \rightarrow 2N_2(g) + O_2(g)$
B	(−)	(−)	???	Rxn is spontaneous at *low* temperatures.	$H_2O(\ell) \rightarrow H_2O(s)$
C	(+)	(+)	???	Rxn is spontaneous at *high* temperatures.	$2NH_3(g) \rightarrow N_2(g) + 3H_2(g)$
D	(+)	(−)	(+)	Rxn is *not* spontaneous at *any* temperature.	$3O_2(g) \rightarrow 2O_3(g)$

Remember: A reaction that is not spontaneous can still be made to occur. In reaction D, for example, $O_3(g)$ (ozone gas) is produced by the action of an electrical discharge, such as lightning, on the $O_2(g)$ present in air.

END-OF-CHAPTER QUESTIONS

Some questions have the symbol "§2" in front of the question number. This symbol means that the question is based on Section II material.

1. What is the maximum number of grams of liquid water at 10°C that can be heated to 30°C by the addition of 84 joules of heat?
   (1) 1.0   (2) 2.0   (3) 20.   (4) 30.

§2   2. How many joules of heat are released when 50 grams of solid magnesium is cooled from 70°C to 60°C?
   (1) 10   (2) 50   (3) 500   (4) 1000

§2   3. If 4 grams of solid aluminum at 1°C absorbs 8 joules of heat, the temperature of the aluminum will change by
   (1) 1 C°   (2) 2 C°   (3) 3 C°   (4) 4 C°

§2   4. The temperature of 50 grams of liquid toluene ($C_6H_5CH_3$) was raised to 50°C by the addition of 900 joules of heat. What was the initial temperature of the toluene?
   (1) 0°C   (2) 20°C   (3) 30°C   (4) 40°C

§2   5. The temperature of solid silver increases from 30.00°C to 40.00°C by the addition of 117.5 joules of heat. What is the mass of the silver?
   (1) 1.00 g   (2) 5.00 g   (3) 10.0 g   (4) 50.0 g

6. The temperature of 100.00 grams of water changes from 16.00°C to 20.00°C. What is the total number of joules of heat energy absorbed by the water?
   (1) 250.0   (2) 400.0   (3) 1000.   (4) 1680.

7. In a reversible reaction, the difference between the activation energy of the forward reaction and the activation energy of the reverse reaction is equal to the
   (1) activated complex   (2) heat of reaction
   (3) potential energy of reactants   (4) potential energy of products

8. A piece of Mg(s) ribbon held in a Bunsen burner flame begins to burn according to the equation

$$2Mg(s) + O_2(g) \rightarrow 2MgO(s)$$

The reaction begins because the reactants
(1) are activated by heat from the Bunsen burner flame
(2) are activated by heat from the burning magnesium
(3) underwent an increase in entropy
(4) underwent a decrease in entropy

9.

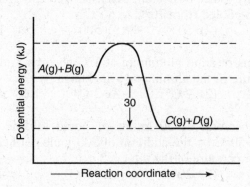

In the diagram above, the reaction

$$A(g) + B(g) \rightarrow C(g) + D(g)$$

has a forward activation energy of 20 kilojoules. What is the activation energy for the reverse reaction?
(1) 10 kJ   (2) 20 kJ   (3) 30 kJ   (4) 50 kJ

§2 **10.** What is the standard heat of formation, in kilojoules per mole of liquid water at 298.15 K?
(1) -285.8 (2) -241.8 (3) $+241.8$ (4) $+285.8$

11. Given the reaction

$$H_2(g) + \frac{1}{2}O_2(g) \rightarrow H_2O(g) + 243 \text{ kJ}$$

If the activation energy for the forward reaction is 168 kilojoules, the activation energy for the reverse reaction, in kilojoules, will be
(1) 163   (2) 168   (3) 243   (4) 411

12. Assume that the potential energy of the products in a chemical reaction is 60 kilojoules. This reaction would be exothermic if the potential energy of the reactants was
(1) 50 kJ   (2) 20 kJ   (3) 30 kJ   (4) 80 kJ

Base your answers to questions 13–15 on the diagram below.

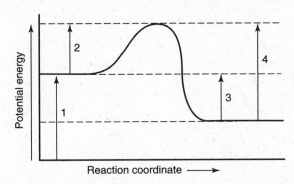

13. The activation energy for the reverse reaction is represented by
(1) 1   (2) 2   (3) 3   (4) 4

14. The heat of reaction (ΔH) is represented by
(1) 1   (2) 2   (3) 3   (4) 4

15. The potential energy of the activated complex is represented by
(1) 1 + 2   (2) 2 + 3   (3) 3 + 4   (4) 4 + 1

§2 **16.** According to Reference Table W-4, how many kilojoules of heat are released when 0.5000 mole of MgO(s) is formed from its elements?
(1) 150.5   (2) 300.8   (3) 601.6   (4) 1204

§2 **17.** According to Reference Table W-4, the formation of 1 mole of which of the following substances *absorbs* the greatest amount of energy?
(1) C_2H_2 (2) C_2H_4 (3) $CuSO_4$ (4) $BaSO_4$

18. Which potential energy diagram represents an exothermic reaction?

(1)
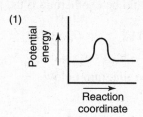

(2)
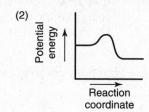

(3)
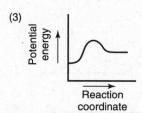

(4)
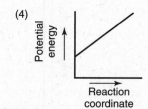

19. When a substance was dissolved in water, the temperature of the water decreased. This process is described as
(1) endothermic, with the release of energy
(2) endothermic, with the absorption of energy
(3) exothermic, with the release of energy
(4) exothermic, with the absorption of energy

20. In a chemical reaction, the difference between the potential energy of the products and the potential energy of the reactants is called the
(1) activation energy   (2) activated complex
(3) kinetic energy   (4) heat of reaction

21. For this reaction:

$$A + B \rightarrow C + \text{heat},$$

the potential energy of the products, as compared with the potential energy of the reactants, is
(1) less and the reaction is exothermic
(2) less and the reaction is endothermic
(3) greater and the reaction is exothermic
(4) greater and the reaction is endothermic

§2 **22.** According to Reference Table W-4, which compound has a higher potential energy than the elements from which it is formed?
(1) aluminum oxide(s)   (2) water(ℓ)
(3) nitrogen(II) oxide(g)   (4) carbon dioxide(g)

§2 **23.** The quantity of heat liberated when 2.00 moles of aluminum oxide is formed under standard conditions is closest to
(1) 419 kJ   (2) 838 kJ   (3) 1680 kJ   (4) 3350 kJ

24. Given the reaction

$$A + B \rightarrow AB + 50 \text{ kJ}$$

If an activation energy of 5 kilojoules is required for the forward reaction, the activation energy of the reverse reaction is
(1) 5 kJ   (2) 45 kJ
(3) 50 kJ   (4) 55 kJ

25. Which of the following best describes exothermic chemical reactions?
(1) They never release heat.
(2) They always release heat.
(3) They are always decomposition reactions.
(4) They are always synthesis reactions.

26. Which type of reaction does this equation represent?

$$N_2(g) + 3H_2(g) \rightarrow 2NH_3(g) + \text{heat}$$

(1) exothermic, with an increase in entropy
(2) exothermic, with a decrease in entropy
(3) endothermic, with an increase in entropy
(4) endothermic, with a decrease in entropy

§2 **27.** A chemical reaction will always occur spontaneously if the reaction has a
(1) negative ΔG (2) positive ΔG
(3) negative ΔH (4) positive ΔH

28. Systems in nature tend to undergo changes toward
(1) lower energy and lower entropy
(2) lower energy and higher entropy
(3) higher energy and lower entropy
(4) higher energy and higher entropy

29. A 1-gram sample of a substance has the greatest entropy when it is in the
(1) solid state   (2) liquid state
(3) crystalline state   (4) gaseous state

§2 **30.** Which change represents an increase in the entropy of a system?
(1) $C_6H_{12}O_6(s) \rightarrow C_6H_{12}O_6(aq)$
(2) $H_2O(\ell) \rightarrow H_2O(s)$
(3) $CO_2(g) \rightarrow CO_2(s)$
(4) $C_2H_5OH(g) \rightarrow C_2H_5OH(\ell)$

31. Which change represents an increase in entropy?
(1) $I_2(s) \rightarrow I_2(g)$ (2) $I_2(g) \rightarrow I_2(\ell)$
(3) $H_2O(g) \rightarrow H_2O(\ell)$ (4) $H_2O(\ell) \rightarrow H_2O(s)$

§2 **32.** A reaction that has a positive ΔG must be
(1) exothermic   (2) endothermic
(3) spontaneous   (4) nonspontaneous

§2 **33.** A chemical reaction releases heat, and the products are in a more disordered state than the reactants. The ΔG for this reaction
(1) must be negative
(2) must be positive
(3) could be negative or positive
(4) could be zero

34. A sample of $H_2O(\ell)$ at 20°C is in equilibrium with its vapor in a sealed container. When the temperature increases to 25°C, the entropy of the system will
(1) decrease   (2) increase   (3) remain the same

35. As 1 gram of $H_2O(\ell)$ changes to 1 gram of $H_2O(s)$, the entropy of the system
(1) decreases   (2) increases   (3) remains the same

Base your answers to questions 36 and 37 on the table below, which represents the energy and entropy changes in four chemical reactions, *A–D*.

Reaction	ΔH	ΔS
A	+	+
B	+	−
C	−	+
D	−	−

§2 **36.** Which reaction *must* be spontaneous?
(1) *A*   (2) *B*   (3) *C*   (4) *D*

134

§2 **37.** Which reaction *must* have a free-energy change (Δ*G*) that is positive?
(1) *A*   (2) *B*   (3) *C*   (4) *D*

38. Given the reaction:

$$S(s) + O_2(g) \rightarrow SO_2(g) + \text{energy}$$

Which diagram best represents the potential energy changes for this reaction?

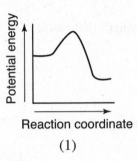

Reaction coordinate
(1)

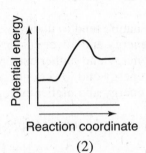

Reaction coordinate
(2)

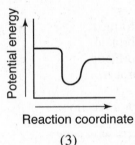

Reaction coordinate
(3)

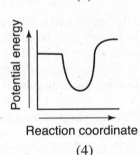

Reaction coordinate
(4)

39. Which phase change represents a *decrease* in entropy?
(1) solid to liquid   (2) gas to liquid
(3) liquid to gas   (4) solid to gas

40. Which change is exothermic?
(1) freezing of water   (2) melting of iron
(3) vaporization of ethanol   (4) sublimation of iodine

41. Which statement correctly describes an endothermic reaction?
   (1) The products have a higher potential energy than the reactants, and the ΔH is negative.
   (2) The products have a higher potential energy than the reactants, and the ΔH is positive.
   (3) The products have a lower potential energy than the reactants, and the ΔH is negative.
   (4) The products have a lower potential energy than the reactants, and the ΔH is positive.

42. Systems in nature tend to undergo changes that result in
   (1) lower energy and lower entropy
   (2) lower energy and higher entropy
   (3) higher energy and lower entropy
   (4) higher energy and higher entropy

43. Given the balanced equation representing a reaction:

$$H_2 \rightarrow H + H$$

What occurs during this reaction?
   (1) Energy is absorbed as bonds are formed.
   (2) Energy is absorbed as bonds are broken.
   (3) Energy is released as bonds are formed.
   (4) Energy is released as bonds are broken.

44. Given the potential energy diagram for a reaction:

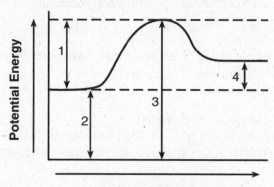

Reaction Coordinate

Which intervals are affected by the addition of a catalyst?
   (1) 1 and 2   (2) 1 and 3   (3) 2 and 4   (4) 3 and 4

Constructed-Response Questions

§2 **1.** The standard heat of formation of C_2H_5OH is -277.6 kilojoule per mole. How much heat will be released when 1.000 gram of this compound is formed from its elements under standard conditions?

§2 **2.** Use the data given below to calculate $\Delta H°$ for this reaction:

$$CaO(s) + CO_2(g) \rightarrow CaCO_3(s)$$

Compound	$\Delta H_f°$(kJ/mol)
CaO (s)	−635.1
CO_2 (g)	−393.5
$CaCO_3$ (s)	−1207.6

§2 **3.** Given the following information:

$$2Cu_2O(s) + O_2(g) \rightarrow 4CuO(s) \quad \Delta H° = -288.0 \text{ kJ}$$
$$CuO(s) + Cu(s) \rightarrow Cu_2O(s) \quad \Delta H° = -11.3 \text{ kJ}$$

Calculate the standard heat of formation of $Cu_2O(s)$.

Base your answers to questions 4 through 6 on the information and potential energy diagram below.

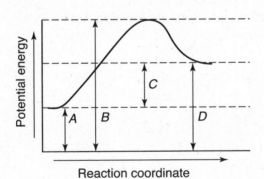

Chemical cold packs are often used to reduce swelling after an athletic injury. The diagram represents the potential energy changes when a cold pack is activated.

4. Which lettered interval on the diagram represents the potential energy of the products?

5. Which lettered interval on the diagram represents the heat of reaction?

6. Identify a reactant listed in Reference Table I that could be mixed with water for use in a chemical cold pack.

Base your answers to questions 7–9 on the information below and on your knowledge of chemistry.

A student made a copper bracelet by hammering a small copper bar into the desired shape. The bracelet had a mass of 30.1 grams and was at a temperature of 21°C in the classroom. After the student wore the bracelet, the bracelet reached a temperature of 33°C. Later, the student removed the bracelet and placed it onto a desk at home, where it cooled from 33°C to 19°C. The specific heat capacity of copper is 0.385 J/g · K.

7. Explain, in terms of heat flow, the change in temperature when the student wore the bracelet.

8. Determine the number of moles of copper in the bracelet.

9. Show a numerical setup for calculating the amount of heat released by the bracelet as it cooled on the desk.

The answers to these questions are found in Appendix 3.

Chapter Six

THE PHASES OF MATTER

KEY IDEAS

This chapter focuses on the gaseous, liquid, and solid phases of matter, as well as changes in phase. In particular, the ideal gas laws and the kinetic-molecular theory are introduced.

KEY OBJECTIVES

At the conclusion of this chapter you will be able to:

- Distinguish among the solid, liquid, and gaseous phases of matter.
- Define the terms *standard atmosphere* and *standard state pressure*.
- Define the term *standard temperature and pressure* (STP).
- Describe the measurement of gas pressure using closed-tube and open-tube manometers.
- Describe and apply the various gas laws to numerical problems.
- Convert between the Celsius and Kelvin temperature scales.
- Calculate the density of a gas at STP.
- Calculate the molar mass of a gas, given its density at STP.
- Solve stoichiometry problems involving gas volumes.
- State the hypotheses of the kinetic-molecular theory (KMT) of gas behavior.
- Define the term *ideal gas*, and the terms *volume*, *pressure*, and *temperature* in relation to the KMT.
- List the conditions under which real gases exhibit most nearly ideal and least ideal behavior.
- Relate vapor pressure to the boiling point of a liquid.
- Solve problems involving gases collected over water.
- Define the terms associated with phase changes.
- Interpret the parts of a heating curve and a cooling curve.
- Solve problems involving changes of phase.
- Interpret phase diagrams.
- Define the terms *triple point*, *critical point*, *critical temperature*, and *critical pressure*.

SECTION I—BASIC (REGENTS-LEVEL) MATERIAL

NYS REGENTS CONCEPTS AND SKILLS

Note: By the time you have finished Section I, you should have mastered the concepts and skills listed below. The Regents chemistry examination will test your knowledge of these items and your ability to apply them.

Concepts are the *basic ideas* that form the body of the Regents chemistry course (what you need to know!).

Skills are the *activities* that demonstrate your mastery of these concepts (how you show that you know them!).

Following each concept or skill is a page reference (given in parentheses) to this chapter.

6.1   Concept:
Temperature is a measurement of the average kinetic energy of the particles in a gas. Temperature is not a form of energy. (Page 150)

6.2   Concept:
The concept of an ideal gas is a model to explain the behavior of gases. A real gas is most like an ideal gas when the real gas is at low pressure and high temperature. (Pages 150–152)

6.3   Concepts:
The kinetic molecular theory (KMT) for an ideal gas states that all gas particles:
(a) are in random, constant, straightline motion;
(b) are separated by great distances relative to their size; the volume of the gas particles is considered negligible;
(c) have no attractive forces between them;
(d) have collisions that may result in the transfer of energy between gas particles, but the total energy of the system remains constant. (Page 154)

6.4   Concept:
The kinetic molecular theory describes the relationships of pressure, volume, temperature, velocity, and frequency and force of collisions among gas molecules. (Page 154)

Skills:
• Explain the gas laws in terms of KMT. (Pages 154–155)
• Solve problems using the combined gas law. (Pages 151–153)

6.5   Concept:
Equal volumes of different gases at the same temperature and pressure contain equal numbers of particles. (Pages 152–153)

Skill:
Convert temperatures between the Celsius and Kelvin scales. (Page 147)

6.6   Concept:
The concepts of kinetic and potential energy can be used to explain physical processes that include fusion (melting), solidification (freezing), vaporization (boiling, evaporation), condensation, sublimation, and deposition. (Pages 158–161)

Skills:
- Interpret heating and cooling curves in terms of changes in kinetic and potential energy, heat of fusion, heat of vaporization, and phase changes. (Pages 159–161)
- Calculate the heat involved in a phase change. (Page 161)

6.1   INTRODUCTION

Matter commonly exists in the solid, liquid, or gaseous phase. The phase of a substance is usually recognized by the characteristics of the substance's shape and volume. Gases have neither definite shape nor definite volume, while solids have both definite shape and definite volume. Liquids do not have definite shape, but they do have definite volume. What phase a particular sample of matter is in depends on the nature of the sample, its temperature, and the pressure exerted on it.

6.2   GASES

As we begin our study of gases, we assume that the sample of gas under discussion behaves *ideally*. While no gas is truly ideal, many samples of gases do exhibit ideal behavior under appropriate conditions. (We will indicate exactly what we mean by the term *ideal behavior* in Section 6.4.) We make this assumption about ideal behavior because, when we do, the "laws" that govern gas behavior become very simple.

To describe the behavior of a sample of an ideal gas, we need to know only four characteristics of the sample: the pressure it exerts, its volume, its temperature, and the number of particles (or numbers of *moles* of particles) it contains. We have already learned how volume, temperature, and number of moles are measured.

Pressure is defined as the force per unit area of surface. Mathematically, this means that, when we apply a force on a surface, we calculate the pressure by *dividing* the force by the area of the surface (pressure = force/area). Pressure gives us a means of describing how a force is distributed over an entire surface. In the English system, the unit of pressure is the *pound per square inch* (lb/in.^{2}). Under normal atmospheric conditions, air exerts a pressure of approximately 14.7 pounds per square inch.

In the SI metric system, the units of pressure are the *pascal* (Pa) and its multiple, the *kilopascal* (kPa). The pascal is a very small unit, and under normal atmospheric conditions air exerts a pressure of approximately 1.01×10^5 pascals (101 kPa).

Currently, there are *two standards* of pressure in the scientific world. The **standard atmosphere** (or *atmosphere*, atm) is defined as *exactly* 101.325 kilopascals; **standard state pressure** is defined as *exactly* 100.000 . . . kilopascals. While the values of standard atmosphere and standard state pressure are not equal, they are very close. Generally, chemists use the *standard atmosphere* in applications involving gas behavior, and the *standard state pressure* in applications involving thermodynamics (Chapter 5).

An older metric unit of pressure, the *millimeter of mercury* (mmHg, or *torr*), is associated with the measurement of air pressure using an instrument known as a *Torricelli barometer*. This instrument consists of a long tube filled with mercury and inverted in a dish containing mercury. After inversion, the mercury in the tube falls to a specific height, and this height (measured in *millimeters*) is a measure of the pressure exerted by the air. One standard atmosphere of pressure supports a column of mercury very nearly equal to 760 millimeters (approximately 30 inches).

The accompanying chart shows the pressure relationships that we will use in this chapter.

$$
\text{1 standard atmosphere (atm)} = \begin{cases} 1.01325 \times 10^5 \text{ Pa [exact]} \\ 101.325 \text{ kPa [exact]} \\ 760 \text{ mmHg [approximate]} \\ 760 \text{ torr [approximate]} \end{cases}
$$

In the study of gas behavior, it is often necessary to refer *jointly* to a standard of pressure and temperature. The term **standard temperature and pressure (STP)** is defined as the ice point of water (273.15 K or 0°C) and 1 standard atmosphere (101.325 kPa or 760 mmHg).

6.3 THE GAS LAWS

Boyle's Law

To investigate the behavior of a gas, we need to change one of its characteristics (pressure, for example) and observe how another characteristic (volume, for example) will change in response. We must be certain, though, that the other two variables (temperature and number of particles, in this case) are not allowed to vary during the investigation. **Boyle's law** is the result of measuring how the volume of a gas varies with pressure (at constant temperature and number of particles). The accompanying table illustrates the results of one such experiment carried out on an ideal gas at 25°C.

Pressure (P) / Pa	Volume (V) / mL	$P \times V$ / (Pa·mL)
500	400	200,000
1000	200	200,000
2000	100	200,000
4000	50	200,000
5000	(Solved in the problem on page 142)	

According to the first four pairs of data given in the table, the volume varies *inversely* with the pressure. In other words, an *increase* in pressure is accompanied by a *decrease* in volume (and vice versa) at constant temperature and number of particles. Moreover, the product of pressure and volume is constant under these conditions. This relationship is known as *Boyle's law*:

> **pressure · volume = constant**
> **(at constant temperature and number of particles)**

If we graph the data given in the table, we produce the curve shown on the next page, known as a *rectangular hyperbola*. This curve is characteristic of inverse relationships.

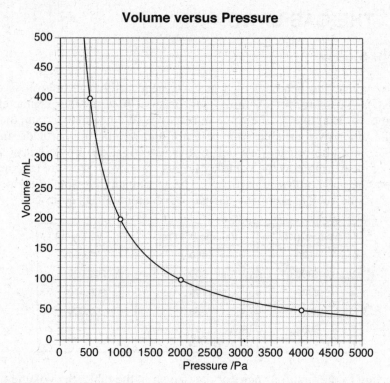

Volume versus Pressure

PROBLEM

According to the graph shown above, what will the volume of the gas be when the pressure is 3000 pascals?

SOLUTION

We locate the point on the curve that corresponds to a pressure of 3000 Pa, and then locate the corresponding value on the volume axis. The volume is approximately 70 mL.

TRY IT YOURSELF

According to the graph, what pressure corresponds to a volume of 150 milliliters?

ANSWER

Approximately 1300 Pa.

Each point on the graph obeys the Boyle's law equation:

$$P \cdot V = \text{constant}$$

If we choose *any* two points, whose respective values for pressure and volume are $P_1 \cdot V_1$ and $P_2 \cdot V_2$, it follows that

$$\boxed{P_1 \cdot V_1 = P_2 \cdot V_2}$$

This is another way of writing Boyle's law. This formula is very useful for solving mathematical problems involving pressure and volume. Note that we can use *any* units of pressure and volume in solving a Boyle's law problem as long as we remain consistent within the problem.

PROBLEM
According to the table on page 143, what will be the volume of the gas when the pressure is 5000 pascals?

SOLUTION
Boyle's law indicates that we can use any other set of values in the table for solving our problem. We will use the second set.

$$P_1 \cdot V_1 = P_2 \cdot V_2$$
$$P_1 = 1000 \text{ Pa}$$
$$V_1 = 200 \text{ mL}$$
$$P_2 = 5000 \text{ Pa}$$
$$V_2 = ???$$
$$1000 \text{ Pa} \cdot 200 \text{ mL} = 5000 \text{ Pa} \cdot V_2$$

$$V_2 = \frac{1000 \text{ Pa} \cdot 200 \text{ mL}}{5000 \text{ Pa}}$$

$$= 40 \text{ mL}$$

PROBLEM
The volume occupied by a gas at STP is 250 liters. At what pressure (in *atmospheres*) will the gas occupy 1500 liters, if the temperature and number of particles remain constant?

SOLUTION
We apply Boyle's law and remember that standard atmospheric pressure equals 1 atm.

$$P_1 \cdot V_1 = P_2 \cdot V_2$$
$$P_1 = 1 \text{ atm}$$
$$V_1 = 250 \text{ L}$$
$$P_2 = ???$$
$$V_2 = 1500 \text{ L}$$
$$1 \text{ atm} \cdot 250 \text{ L} = P_2 \cdot 1500 \text{ L}$$

$$P_2 = \frac{1 \text{ atm} \cdot 250 \text{ L}}{1500 \text{ L}}$$

$$= 0.17 \text{ atm}$$

TRY IT YOURSELF
The volume of an ideal gas is 32. liters at a pressure of 10. kilopascals. What will the volume be at a pressure of 2.0 kilopascals, temperature and number of particles remaining constant?

ANSWER
160 L

Charles's Law and the Kelvin Temperature Scale

To investigate the relationship between the temperature and the volume of an ideally behaving gas, we must hold the pressure and the number of particles constant. We begin at 0°C and measure the volumes of two samples of an ideally behaving gas. We then vary the temperature of each sample and measure the respective volumes of the samples. The accompanying table provides the data for these investigations.

Temperature/°C	Experiment 1 Volume/mL	Experiment 2 Volume/mL
−100	173	346
−50	223	446
−20	253	506
−10	263	526
0	273	546
+10	283	566
+20	293	586
+50	323	646
+100	373	746

From the table, we can draw two conclusions:

1. The volume of the gas increases (decreases) with increasing (decreasing) temperature.
2. The change in volume with change in temperature is regular: In experiment 1 the volume changes by 1 milliliter for each 1° change (1 mL/C°); in experiment 2 the volume changes by 2 milliliters for each 1° change (2 mL/C°).

A graph of these results follows.

Volume versus Temperature

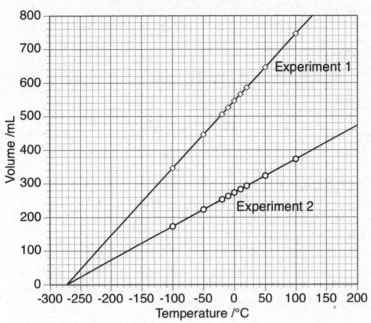

When we inspect the graph, we note that the two experiments produce two straight lines that converge to a point. This point represents a volume of 0 milliliter. By inspecting the table above, we can deduce that this value will be reached—in both experiments—at approximately −273°C. What is truly amazing is that *every* similar experiment will yield a straight-line graph that drops to zero volume at approximately −273°C! Obviously, there is something very special about this temperature. Moreover, it is the lowest temperature possible because no gas, ideally behaving or not, can have a negative volume. The precise value of this lowest temperature is –273.15°C.

Scientists found it very useful to define a new temperature scale in which this lowest possible temperature (−273.15°C) was given a value of 0 and was referred to as **absolute zero**. Today, this temperature scale is called the **Kelvin** or **absolute** scale, and it is related to the Celsius scale by this equation:

$$T_C = T_K - 273.15$$

A simplified version of this equation is given in Reference Table T in Appendix 1:

$$K = °C + 273$$

We can now assume that the value of absolute zero is −273°C. With this relationship, −273°C is assigned the value 0 K, while 0°C equals 273 K and 100°C equals 373 K. Note that there are *no* negative Kelvin temperatures.

PROBLEM
What Kelvin temperature corresponds to 35°C?

SOLUTION

$$K = °C + 273$$
$$= 35°C + 273$$
$$= 308 \text{ K}$$

TRY IT YOURSELF
What Celsius temperature corresponds to 260 K?

ANSWER
−13°C

If we adopt the Kelvin scale as our scale of choice, the relationship between the volume of a gas and its temperature is simplified, as shown in this modification of the preceding graph:

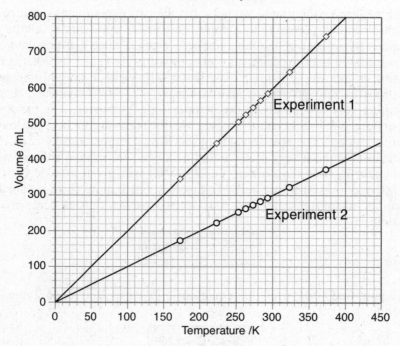

Volume versus Temperature

148

The simple relationship that can be drawn from this graph is that the volume of an ideally behaving gas is *directly proportional* to the Kelvin temperature:

$$\frac{\text{volume}}{\text{temperature}} = \text{constant}$$

(at constant pressure and number of particles)

We can write this relationship, which is known as **Charles's law**, in its equivalent form:

$$\frac{V_1}{T_1} = \frac{V_2}{T_2}$$

where the temperature is *always* specified in the Kelvin scale.

PROBLEM

The volume of an ideally behaving gas is 300 liters at 227°C. What volume will the gas occupy at 27°C, pressure and number of particles remaining constant?

SOLUTION

$$\frac{V_1}{T_1} = \frac{V_2}{T_2}$$

$$V_1 = 300 \text{ L}$$

$$T_1 = 227°C + 273 = 500 \text{ K}$$

$$V_2 = \text{???}$$

$$T_2 = 27°C + 273 = 300 \text{K}$$

$$\frac{300 \text{ L}}{500 \text{ K}} = \frac{V_2}{300 \text{ K}}$$

$$V_2 = \frac{300 \text{ L} \cdot 300 \text{ K}}{500 \text{ K}}$$

$$= 180 \text{ L}$$

TRY IT YOURSELF
The volume of an ideally behaving gas is 50 cubic centimeters at 250 K. At what temperature will the volume of the gas be 200 cubic centimeters, pressure and number of particles remaining constant?

ANSWER
1000 K

Relating Pressure and Temperature (Gay-Lussac's Law)

At constant volume and number of particles, the relationship between the pressure and the Kelvin temperature of an ideally behaving gas is similar to that given by Charles's law:

$$\frac{\text{pressure}}{\text{temperature}} = \text{constant}$$

(at constant volume and number of particles)

Or, in equivalent form:

$$\frac{P_1}{T_1} = \frac{P_2}{T_2}$$

This relationship is known as **Gay-Lussac's law**.

PROBLEM
The pressure exerted by an ideally behaving gas is 7.00 kilopascals at 200. K. What pressure does the gas exert at 500. K, volume and number of particles remaining constant?

SOLUTION

$$\frac{P_1}{T_1} = \frac{P_2}{T_2}$$

$$P_1 = 7.00 \text{ kPa}$$

$$T_1 = 200. \text{ K}$$

$$P_2 = ???$$

$$T_2 = 500 \text{ K}$$

$$\frac{7.00 \text{ kPa}}{200. \text{ K}} = \frac{P_2}{500 \text{ K}}$$

$$P_2 = \frac{7.00 \text{ kPa} \cdot 500 \cancel{K}}{200. \cancel{K}}$$

$$= 17.5 \text{ kPa}$$

TRY IT YOURSELF
The pressure exerted by an ideally behaving gas is 50,000 pascals at a temperature of 327°C. At what temperature, in degrees Celsius, will the pressure exerted by the gas be 25,000 pascals, volume and number of particles remaining constant? (*Hint:* Remember that Gay-Lussac's law requires *Kelvin* temperature.)

ANSWER
27°C

The Combined Gas Law

It is possible to combine the three laws for ideally behaving gases into a single relationship:

$$\frac{\text{pressure} \cdot \text{volume}}{\text{temperature}} = \text{constant}$$

(at constant number of particles)

or, in equivalent form:

$$\frac{P_1 \cdot V_1}{T_1} = \frac{P_2 \cdot V_2}{T_2}$$

This **combined gas law** replaces the three individual laws—Boyle's, Charles's, and Gay-Lussac's—for ideally behaving gases. For example, if the pressure is held constant in a given problem, then $P_1 = P_2$ and the pressure term can be eliminated from the equation, leaving

$$\frac{V_1}{T_1} = \frac{V_2}{T_2} \qquad \text{(Charles's law)}$$

151

The additional benefit we obtain from using the combined law lies in the fact that it enables us to solve problems in which pressure, volume, and temperature may be changing.

PROBLEM
An ideally behaving gas occupies 500 milliliters at STP. What volume does it occupy at 546 K and 0.5 atmosphere, number of particles remaining constant?

SOLUTION

$$\frac{P_1 \cdot V_1}{T_1} = \frac{P_2 \cdot V_2}{T_2}$$

$$P_1 = 1 \text{ atm}$$
$$V_1 = 500 \text{ mL}$$
$$T_1 = 273 \text{ K}$$
$$P_2 = 0.5 \text{ atm}$$
$$V_2 = \text{ ???}$$
$$T_2 = 546 \text{ K}$$

$$\frac{1 \text{ atm} \cdot 500 \text{ mL}}{273 \text{ K}} = \frac{0.5 \text{ atm} \cdot V_2}{546 \text{ K}}$$

$$V_2 = \frac{1 \text{ atm} \cdot 500 \text{ mL} \cdot 546 \text{ K}}{273 \text{ K} \cdot 0.5 \text{ atm}}$$

$$= 2000 \text{ mL}$$

TRY IT YOURSELF
An ideally behaving gas occupies 12 cubic decimeters at 0.5 atmosphere and 300 K. At what temperature will this gas occupy 6 cubic decimeters at 0.25 atmosphere, number of particles remaining constant?

ANSWER
75 K

Relating Volume and Moles (Avogadro's Law)

We now ask this question: How does the number of gas particles in a container affect the volume of a gas when the pressure and temperature are held constant? Imagine that we have a container whose volume is 5.0 liters and that contains n moles of ideally behaving gas particles at STP. Now suppose we have a second container that is identical to the first in every respect. If we join the containers, as shown in the diagram, it is obvious that we have a volume of 10 liters and $2n$ moles of gas particles.

Container 1	Container 2
5.0 liters n moles at STP	5.0 liters n moles at STP

Joined Containers

10. liters $2n$ moles at STP

Amadeo Avogadro first proposed an hypothesis about gases in 1811: *At the same temperature and pressure, equal volumes of any (ideally behaving) gas contain equal numbers of particles*. Therefore, each of the three containers in the accompanying diagram holds the same number, n, of particles, even though there is a different gas in each container.

Gas: CO	Gas: H_2	Gas: Ar
Pressure: 10. kPa	Pressure: 10. kPa	Pressure: 10. kPa
Volume: 5.0 L	Volume: 5.0 L	Volume: 5.0 L
Temperature: 800 K	Temperature: 800 K	Temperature: 800 K
Number of particles: n	Number of particles: n	Number of particles: n

Avogadro's law states also that the volume of an ideally behaving gas is *directly proportional* to the number of gas particles present *if the temperature and pressure are held constant*. This relationship, known as *Avogadro's law*, is written as follows:

$$\frac{\text{volume}}{\text{number of particles}} = \text{constant}$$

(at constant temperature and pressure)

or, in equivalent form:

$$\frac{V_1}{n_1} = \frac{V_2}{n_2}$$

This basic law of nature follows from the observation that all the gas laws we have studied do *not* depend on the *identity* of the gas. The only requirement is that the sample of gas behaves ideally.

6.4   THE KINETIC-MOLECULAR THEORY (KMT) OF GAS BEHAVIOR

So far, we have described gas behavior in terms of *experimental laws*; that is, laws that can be investigated in the laboratory. We have assumed that all of the gases behaved ideally, but we have never actually defined the term *ideal gas*.

The KMT Model of an Ideal Gas

Now we take a different approach. We propose a model (the **kinetic-molecular theory** or **KMT model**) of an *ideal* gas, and we manipulate the model according to the basic laws of (shudder!) mathematics and physics. Then we look at the results. (We will not do all of the fancy math; we will be content to accept the conclusions.)

An **ideal gas** is a collection of particles that:

- Have mass but negligible volume.
- Move randomly in straight lines.
- Are *not* subject to any attractive or repulsive forces (except during collisions with each other or with the walls of their container).
- Collide in a *perfectly elastic* fashion; that is, no energy is lost during a collision.

The KMT model of gas behavior enables us to draw three conclusions:

1. The **volume** occupied by an ideal gas is essentially the volume of its container.
2. The **pressure** exerted by an ideal gas is related to the number of collisions that the particles make with the walls of the container in a given amount of time.
3. The **absolute (Kelvin) temperature** of an ideal gas is directly proportional to the *average* kinetic energy of the gas particles. (In other words, if the absolute temperature is doubled, the average kinetic energy of the particles is also doubled.)

Astounding as it seems, *all* of the gas laws we have developed in this chapter can be derived as a direct result of the KMT model. Therefore, we can conclude that the KMT is a well-constructed model for explaining the behavior of ideal gases.

PROBLEM

Use the KMT to explain why reducing the volume of a gas at constant temperature increases the pressure that the gas exerts on the walls of the container.

SOLUTION

Refer to the accompanying diagram.

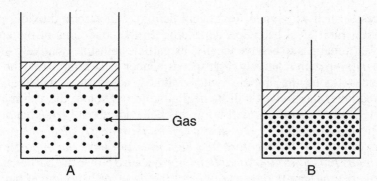

In the cylinder with the reduced volume (cylinder *B*), the molecules of the gas have less space in which to move. As a result, the molecules collide more frequently with the walls of the container. Since pressure depends on the frequency of these collisions, the pressure exerted by the gas increases.

PROBLEM

Use the KMT to explain why lowering the temperature of a gas at constant volume decreases the pressure that the gas exerts on the walls of the container.

SOLUTION

Refer to the accompanying diagram.

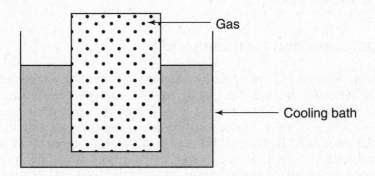

As the temperature of the gas is lowered, the average kinetic energy of the gas molecules decreases. Since the molecules are, on the average, traveling more slowly, they collide less frequently with the walls of the container and the pressure exerted on the walls decreases.

The Behavior of Real Gases

Real gases, such as oxygen, hydrogen, helium, and carbon dioxide, may or may not exhibit ideal behavior, depending on the conditions of the environment. For a real gas to behave ideally, its particles must be relatively far apart and be moving with relatively high speeds. Under these conditions, the attractive forces among the particles will be almost nonexistent, and the volume of the particles themselves will be negligible in relation to the volume of the container. What conditions will bring this behavior about? *High temperature and low pressure enable a real gas to exhibit most nearly ideal behavior*.

Conversely, *low temperature and high pressure have the opposite effect: the real gas fails to exhibit ideal behavior* because either the particles attract each other significantly or the volume of the particles is no longer negligible in relation to the volume of the container. As a consequence, gases under these conditions do not obey the gas laws completely and chemists have had to devise other relationships to correct for this nonideal behavior.

6.5 LIQUIDS

Liquids exist because particles of matter do *not* exhibit ideal behavior: they have a measurable volume and they exert forces on one another. When the temperature of a gas is sufficiently low and the pressure is sufficiently high, the gas becomes a liquid. The motion of the liquid particles becomes more restricted (the particles cannot move freely through space). As a result, the volume of the liquid becomes definite, but the liquid continues to take the shape of its container.

Evaporation and Condensation

According to the KMT, as applied to liquids, the absolute temperature of a liquid is related to the *average* kinetic energy of the particles of the liquid. However, some of the particles have enough kinetic energy to overcome the attractive forces of their neighboring particles, and they are able to escape from the surface of the liquid. We call this process **evaporation** because the particles that escape enter the *vapor* or gas phase above the liquid. The reverse process, called **condensation**, also occurs: particles of vapor with lower kinetic energies are recaptured by the liquid.

The rate of evaporation depends on three factors:

1. The nature of the liquid (ethanol evaporates more quickly than water).
2. The temperature of the liquid (hot water evaporates more quickly than cold water).
3. The surface area of the liquid (a puddle of water evaporates more quickly than water in a bottle).

Vapor Pressure

If we place a liquid in a sealed container, the vapor produced by the evaporating liquid will exert pressure on the surface of the liquid. Eventually, this pressure will seem to prevent the liquid from evaporating further. Actually, the *rate* of evaporation becomes equal to the *rate* of condensation: a condition we call **dynamic equilibrium**. The pressure that exists between the vapor and the liquid is known as the **equilibrium vapor pressure** (or simply the **vapor pressure**). Vapor pressure depends on the nature of the liquid and its temperature.

In the accompanying graph of vapor pressure as a function of temperature for five liquids, note that the vapor pressure rises slowly at low temperatures and climbs more rapidly at higher temperatures. At any given temperature, acetone has a higher vapor pressure than ethanol. As a result, the evaporation rate for acetone is greater than the rate for ethanol. A similar graph is shown in Reference Table H in Appendix 1.

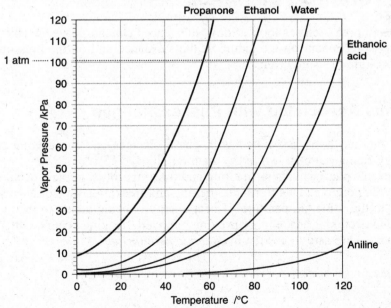

Vapor Pressure versus Temperature

If a liquid is heated in an open container, its vapor pressure will rise until it equals the pressure of the atmosphere above the liquid. At this point the liquid will begin to boil, and the temperature at which this occurs is known as the *boiling point* of the liquid. A liquid can have any number of boiling points, depending on the value of the pressure above the liquid.

When the atmospheric pressure is 1 atmosphere, the boiling temperature is known as the **normal boiling point** of the liquid. If we refer to the previous graph, we can see that propanone has the lowest normal boiling point (56°C) because it reaches a vapor pressure of 760 kilopascals at a lower temperature than any of the other liquids. According to the graph, the normal boiling points of ethanol, water, and ethanoic acid are 78°C, 100°C, and 118°C, respectively. The normal boiling point of aniline (184°C) is *not* shown on the graph because its vapor pressure rises too slowly as temperature is increased.

6.6 SOLIDS

As the temperature of a liquid is lowered, the forces of attraction between the particles become stronger and the particles begin to arrange themselves in an orderly fashion. The motion of the particles becomes severely restricted (largely vibrations), and the substance is said to be in the solid phase. All true solids have crystalline structure, and we say that the particles occupy positions within the crystal *lattice* of the substance. Solids and their crystal systems are studied in more advanced courses in chemistry.

6.7 CHANGE OF PHASE

We now take a closer look at how substances change phase. A change of phase is dependent on the nature of the substance and on the pressure and temperature of its environment.

Terms Associated with Phase Changes

The various terms associated with phase changes and the accompanying energy changes are shown in the diagram on page 159.

Reverse phase changes—melting and freezing, for example—occur at the same temperature. Therefore, the melting and freezing points of a substance refer to the same temperature. If heat energy is being added to the system, melting occurs; if heat energy is being removed, freezing occurs. Otherwise, the phases are said to coexist in *dynamic equilibrium*.

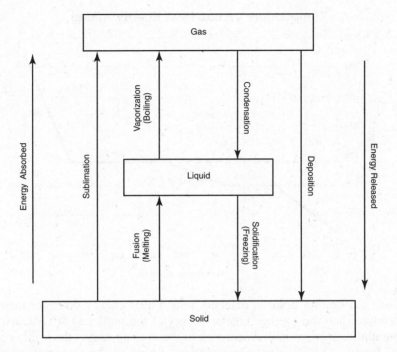

Sublimation is the most peculiar phase change because no liquid phase is encountered; the solid passes directly into the gas phase. Substances that sublime have relatively high vapor pressures (compared to the atmospheric pressure), as well as relatively weak forces of attraction between the particles of the solid. Dry ice [CO_2(s)] sublimes readily at room temperature and pressure. *Deposition*, the reverse of sublimation, is demonstrated when a carbon dioxide fire extinguisher is released. The rapid cooling of the carbon dioxide gas causes crystals of dry ice to form.

Heating and Cooling Curves

Suppose we take a solid that does not sublime, gradually add heat energy to it, and note how the temperature of the system changes as time progresses. If we graphed the results, we would produce a curve similar to the one shown that follows the next paragraph.

As heat energy is added, the temperature of the solid rises, indicating that the heat energy is being used to increase the average kinetic energy of the particles. After a while, however, the temperature remains constant, even though heat energy is still being added to the system. At this point the phase change of melting is occurring. According to the heating curve, the melting point of this solid is 100°C. The energy is being stored as potential energy. When melting has been completed, the temperature of the liquid begins to rise again as the average kinetic energy of the particles is increasing.

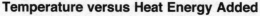

Temperature versus Heat Energy Added

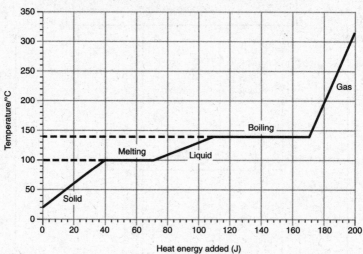

When the boiling point is reached, added heat energy does not raise the temperature (and the average kinetic energy of the particles) but is converted to potential energy instead. According to the heating curve, the boiling point of this liquid is 140°C. At the completion of boiling, the temperature of the gas again rises with the addition of heat energy.

If we remove heat energy from a gas, we obtain the cooling curve given below. Note that it is the reverse of the heating curve above.

Temperature versus Heat Energy Added

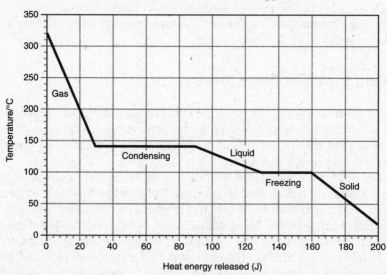

Inspection of the cooling curve confirms that condensation takes place at the boiling point and freezing occurs at the melting point. The only change is that heat energy is being removed from the system.

During melting, a specific amount of heat energy was added to the solid in order to convert it to a liquid. This energy, known as the *heat of fusion* (H_f), is sometimes called a *latent* (hidden) heat because its addition is not accompanied by a temperature change. Heat of fusion is measured in joules per gram (J/g). According to Reference Table B in Appendix 1 of this book, the heat of fusion of ice at 0°C is 333.6 joules per gram. This means that 333.6 joules of heat energy is needed to convert each gram of ice into liquid water. Conversely, during freezing, 333.6 joules of heat energy is released for each gram of water that is converted into ice.

Similarly, during boiling, a specific amount of heat energy, known as the *heat of vaporization* (H_v), is needed to convert a liquid into a gas. The heat of vaporization of water at 100°C is 2259 joules per gram. Therefore, as each gram of water is converted into steam, 2259 joules of heat energy is absorbed by the system. As each gram of steam condenses to liquid water, 2259 joules of heat energy is released.

PROBLEM
How much heat energy is needed to melt 6.00 grams of solid water (ice) at its melting point?

SOLUTION
We find the heat of fusion of water in Reference Table B and use this value and the factor-label method to solve the problem.

$$\Delta H = 6.00 \text{ g} \cdot \frac{333.6 \text{ J}}{1 \text{ g}}$$

$$= 2.00 \times 10^3 \text{ J}$$

(Note the use of the symbol ΔH to represent the quantity of the heat absorbed or released.)

SECTION II—ADDITIONAL MATERIAL

6.1A MEASURING GAS PRESSURE IN THE LABORATORY

In laboratories, the pressure of a gas enclosed in a container is usually measured by a device called a **manometer**, which comes in two forms: closed-tube and open-tube. The operation of a manometer is similar to that of a barometer. A bent tube is filled with a liquid, such as mercury, and the tube is connected to the container.

Closed-tube manometers are usually used when the pressure of the enclosed gas is significantly less than atmospheric pressure. The accompanying diagram illustrates a closed-tube manometer.

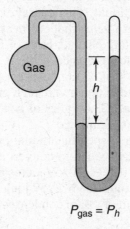

$$P_{gas} = P_h$$

The pressure of the enclosed gas is simply pressure represented by the difference in the heights (h) of the two liquid levels:

$$P_{gas} = P_h$$

Open-tube manometers are used when the pressure of the enclosed gas is at or near atmospheric pressure. The accompanying diagram shows three open-tube manometer arrangements, (a), (b), and (c):

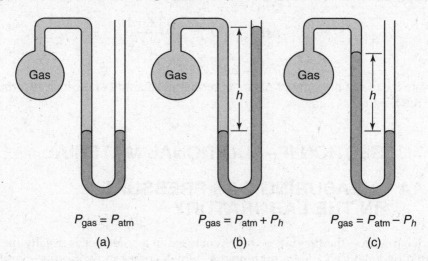

$$P_{gas} = P_{atm} \qquad P_{gas} = P_{atm} + P_h \qquad P_{gas} = P_{atm} - P_h$$

(a) (b) (c)

When the pressure of the enclosed gas is *equal* to the atmospheric pressure, both liquid levels are *equal*, as in arrangement (a):

$$P_{gas} = P_{atm}$$

When the gas pressure is *greater* than the atmospheric pressure, the liquid level nearest the enclosed gas is *lower* than the level in the open tube, as in arrangement (b):

$$P_{gas} = P_{atm} + P_h$$

When the gas pressure is *less* than the atmospheric pressure, the liquid level nearest the enclosed gas is *higher* than the level in the open tube, as in arrangement (c):

$$P_{gas} = P_{atm} - P_h$$

6.2A   THE IDEAL (UNIVERSAL) GAS LAW

We are now ready to combine all of the preceding gas laws into a single, universal relationship that can be written as follows:

$$\frac{\text{pressure} \cdot \text{volume}}{\text{number of particles} \cdot \text{temperature}} = \text{constant}$$

Mathematically, the **ideal gas law** can be expressed as:

$$\frac{P \cdot V}{n \cdot T} = R$$

or, in its more usual form:

$$P \cdot V = n \cdot R \cdot T$$

where R is a constant (called the **molar gas constant**) that depends on the units of pressure and volume. If the pressure is measured in atmospheres and the volume in liters, then R has the value 0.0821 liter · atmospheres per mole · kelvin (0.0821 L · atm/mol · K). In the SI system, where pressure is measured in pascals and volume in cubic meters, R has the value 8.31 joules per mole · kelvin (8.31 J/mol · K).

PROBLEM
How many moles of an ideally behaving gas occupy 400 liters at 0.821 atmosphere and 200 K?

SOLUTION

$$P \cdot V = n \cdot R \cdot T$$
$$P = 0.821 \text{ atm}$$
$$V = 400 \text{ L}$$
$$n = ???$$
$$R = 0.0821 \text{ L} \cdot \text{atm/mol} \cdot \text{K}$$
$$T = 200 \text{ K}$$
$$0.821 \text{ atm} \cdot 400 \text{ L} = n \cdot 0.0821 \text{ L} \cdot \text{atm/mol} \cdot \text{K} \cdot 200 \text{ K}$$
$$n = \frac{0.821 \text{ atm} \cdot 400 \text{ L}}{0.0821 \text{ L} \cdot \text{atm}/\text{mol} \cdot \text{K} \cdot 200 \text{ K}}$$
$$= 20 \text{ moles of gas}$$

PROBLEM

What volume will an ideally behaving gas containing 1.00 mole of particles occupy at STP?

SOLUTION

$$P \cdot V = n \cdot R \cdot T$$
$$P = 1.00 \text{ atm}$$
$$V = ???$$
$$n = 1.00 \text{ mol}$$
$$R = 0.0821 \text{ L} \cdot \text{atm/mol} \cdot \text{K}$$
$$T = 273 \text{ K}$$
$$1.00 \text{ atm} \cdot V = 1.00 \text{ mol} \cdot 0.0821 \text{ L} \cdot \text{atm/mol} \cdot \text{K} \cdot 273 \text{ K}$$
$$V = \frac{1.00 \text{ mol} \cdot 0.0821 \text{ L} \cdot \text{atm}/\text{mol} \cdot \text{K} \cdot 273 \text{ K}}{1.00 \text{ atm}}$$
$$V = 22.4 \text{ L}$$

This volume, 22.4 liters, appears so often in chemistry that it is given a special name—the **molar volume** *at STP* (denoted as V_m); its units are liters per mole (L/mol).

PROBLEM

What is the volume occupied by 3.00 moles of nitrogen gas (N_2) at STP?

SOLUTION

$$3.00 \text{ mol } N_2 \left(\frac{22.4 \text{ L } N_2}{1 \text{ mol } N_2} \right) = 67.2 \text{ L}$$

TRY IT YOURSELF
What volume is occupied by 8.00 grams of helium gas (He) at STP?

ANSWER
44.8 L

6.3A   THE DENSITY OF AN IDEAL GAS AT STP

Since the mole is associated with a mass (molar mass) and a gas volume (molar volume) at STP, we can calculate the gas *density* for any substance that is assumed to behave as an ideal gas at STP. We recall that density is calculated by dividing the mass of a substance by its volume. Therefore, we can calculate this special density from the following relationship:

$$d_{STP} = \frac{\mathcal{M}}{V_m}$$

where V_m is 22.4 liters per mole at STP.

PROBLEM
Calculate the density of CO gas at STP.

SOLUTION
The molar mass of CO is 28.01 g/mol. Therefore:

$$d_{STP} = \frac{28.01\,\frac{g}{mol}}{22.4\,\frac{L}{mol}} = 1.25\,\frac{g}{L}\,(\text{at STP})$$

TRY IT YOURSELF
Calculate the density of O_2 gas at STP.

ANSWER
1.43 g/L

Molar Mass from Gas Density

Since the molar mass of a gaseous substance can be used to calculate the density of the substance at STP, we can reverse the process; that is, we can calculate the molar mass from the gas density at STP. This is accomplished by multiplying the density by V_m (22.4 L/mol at STP), as given by the following relationship:

$$\mathcal{M} = d_{STP} \cdot V_m$$

PROBLEM

The density of a gas is 1.96 grams per liter at STP. Calculate its molar mass.

SOLUTION

$$\frac{1.96 \text{ g}}{\cancel{L}} \cdot \frac{22.4 \cancel{L}}{\text{mol}} = 43.9 \text{ g/mol}$$

TRY IT YOURSELF

The density of a gas is 2.59 grams per liter at STP. Calculate its molar mass.

ANSWER

58.0 g/mol

Determination of Molecular Formula from Gas Density

If the empirical formula and the density of a gaseous substance at STP are known, the molecular formula of the substance is easily calculated.

PROBLEM

The empirical formula of a gaseous substance is CH, and its density at STP is 3.48 grams per liter. Calculate the molecular formula of this substance.

SOLUTION

The molar mass of the substance is calculated from this relationship:

$$d_{STP} \cdot 22.4 \frac{L}{\text{mol}} = \mathcal{M}$$

$$\mathcal{M} = \frac{3.48 \text{ g}}{\cancel{L}} \cdot \frac{22.4 \cancel{L}}{\text{mol}} = 78.0 \text{ g/mol}$$

Since the empirical formula is CH, the *empirical* molar mass is 13.0 g/mol.

The molecular formula *must* be a multiple of the empirical formula. We can find this multiple by dividing the two molar masses:

$$\frac{78.0 \text{ g/mol}}{13.0 \text{ g/mol}} = 6.00$$

Therefore, the molecular formula is *six times* the empirical formula:

$$\text{Molecular formula} = 6 \cdot CH = C_6H_6$$

6.4A   GASES AND CHEMICAL REACTIONS

Mass-Volume Gas Problems

PROBLEM
In the equation $2CO(g) + O_2(g) \rightarrow 2CO_2(g)$, how many liters of $CO_2(g)$ at STP are produced by the reaction of 64.0 grams of $O_2(g)$?

SOLUTION
This type of problem is also solved by using FLM. We need the connections among mass, volume, and moles. Here is the solution map:

$$64.0 \text{ g } O_2 \xrightarrow{} \text{mol } O_2 \xrightarrow[\text{coefficients}]{\text{equation}} \text{mol } CO_2 \xrightarrow{} \text{L } CO_2$$

The solution is as follows:

$$\text{L } CO_2 = 64.0 \text{ g } O_2 \cdot \frac{1 \text{ mol } O_2}{32.0 \text{ g } O_2} \cdot \frac{2 \text{ mol } CO_2}{1 \text{ mol } O_2} \cdot \frac{22.4 \text{ L } CO_2}{1 \text{ mol } CO_2} = 89.6 \text{ L } CO_2$$

TRY IT YOURSELF
In the equation $2CO(g) + O_2(g) \rightarrow 2CO_2(g)$, how many liters of $O_2(g)$ at STP are able to react with 28.0 grams of $CO(g)$?

ANSWER
11.2 L

Volume-Volume Gas Problems

PROBLEM
In the equation $4NH_3(g) + 5O_2(g) \rightarrow 4NO(g) + 6H_2O(l)$, how many liters of $NH_3(g)$ at STP are needed to react with 200. liters of $O_2(g)$ at STP?

SOLUTION

To solve this problem, we need to convert between volume and numbers of moles. Our solution map is as follows:

$$200. \text{ L O}_2 \xrightarrow{} \text{mol O}_2 \xrightarrow[\text{coefficients}]{\text{equation}} \text{mol NH}_3 \xrightarrow{} \text{L NH}_3$$

The solution is given below:

$$\text{L NH}_3 = 200. \ \cancel{\text{L O}_2} \cdot \frac{1 \ \cancel{\text{mol O}_2}}{22.4 \ \cancel{\text{L}}} \cdot \frac{4 \ \cancel{\text{mol NH}_3}}{5 \ \cancel{\text{mol O}_2}} \cdot \frac{22.4 \ \text{L NH}_3}{1 \ \cancel{\text{mol NH}_3}} = 160. \ \text{L NH}_3$$

Volume-volume problems are particularly easy to solve because the factor "22.4" *always* cancels in the calculations. (Why?)

TRY IT YOURSELF

In the equation $4NH_3(g) + 5O_2(g) \rightarrow 4NO(g) + 6H_2O(\ell)$, how many liters of $NH_3(g)$ at STP are needed to produce 1000 liters of $NO(g)$ at STP?

ANSWER

1000 L

6.5A   DALTON'S LAW OF PARTIAL PRESSURES

Imagine that we have a container with a *mixture* of three ideally behaving gases, A, B, and C. We know that the gases would exert a combined pressure on the walls of the container (P_{total}). What we would like to know is how this combined pressure is related to the individual pressures (P_A, P_B, and P_C) of the three gases in the container. These individual pressures are called *partial pressures*. A partial pressure is the pressure that a single gas would exert *if it were the only gas in the container*.

Dalton discovered that a simple relationship exists between the total pressure of a gas mixture and the partial pressures of its components. **Dalton's law of partial pressures** states that the *sum* of the partial pressures equals the total pressure exerted on the container. For our example, this relationship is as follows:

$$P_{total} = P_A + P_B + P_C$$

The total pressure of the three gas components in the mixture described above is 55.0 kilopascals. If P_A = 20.0 kilopascals and P_B = 7.5 kilopascals, what is the partial pressure of gas C (P_C)?

SOLUTION

$$P_{total} = P_A + P_B + P_C$$
$$P_C = P_{total} - P_A - P_B$$
$$= 55.0 \text{ kPa} - 20.0 \text{ kPa} - 7.5 \text{ kPa}$$
$$= 27.5 \text{ kPa}$$

TRY IT YOURSELF

Four gases in a tank have respective partial pressures of 0.20 atmosphere, 0.35 atmosphere, 0.15 atmosphere, and 0.60 atmosphere. What is the total gas pressure inside the tank?

ANSWER

1.30 atm

The simple relationship expressed in Dalton's law is a direct result of the fact that each gas in a mixture occupies the *entire* volume of the container. For example, if we sample the air in various parts of a room, we expect to find nitrogen, oxygen, and other gases in every sample we take. Since the volume and temperature are the same for each of the components in a gas mixture, it follows that the partial pressure of each component is directly related to the *number of moles* of that component present in the mixture. There is a simple relationship among the partial pressure of any component in a gas mixture, the total pressure of the mixture, and the number of moles of gas present:

$$P_i = \left(\frac{n_i}{n_{total}}\right) \cdot P_{total}$$

where P_i and n_i represent the partial pressure and number of moles, respectively, of any component in the mixture, and P_{total} and n_{total} represent the pressure and number of moles, respectively, of the *entire mixture*.

PROBLEM

A tank contains 1.5 moles of gas A, 2.0 moles of gas B, and 2.5 moles of gas C. If the total pressure of the gas mixture is 0.80 atmosphere, what is the partial pressure of gas A?

SOLUTION

$$P_A = \left(\frac{n_A}{n_{\text{total}}}\right) \cdot P_{\text{total}}$$

$$= \left(\frac{1.5 \text{ mol}}{1.5 \text{ mol} + 2.0 \text{ mol} + 2.5 \text{ mol}}\right) \cdot 0.80 \text{ atm}$$

$$= \left(\frac{1.5 \text{ mol}}{6.0 \text{ mol}}\right) \cdot 0.80 \text{ atm}$$

$$= 0.20 \text{ atm}$$

TRY IT YOURSELF
Calculate the partial pressure of gas C in the problem given above.

ANSWER
0.33 atm

6.6A   GRAHAM'S LAW OF EFFUSION (DIFFUSION)

Effusion is the process by which gases escape through small openings or *pores* in materials. We have all had the experience of buying a balloon filled with helium and discovering that, one day later, the helium has escaped and the balloon has lost its firmness. This is an example of a gas effusing.

When gases are brought into contact, they mix with one another. This process is known as *diffusion*.

In 1832 the Scottish chemist Thomas Graham investigated the effusion and diffusion of gases and discovered that both processes obey the same simple law known as **Graham's law**: At a given temperature and pressure, the greater the density of a gas, the *slower* the gas effuses (and diffuses). Helium and hydrogen, the least dense gases, effuse (and diffuse) most rapidly, while gases with greater densities, such as oxygen and sulfur dioxide, effuse (and diffuse) more slowly.

Graham's law relates the relative rates of effusion (and diffusion) of gases to their molar masses: the greater the molar mass of a gas, the *slower* it effuses (and diffuses). Gases with small molar masses effuse (and diffuse) most rapidly, while gases with large molar masses effuse (and diffuse) more slowly.

Mathematically, *Graham's law* is expressed as a comparison between two ideally behaving gases, A and B, at a specified temperature and pressure. Their rates of effusion (and diffusion) are represented by R_A and R_B, and their molar masses by $\mathcal{M}_A$ and $\mathcal{M}_B$:

$$\frac{R_A}{R_B} = \sqrt{\frac{\mathcal{M}_B}{\mathcal{M}_A}}$$

PROBLEM

Calculate the rate of effusion (and diffusion) of hydrogen gas ($\mathcal{M} = 2.0$ g/mol) relative to that of oxygen gas ($\mathcal{M} = 32.0$ g/mol) at the same temperature and pressure.

SOLUTION

$$\frac{R_{H_2}}{R_{O_2}} = \sqrt{\frac{\mathcal{M}_{O_2}}{\mathcal{M}_{H_2}}}$$

$$= \sqrt{\frac{32.0 \text{ g/mol}}{2.0 \text{ g/mol}}} = \sqrt{\frac{16.0}{1.0}} = \frac{4.0}{1.0}$$

This result means that, under the same conditions of temperature and pressure, hydrogen gas will effuse (and diffuse) four times faster than oxygen gas.

TRY IT YOURSELF

Compare the rate of effusion (and diffusion) of O_2 gas ($\mathcal{M} = 32.0$ g/mol) relative to that of SO_2 gas ($\mathcal{M} = 64.0$ g/mol) at STP.

ANSWER

O_2 will effuse (and diffuse) 1.41 times faster than SO_2.

6.7A GASES COLLECTED OVER WATER

When a gas, such as hydrogen, is collected over water, there are actually *two* gases present in the container: hydrogen and water vapor. According to Dalton's law of partial pressures, both gases occupy the entire volume of the container but each gas exerts its own partial pressure on the container. The pressure of the water vapor may be determined by referring to Table X in Appendix 2. The pressure of the hydrogen is found by *subtracting* the vapor pressure of the water from the total pressure of the gas mixture.

PROBLEM

Hydrogen is collected in a 250-milliliter container over water at 30°C. The total pressure is 101.246 kilopascals.
(a) What is the pressure of the "dry" hydrogen gas?
(b) What is the volume occupied by the water vapor?

SOLUTIONS

(a) According to Table X in Appendix 2, the vapor pressure of water at 30°C is 4.246 kPa. According to Dalton's law:

$$P_{H_2} + P_{H_2O} = P_{total}$$
$$P_{H_2} = P_{total} - P_{H_2O}$$
$$= 101.246 \text{ kPa} - 4.246 \text{ kPa}$$
$$= 97.000 \text{ kPa}$$

(b) The volume of the water vapor is 250 mL (as is the volume of the hydrogen gas).

TRY IT YOURSELF

When hydrogen gas is collected over water at 20°C, it is found that the partial pressure of the hydrogen is 96.349 kilopascals. At what (total) pressure was the gas collected?

ANSWER

98.688 kPa

PROBLEM

Five hundred milliliters of a gas is collected over water at 298 K (25°C). The atmospheric pressure is 101.169 kilopascals. What would the volume of the "dry" gas be at STP?

SOLUTION

This problem involves Dalton's law and the combined gas law. When the gas is collected, the *total* gas pressure in the container becomes equal to the atmospheric pressure (101.169 kPa). The volume of the "dry" gas in the original container is (as always) 500 mL.

The partial pressure of the dry gas is 98.000 kPa (101.169 kPa − 3.169 kPa). We now apply the combined gas law:

$$\frac{P_1 \cdot V_1}{T_1} = \frac{P_2 \cdot V_2}{T_2}$$

$$\frac{98.000 \text{ kPa} \cdot 500 \text{ mL}}{298 \text{ K}} = \frac{101.325 \text{ kPa} \cdot V_2}{273 \text{ K}}$$

$$V_2 = \frac{98.000 \text{ kPa} \cdot 500 \text{ mL} \cdot 273 \text{ K}}{298 \text{ K} \cdot 101.325 \text{ kPa}}$$

$$= 443 \text{ mL}$$

TRY IT YOURSELF
A 940.0-milliliter sample of gas is collected over water at 40.0°C. The total pressure of the gas is 85.592 kilopascals. Calculate the volume of the "dry" gas at 25.0°C and 72.661 kilopascals.

ANSWER
963.3 mL

6.8A ADDITIONAL FUSION AND VAPORIZATION PROBLEMS

Reference Table W-2 in Appendix 2 lists the heats of fusion and vaporization of selected substances. (For convenience, the melting points and boiling points of these substances are also included.)

When we inspect the table, we see that the heat of vaporization of a substance is always considerably larger than the corresponding heat of fusion. We may conclude that more energy is needed to convert a liquid into a gas than to convert a solid into a liquid. This fact can be explained by noting that a gas is a collection of particles that have little or no attractions among them. The large energy absorbed during the boiling process is used to free the particles of a liquid from their mutual attractions.

TRY IT YOURSELF

How much heat energy is absorbed when 20.0 grams of solid ethanol is converted to liquid at its melting point?

ANSWER
2080 J

PROBLEM

How many grams of oxygen gas will release 639 joules of heat energy as the gas condenses at its boiling point?

SOLUTION
The heat of vaporization of oxygen, as obtained from Reference Table W-2, is 213 J/g. Also:

$$x\,g = \frac{639\ \cancel{J}}{213\ \cancel{J}}$$

$$x = 3\ g$$

$$639\ J = x\,g \cdot \frac{213\ J}{1\ g}$$

$$x = 3.00\ g$$

173

TRY IT YOURSELF

How much heat energy is released by 10.0 grams of liquid copper as it solidifies at its melting point?

ANSWER
$\Delta H = 1340$ J

6.9A PHASE DIAGRAMS

Every phase change (solid $\rightleftarrows$ liquid, liquid $\rightleftarrows$ gas, solid $\rightleftarrows$ gas) occurs at a specific combination of temperature and pressure. For example, the phase change between ice and liquid water occurs at 0°C when the pressure is 1 atmosphere. The *direction* of the phase change depends on whether heat energy is being added or removed from the system. In our example, liquid water will freeze if heat energy is being *removed* from the ice-water system.

If *no* heat energy is added or removed, the two phases will remain in a state of *dynamic equilibrium* at that temperature and pressure. In our example, the rate of melting ice will be exactly equal to the rate of freezing water and the quantities of ice and water will remain constant over time.

We can summarize all of these relationships graphically by using a *phase diagram*. A general phase diagram follows.

Pressure versus Temperature

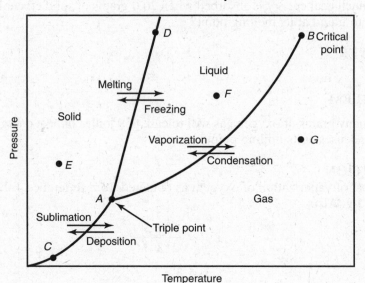

Each line represents the pressure-temperature combination at which two phases are in dynamic equilibrium. For example, line *CA*, the sublimation-deposition line, represents the conditions under which the solid and gas phases can be in dynamic equilibrium. At any combination of temperature and pressure that does *not* lie on one of the lines (points *E*, *F*, and *G*, for example) *only one phase* can exist.

The phase diagram includes two points of special interest:

1. Point *A* is known as the **triple point**. It is the only point at which all three phases (solid, liquid, and gas) can coexist in dynamic equilibrium.
2. Point *B* is known as the **critical point**. The temperature and pressure at the critical point are known, respectively, as the **critical temperature** and **critical pressure**. Above the critical temperature, only the gas phase can exist, no matter how high the pressure, because the kinetic energies of the gas particles are too high to allow the formation of a liquid.

The accompanying table lists the critical temperatures and pressures for four substances. We note that the critical pressures are generally quite large.

Substance	Critical Temperature/°C	Critical Pressure/atm
Ammonia (NH_3)	132.6	111.5
Carbon dioxide (CO_2)	31.3	73.0
Nitrogen (N_2)	−146.9	33.5
Water (H_2O)	374.6	217.7

The accompanying phase diagrams for (a) water and (b) carbon dioxide are *not* drawn to scale.

Pressure versus Temperature

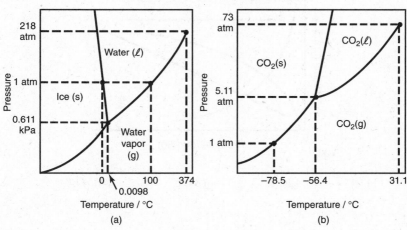

(a)   (b)

In the phase diagram for water, we note that the solid-liquid line tilts toward the left. This is a direct result of the fact that liquid water is more dense than ice at temperatures close to 0°C. It is also the reason why ice melts when the pressure is increased sufficiently. The triple point of water (0.611 kPa, 0.0098°C) provides the basis for defining the Kelvin temperature scale. *This point is assigned an exact value of 273.16 K.*

The phase diagram for carbon dioxide has the same appearance as the general phase diagram shown above because solid carbon dioxide is more dense than the liquid phase. We also note that, at a constant pressure of 1 atmosphere, carbon dioxide passes directly from the solid to the gas phase as the temperature is increased; the liquid phase does not even come into existence until the pressure reaches 5.11 atmospheres.

END-OF-CHAPTER QUESTIONS

Some questions have the symbol "§2" in front of the question number. This symbol means that the question is based on Section II material.

§2 **1.** A 1-liter flask contains two gases at a total pressure of 3.0 atmospheres. If the partial pressure of one of the gases is 0.5 atmosphere, the partial pressure of the other gas must be
(1) 1.0 atm   (2) 2.5 atm   (3) 1.5 atm   (4) 0.50 atm

2. At constant temperature, which line in the graph best shows the relationship between the volume of an ideal gas and its pressure?

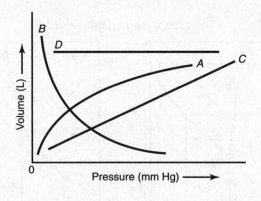

(1) *A*   (2) *B*   (3) *C*   (4) *D*

§2 **3.** A container that is filled with 1.00 mole of oxygen and 2.00 moles of hydrogen has a total pressure of 75.0 kilopascals. What is the partial pressure of the oxygen?
(1) 10.0 kPa   (2) 20.0 kPa
(3) 25.0 kPa   (4) 50.0 kPa

4. At STP, 44.8 liters of CO_2 contains the same number of molecules as
(1) 1.00 mole of He   (2) 2.00 moles of Ne
(3) 0.500 mole of H_2 (4) 4.00 moles of N_2

5. The pressure of 200. milliliters of a gas at constant temperature is changed from 380. torr to 760. torr. The new volume of the gas is
(1) 100. mL   (2) 200. mL
(3) 400. mL   (4) 800. mL

§2 **6.** At STP, the volume occupied by 32 grams of a gas is 11.2 liters. The molar mass of this gas is closest to
(1) 8.0 g/mol   (2) 16 g/mol
(3) 32 g/mol   (4) 64 g/mol

§2 **7.** In a laboratory experiment, hydrogen gas is collected over water. If the atmospheric pressure is 101.645 kPa and the room temperature is 22°C, the partial pressure of the hydrogen is
(1) 104.294 kPa   (2) 103.605 kPa
(3) 101.645 kPa   (4) 99.000 kPa

§2 **8.** If 2.5 moles of a gas occupy 30. liters at a certain temperature and pressure, what volume will 7.5 moles of the gas occupy at the same temperature and pressure?
(1) 10. L   (2) 30. L   (3) 90. L   (4) 180 L

§2 Base your answers to questions 9–11 on the accompanying graph and on the following information:

The graph represents an experiment in which the volumes of four cylinders (*A*, *B*, *C*, and *D*) of the same ideal gas were measured at different temperatures. Each cylinder contained 1.0 mole of this gas. During the experiment, the pressure of each cylinder was not allowed to change.

Volume versus Temperature

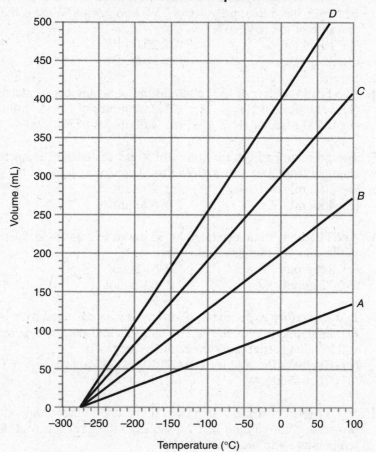

§2 **9.** In which cylinder was the pressure of the gas least?
  (1) *A*   (2) *B*   (3) *C*   (4) *D*

§2 **10.** At 546 K, in which cylinder would the gas occupy a volume of 400 milliliters?
  (1) *A*   (2) *B*   (3) *C*   (4) *D*

§2 **11.** For each 1° decrease in temperature, the volume occupied by the gas in cylinder *A* decreased by

  (1) $\dfrac{1}{273}$ mL (2) $\dfrac{100}{273}$ mL (3) $\dfrac{273}{100}$ mL (4) $\dfrac{273}{1}$ mL

12. At which temperature does a water sample have the highest average kinetic energy?
(1) 0°C   (2) 100°C   (3) 0 K   (4) 100 K

13. Which gas has properties that are most similar to those of an ideal gas?
(1) N_2 (2) O_2 (3) He   (4) Xe

14. When the average kinetic energy of a gaseous system is increased, the average molecular velocity of the system
(1) increases and the molecular mass increases
(2) decreases and the molecular mass increases
(3) increases and the molecular mass remains the same
(4) decreases and the molecular mass remains the same

15. Under which conditions does a real gas behave most like an ideal gas?
(1) at high temperatures and low pressures
(2) at high temperatures and high pressures
(3) at low temperatures and low pressures
(4) at low temperatures and high pressures

16. Which temperature represents absolute zero?
(1) 0 K   (2) 0°C   (3) 273 K   (4) 273°C

17. Which of the following substances is made up of particles with the highest average kinetic energy?
(1) Fe(s) at 35°C   (2) $Br_2(\ell)$ at 20°C
(3) $H_2O(\ell)$ at 30°C   (4) $CO_2(g)$ at 25°C

18. When a sample of a gas is heated at constant pressure, the average kinetic energy of its molecules
(1) decreases, and the volume of the gas increases
(2) decreases, and the volume of the gas decreases
(3) increases, and the volume of the gas increases
(4) increases, and the volume of the gas decreases

§2 **19.** Which gas has, under high pressure and low temperature, behavior closest to that of an ideal gas?
(1) $H_2(g)$ (2) $O_2(g)$ (3) $NH_3(g)$ (4) $CO_2(g)$

20. Compared to the average kinetic energy of 1 mole of water at 0°C, the average kinetic energy of 1 mole of water at 298 K is
(1) the same, and the number of molecules is the same
(2) the same, but the number of molecules is greater
(3) greater, and the number of molecules is greater
(4) greater, but the number of molecules is the same

21. When the vapor pressure of a liquid in an open container equals the atmospheric pressure, the liquid will
(1) freeze   (2) crystallize   (3) melt   (4) boil

22. As the temperature of a liquid increases, its vapor pressure
(1) decreases   (2) increases   (3) remains the same

23. Which sample of water has the greatest vapor pressure?
(1) 100 mL at 20°C   (2) 200 mL at 25°C
(3) 20 mL at 30°C   (4) 40 mL at 35°C

§2 **24.** The temperature of a sample of water is changed from 60°C to 70°C. According to Reference Table X, the change in the vapor pressure of the water is
(1) 11.25 kPa   (2) 19.93 kPa   (3) 31.18 kPa   (4) 51.11. kPa

§2 **25.** Water will boil at a temperature of 40°C when the pressure on its surface is
(1) 1.938 kPa   (2) 2.339 kPa   (3) 7.381 kPa   (4) 101.3 kPa

26. If 1.00 mole of $H_2(g)$ at STP is compared to 1.00 mole of He(g) at STP, the volumes of the gases will be found to be
(1) equal, and their masses unequal
(2) equal, and their masses equal
(3) unequal, and their masses unequal
(4) unequal, and their masses equal

27. The graph below shows the equilibrium vapor-pressure curves of liquids *A, B, C,* and *D*.

Pressure versus Temperature

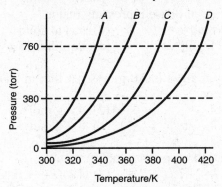

Which liquid has the lowest normal boiling point?
(1) *A*　　　　(2) *B*　　　　(3) *C*　　　　(4) *D*

28. Which sample contains particles arranged in a regular geometric pattern?
(1) $CO_2(\ell)$　(2) $CO_2(s)$　　(3) $CO_2(g)$　　(4) $CO_2(aq)$

29. The particles in a typical crystalline solid are arranged
(1) randomly and relatively far apart
(2) randomly and relatively close together
(3) regularly and relatively far apart
(4) regularly and relatively close together

30. The accompanying graph represents changes of state for an unknown substance. What is the boiling temperature of the substance?

Temperature versus Time

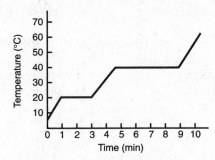

(1) 0°C　　　　(2) 20°C　　　　(3) 70°C　　　　(4) 40°C

31. What is the approximate number of joules of heat needed to change 150 grams of ice to liquid water at 0°C?
(1) 50,000   (2) 90,000   (3) 130,000   (4) 230,000

32. Which change of phase represents fusion?
(1) gas to liquid   (2) gas to solid
(3) solid to liquid   (4) liquid to gas

33. The density of a gas is 2.0 grams per liter at STP. Its molar mass is approximately
(1) 67 g/mol   (2) 45 g/mol
(3) 22 g/mol   (4) 8.0 g/mol

34. Which equation represents sublimation?
(1) $NH_3(g) \rightarrow NH_3(\ell)$ (2) $H_2O(\ell) \rightarrow H_2O(g)$
(3) $HCl(g) \rightarrow HCl(aq)$ (4) $CO_2(s) \rightarrow CO_2(g)$

§2 **35.** According to Reference Table W-2 in Appendix 2, which of the following solids requires the most heat energy to change it to a liquid at its melting point?
(1) aluminum   (2) copper   (3) lead   (4) water

§2 **36.** The temperature above which a gas can no longer be liquefied is known as the
(1) fusion point   (2) vaporization point
(3) critical point   (4) triple point

37. At STP, 32 grams of O_2 will occupy the same volume as
(1) 64 g of H_2 (2) 32 g of SO_2
(3) 8.0 g of CH_4 (4) 4.0 g of He

38. Which change of phase is exothermic?
(1) $H_2O(s) \rightarrow H_2O(g)$ (2) $CO_2(s) \rightarrow CO_2(\ell)$
(3) $H_2S(g) \rightarrow H_2S(\ell)$ (4) $NH_3(\ell) \rightarrow NH_3(g)$

39. Which term represents the change of a substance from the liquid phase to the gaseous phase?
(1) condensation   (2) vaporization
(3) sublimation   (4) fusion

40. The number of joules per gram required to melt ice at its melting point is called
(1) sublimation
(2) vapor pressure
(3) heat of vaporization
(4) heat of fusion

§2 **41.** Which gas is most dense at STP?
(1) $CO(g)$ (2) $NO(g)$ (3) $N_2(g)$ (4) $O_2(g)$

§2 **42.** Given the reaction

$$2C_2H_6(g) + 7O_2(g) \rightarrow 4CO_2(g) + 6H_2O(g)$$

At STP, what is the total volume of $CO_2(g)$ formed when 6.0 liters of $C_2H_6(g)$ is completely oxidized?
(1) 24 L   (2) 12 L   (3) 6.0 L   (4) 4.0 L

§2 **43.** Given this reaction at STP:

$$N_2(g) + 3H_2(g) \rightarrow 2NH_3(g)$$

What is the total number of liters of NH_3 formed when 20 liters of N_2 reacts completely?
(1) 10   (2) 20   (3) 30   (4) 40

§2 **44.** Given this balanced equation:

$$3Fe + 4H_2O \rightarrow Fe_3O_4 + 4H_2$$

What is the total number of liters of H_2 produced at STP when 36.0 grams of H_2O is consumed?
(1) 22.4   (2) 33.6   (3) 44.8   (4) 89.6

45. According to Reference Table H in Appendix 1, what is the vapor pressure of propanone at 45°C?
(1) 22 kPa   (2) 33 kPa   (3) 70 kPa   (4) 98 kPa

46. The freezing point of bromine is
(1) 539°C   (2) 7°C   (3) –539°C   (4) –7°C

47. The graph below represents the heating curve of a substance that starts as a solid below its freezing point.

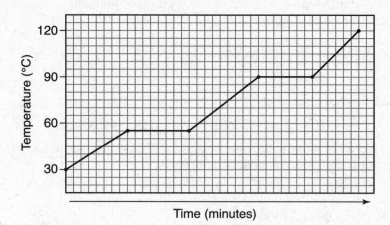

Time (minutes)

What is the freezing point of this substance?
(1) 30°C   (2) 55°C   (3) 90°C   (4) 120°C

48. An increase in the average kinetic energy of a sample of copper atoms occurs with an increase in
(1) concentration
(2) temperature
(3) pressure
(4) volume

49. Which statement describes particles of an ideal gas, based on the kinetic-molecular theory?
(1) Gas particles are separated by distances smaller than the size of the gas particles.
(2) Gas particles do not transfer energy to each other when they collide.
(3) Gas particles have no attractive forces between them.
(4) Gas particles move in predictable, circular movements.

50. Compared to a 1.0-gram sample of chlorine gas at standard pressure, a 1.0-gram sample of solid aluminum at standard pressure has
(1) a lower melting point
(2) a higher boiling point
(3) a lower density
(4) a greater volume

51. At STP, a 1-liter sample of Ne(g) and a 1-liter sample of Kr(g) have the same
    (1) mass
    (2) density
    (3) number of atoms
    (4) number of electrons

52. A rigid cylinder with a movable piston contains a sample of gas. At 300. K, this sample has a pressure of 240. kilopascals and a volume of 70.0 milliliters. What is the volume of this sample when the pressure is changed to 150. K and the pressure is changed to 160. kilopascals?
    (1) 35.0 mL   (2) 52.5 mL   (3) 70.0 mL   (4) 105 mL

Constructed-Response Questions

1. A sample of gas in a container at 30.0°C exerts a pressure of 700. torr. If the temperature of the container is raised to 200.°C at constant volume, what pressure will the gas exert?

2. A sample of gas at 30.00°C and standard pressure occupies a volume of 250.0 milliliters. Calculate the temperature, in degrees Celsius, at which the gas will occupy a volume of 1000. milliliters at standard pressure.

§2 3. A sample of oxygen gas is collected over water at 27°C. If the gas is collected under a pressure of 101.567 kPa, what is the partial pressure of the oxygen gas in the collection vessel?

§2 4. A 750.-milliliter sample of hydrogen gas is collected over water at 30°C. If the gas is collected under a pressure of 98.659 kPa, what volume will the *dry* hydrogen gas occupy at STP?

§2 5. Calculate the relative rates of effusion of $N_2(g)$ and $CO_2(g)$.

§2 6. Calculate the density of CH_4 gas at 303 K and a pressure of 0.658 atmosphere. (*Hint:* Assume that you have 1.00 mole of CH_4, and use $PV = nRT$ to calculate the volume of the gas.)

§2 **7.** Given the equation

$$Ni\ (s) + 4CO\ (g) \rightarrow Ni(CO)_4\ (g):$$

(a) How many liters of CO(g) at STP are needed to react with 15 grams of Ni(s)?

(b) How many liters of $Ni(CO)_4$ (g) at STP will be produced by the reaction of 1000. liters of CO(g) at STP with excess Ni(s)?

Base your answers to questions 8 and 9 on the diagram below, which shows a piston confining a gas in a cylinder.

8. Using the set of axes below, sketch the general relationship between the pressure and the volume of an ideal gas at constant temperature.

9. The volume of gas in the cylinder is 6.2 milliliters. The pressure of the gas is 1.4 atmospheres. The piston is then pushed in until the volume of gas is 3.1 milliliters while the temperature remains constant. Calculate the pressure of the gas, in atmospheres, after the change in volume.

Base your answers to questions 10 and 11 on the information below and on your knowledge of chemistry. A beaker contains a liquid sample of a molecular substance. Both the beaker and the liquid are at 194 K. The graph below represents the relationship between temperature and time as the beaker and its contents are cooled for 12 minutes in a refrigerated chamber.

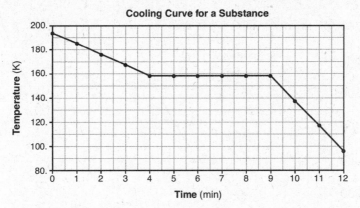

Cooling Curve for a Substance

10. State what happens to the average kinetic energy of the molecules of the sample during the first 3 minutes.

11. Identify the physical change occurring during the time interval minute 4 to minute 9.

Base your answers to questions 12 and 13 on the information below and on your knowledge of chemistry. The diagram below represents a cylinder with a movable piston. The cylinder contains 1.0 liter of oxygen gas at STP. The movable piston is pushed downward at constant temperature until the volume of the $O_2(g)$ is 0.50 liter.

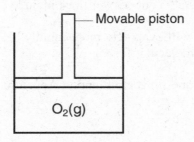

Movable piston

$O_2(g)$

12. Determine the new pressure of the $O_2(g)$ in the cylinder in atmospheres.

13. State the effect on the frequency of gas molecule collisions when the movable piston is pushed farther downward into the cylinder.

Base your answers to questions 14–16 on the graph below, which shows the vapor pressure curves for liquids A and B.

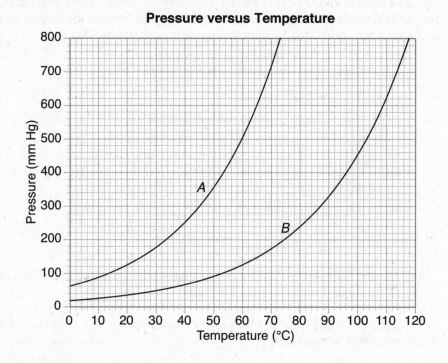

Pressure versus Temperature

14. What is the vapor pressure of liquid A at 70°C? Your answer must include correct units.

15. At what temperature does liquid B have the same vapor pressure as liquid A at 70°C? Your answer must include correct units.

16. Which liquid will evaporate more rapidly? Explain your answer in terms of intermolecular forces.

The answers to these questions are found in Appendix 3.

Chapter Seven
NUCLEAR CHEMISTRY

This chapter focuses on the structure and reactivity of the atomic nucleus. Applications of radioactivity are discussed, as is energy production from nuclear fission and fusion. The chapter concludes with a brief discussion of fundamental interactions and particles.

KEY OBJECTIVES
At the conclusion of this chapter you will be able to:
- List the common nuclear particles, and write their symbols.
- Balance nuclear equations.
- Define the terms *nuclide, natural radioactivity*, *transmutation*, and *radioactive decay*.
- Describe the various types of radioactive decay.
- Explain how radioactive uranium–238 decays to the stable nuclide lead–206.
- Indicate how radioactive emanations can be separated and detected.
- Define the term *half-life*, and solve simple half-life problems.
- Describe the uses of radioactive isotopes.
- Describe how nuclear reactions can be induced, and indicate the role of accelerators in this process.
- Define the term *nuclear fission*, and describe how energy can be obtained from fission.
- List the parts of a fission reactor, and describe their functions.
- Indicate the safety procedures used in the disposal of radioactive wastes.
- Define the term *nuclear fusion*, and explain the role of this process in energy production.
- Describe the four fundamental interactions (forces) present in the universe.
- Describe the role that quarks play in the composition of protons, neutrons, and other nuclear particles.

SECTION I—BASIC (REGENTS-LEVEL) MATERIAL

NYS REGENTS CONCEPTS AND SKILLS

Note: By the time you have finished Section I, you should have mastered the concepts and skills listed below. The Regents chemistry examination will test your knowledge of these items and your ability to apply them.

Concepts are the *basic ideas* that form the body of the Regents chemistry course (what you need to know!).

Skills are the *activities* that demonstrate your mastery of these concepts (how you show that you know them!).

Following each concept or skill is a page reference (given in parentheses) to this chapter.

7.1 Concept:
- The stability of an isotope is based on the ratio of neutrons and protons in its nucleus. (Page 193)
- Although most nuclei are stable, some are unstable and decay spontaneously, emitting radiation. (Page 193)

7.2 Concept:
Each radioactive isotope has a specific mode and rate of decay (half-life). (Pages 193–197)

Skill:
Given the appropriate data or radioactive decay curve, calculate the:
  (a) initial amount,
  (b) fraction remaining,
  (c) half-life
of a radioactive isotope. (Pages 195–196)

7.3 Concepts:
- A change in the nucleus of an atom that converts the atom from one element to another is called transmutation. (Page 193)
- Transmutation can occur naturally or can be induced by bombardment of the nucleus by high-energy particles. (Pages 193, 197)

7.4 Concepts:
- Spontaneous decay can involve the release of:
  (a) alpha particles,
  (b) beta particles,
  (c) positrons,
  (d) gamma radiation

from the nucleus of an unstable isotope. (Pages 193–194)
- These emissions differ in mass, charge, ionizing power, and penetrating power. (Pages 193–194)

Skills:
- Determine the decay mode of a nuclear reaction. (Pages 193–194)
- Write nuclear equations that represent alpha and beta decay. (Pages 193–194)

7.5 Concept:
Nuclear reactions include:
    (a) natural and artificial transmutation (Pages 193, 197);
    (b) fission (Page 198);
    (c) fusion (Pages 205–206).

Skill:
Compare and contrast fission and fusion reactions. (Pages 198, 205–206)

7.6 Concept:
Both benefits and risks are associated with fission and fusion reactions. (Pages 198, 205–206)

7.7 Concept:
Nuclear reactions can be represented by equations that include:
    (a) symbols that represent atomic nuclei (with mass numbers and atomic numbers) (Page 192);
    (b) subatomic particles (with mass numbers and charges) (Pages 192–193);
    (c) emissions such as alpha, beta, and gamma radiation. (Pages 193–194, 199–200)

Skills:
- Complete nuclear equations. (Page 199)
- Predict missing particles from nuclear equations. (Page 199)

7.8 Concept:
The energy released in a nuclear reaction (fission or fusion) comes from the fractional amount of mass converted into energy. (Page 198)

7.9 Concept:
The energy released during nuclear reactions is much greater than the energy released during chemical reactions. (Page 198)

7.10 Concept:
There are inherent risks associated with radioactivity and the use of radioactive isotopes. Risks can include:

(a) biological exposure (Page 205);
(b) longterm storage and disposal of isotopes (Page 205);
(c) nuclear accidents (Page 205).

7.11 Concept:
Radioactive isotopes have many beneficial uses. Radioactive isotopes are used in research, medicine, and industry for such applications as:
(a) radioactive dating (Pages 196–197);
(b) tracing chemical and biological processes (Pages 196–197);
(c) industrial measurement (Pages 196–197);
(d) nuclear power (Page 198);
(e) detection and treatment of disease (Pages 196–197).

Skill:
Identify specific uses of common radioactive isotopes such as:
(a) I-131 in diagnosing and treating thyroid disorders (Pages 1960–197);
(b) Co-60 in treating cancer (Pages 196–197);
(c) the C-14 to C-12 ratio to date organic material (Pages 196–197);
(d) the U-238 to Pb-206 ratio to date geological formations (Pages 196–197).

7.1 NUCLEAR PARTICLES

In Chapter 2 we discussed and defined the important terms *nucleon, atomic number, mass number,* and *isotope.* In addition, we used a symbol such as $^{14}_{6}C$ to describe a particular atomic nucleus (or **nuclide**). You should review this material now.

As we study nuclear chemistry, we will see that certain nuclear particles appear frequently in nuclear processes. For this reason, the special names and symbols are listed in Reference Table O in Appendix 1.

7.2 NUCLEAR EQUATIONS

A nuclear reaction is a change that occurs within or among atomic nuclei and is represented by a *nuclear equation* such as the following:

$$^{15}_{7}N + ^{1}_{1}H \rightarrow ^{12}_{6}C + ^{4}_{2}He$$

If we examine this equation carefully, we note that the sum of the atomic numbers on the left side $(7 + 1)$ equals the sum of the atomic numbers on the right side $(6 + 2)$. This equality demonstrates the fact that electric charge must be conserved in a nuclear reaction. Similarly, the sum of the mass numbers on the left side of the equation $(15 + 1)$ equals the sum of the mass numbers on the right side $(12 + 4)$.

This nuclear equation is considered to be *balanced* because both charge and mass number are conserved. All of the nuclear equations that we write will be balanced equations.

7.3   NATURAL RADIOACTIVITY AND RADIOACTIVE DECAY

At the end of the nineteenth century, the French scientist Henri Becquerel discovered the phenomenon of **natural radioactivity**, that is, the spontaneous disintegration of an atomic nucleus. When the nucleus disintegrates, the process is accompanied by the emission of subatomic particles and/or photons such as X-rays. During this process, one element may change to another; this change is known as **transmutation**.

A large number of elements have radioactive isotopes, such as hydrogen-3 (tritium) and carbon-14, in addition to their stable isotopes, such as hydrogen-1 and carbon-12. However, *elements whose atomic numbers are greater than 83 have no known stable isotopes*.

The earliest radioactive emanations were identified as positively charged alpha particles, negatively charged beta particles, and uncharged gamma radiation that is similar to X-rays but far more energetic. (These are listed in Reference Table O in Appendix 1.) Each of these emanations is different in mass (alpha particles being the most massive) and penetrating power (gamma radiation is the most penetrating; alpha particles are the least).

Why is a particular nuclide stable or radioactive? The answer is related, in part, to the *ratio of neutrons to protons* within the nucleus of the nuclide. Radioactive nuclei break down naturally by a series of processes, known collectively as **radioactive decay** or **natural transmutation**, in order to become more stable; radioactive decay is an attempt to "correct" the ratio of neutrons to protons. However, stability may not occur immediately; a series of decay reactions may be required before a stable nucleus is finally produced.

Alpha Decay

The following nuclear reaction:

$$^{238}_{92}U \rightarrow \, ^{234}_{90}Th + \, ^{4}_{2}He$$

is an example of *alpha decay*. The uranium-238 nucleus (the *parent nucleus*) breaks down to produce a thorium-234 nucleus (the *daughter nucleus*) and a helium-4 nucleus (an *alpha particle*).

Whenever alpha decay occurs, the atomic number of the daughter nucleus (as compared with that of its parent) is *decreased* by 2 and its mass number is *decreased* by 4. Many heavier radioactive nuclei, especially those with atomic numbers greater than 83, undergo alpha decay as a way of reducing the numbers of protons and neutrons present.

PROBLEM
The nuclide $^{220}_{87}$Fr undergoes alpha decay. Write a balanced equation that illustrates this process.

SOLUTION
$$^{220}_{87}\text{Fr} \rightarrow {}^{216}_{85}\text{At} + {}^{4}_{2}\text{He}$$

TRY IT YOURSELF
When a certain radioactive nuclide undergoes alpha decay, $^{218}_{84}$Po is formed as a daughter nucleus. Identify the parent nuclide.

ANSWER
$^{222}_{86}$Rn

Beta (−) Decay

Certain nuclei undergo radioactive decay and produce an *electron* [a *beta* (−) *particle*] in the reaction. In *beta* (−) *decay,* the atomic number of the daughter nucleus is *increased* by 1 while the mass number remains *unchanged*. Beta (−) decay serves to *lower* the neutron to proton ratio. The following equation illustrates the process of beta (−) decay:

$$^{234}_{90}\text{Th} \rightarrow {}^{234}_{91}\text{Pa} + {}^{0}_{-1}\text{e}$$

PROBLEM
The nuclide $^{32}_{15}$P undergoes beta (−) decay. Write a balanced nuclear equation that illustrates this process.

SOLUTION
$$^{32}_{15}\text{P} \rightarrow {}^{32}_{16}\text{S} + {}^{0}_{-1}\text{e}$$

TRY IT YOURSELF
Write a balanced nuclear equation that shows how the daughter nucleus $^{14}_{7}$N is produced by beta (−) decay.

ANSWER
$$^{14}_{6}\text{C} \rightarrow {}^{14}_{7}\text{N} + {}^{0}_{-1}\text{e}$$

7.4 HALF-LIFE

Another method of measuring radioactivity involves the use of a quantity called *half-life*. The **half-life** is the time required for a substance to decay to one-half of its initial value. (Half-lives are not limited to radioactive substances; they are used also to measure the persistence of medicines in the human body and of pesticides in soil.)

As an example, consider the radioactive isotope (*radioisotope*, for short) iodine-131, whose half-life is approximately 8 days. If we have a 16-milligram sample of this isotope, its decay over a period of time occurs as follows:

$$16 \text{ mg} \xrightarrow{8d} 8 \text{ mg} \xrightarrow{8d} 4 \text{ mg} \xrightarrow{8d} 2 \text{ mg} \xrightarrow{8d} 1 \text{ mg} \dots$$

After 32 days of decay, 1 milligram of iodine-131 remains unchanged. The other 15 milligrams have not simply disappeared; they have been transformed into other substances. Moreover, the isotope will *never* decay to 0 milligram; it will simply come closer and closer to this value.

The half-life is constant: it cannot be altered by changes in pressure, temperature, or chemical combination. It depends solely on the nature of the radioactive nucleus. Half-lives range from fractions of a second (fermium-244, half-life = 3.3×10^{-3} s) to *billions* of years (uranium-238, half-life = 4.51×10^9 y). Reference Table N in Appendix 1 gives the half-lives of a number of radioactive isotopes.

The accompanying graph shows how a 16-milligram sample of iodine-131 decays over a period of time. The shape of this graph is characteristic of *all* radioactive decay processes.

Mass of Iodine 131 versus Time

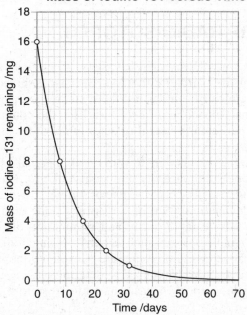

PROBLEM

According to the graph shown above, approximately how much of the radio-isotope iodine-131 remains after 20 days of decay?

SOLUTION

Inspection of the graph shows that approximately 3 mg of the radioisotope remains.

TRY IT YOURSELF

According to the graph, how long must the radioisotope iodine-131 decay so that 12 milligrams remain unchanged?

ANSWER

Approximately 3 d

PROBLEM

According to Reference Table N in Appendix 1, how much of a 100.-micro-gram (μg) sample of nitrogen-16 will remain after 28.8 seconds of decay?

SOLUTION

According to Reference Table N, the half-life of nitrogen-16 is 7.2 s. Since 28.8 s represents four half-lives (28.8/7.2), the decay is as follows:

$$100. \, \mu g \xrightarrow{7.2 \, s} 50.0 \, \mu g \xrightarrow{7.2 \, s} 25.0 \, \mu g \xrightarrow{7.2 \, s} 12.5 \, \mu g \xrightarrow{7.2 \, s} 6.25 \, \mu g$$

After 28.8 s, 6.25 μg of nitrogen-16 will remain.

7.5 USES OF RADIOISOTOPES

Radioactive isotopes can be used in a variety of applications that depend on the properties of the particular isotope. Because radioisotopes (such as carbon-14) are chemically similar to their stable counterparts (such as carbon-12), they can be used as *tracers* to follow the course of a reaction. This application is particularly useful in determining how certain biochemical reactions occur.

Radioisotopes are used also in a variety of medical applications that depend on the isotope's radioactivity and short half-life. These applications include medical diagnoses, such as the use of iodine-131 in uncovering thyroid disorders and technetium-99m in performing bone scans. A short half-life is necessary to ensure rapid decay and elimination of the radioisotope from the body. Also, radioisotopes such as cobalt-60 and iodine-131 are used in treating certain cancerous tumors that are considered inoperable.

Certain radioisotopes have been used to determine the age of a sample of material. For example, carbon-14 has a half-life close to 5,700 years. This radioisotope is produced in the upper atmosphere and is incorporated into carbon dioxide (CO_2) along with nonradioactive carbon-12. Plants, such as trees, take in the CO_2 and incorporate it into the body of the plant (wood, in the case of a tree). As long as the tree is alive, the ratio of carbon-14 to carbon-12 remains constant. When the tree dies or is cut down, however, the carbon-14 decays and is not replaced. Therefore, the ratio of carbon-14 to carbon-12 becomes smaller as time progresses. By comparing the ratio of carbon-14 to carbon-12 in a wooden bowl to the ratio of these isotopes in the atmosphere, scientists can determine the age of the bowl with considerable accuracy. To determine the ages of geological formations, radioactive isotopes with much longer half-lives, such as uranium-238 and lead-206, are needed for the ratios.

7.6 INDUCED NUCLEAR REACTIONS

All of the nuclear reactions we have studied so far have been *natural transmutations*. We now turn our attention to nuclear changes that have been produced artificially, or *induced*. To induce a nuclear reaction, a target nucleus is bombarded with a nuclear particle.

Artificial Transmutation

The first induced nuclear reactions used alpha particles (because of their large masses) as the bombarding particles. In 1919, Rutherford bombarded nitrogen-14 nuclei with alpha particles to produce a reaction in which the nitrogen-14 was artificially changed or *transmuted* into oxygen-17:

$$^{14}_{7}N + {}^{4}_{2}He \rightarrow {}^{17}_{8}O + {}^{1}_{1}H$$

This process is known as **artifical transmutation**.

In another artificial transmutation, English physicist Sir James Chadwick bombarded beryllium-9 and identified a stream of uncharged particles that we now call *neutrons*:

$$^{9}_{4}Be + {}^{4}_{2}He \rightarrow {}^{1}_{0}n + {}^{12}_{6}C$$

Two years later, in 1934, French physicists Frédéric Joliot-Curie and Irène Joliot-Curie bombarded aluminum-27 and produced the first artificially radioactive isotope, phosphorus-30:

$$^{27}_{13}Al + {}^{4}_{2}He \rightarrow {}^{1}_{0}n + {}^{30}_{15}P$$

The phosphorus-30 undergoes positron decay (see page 194) and forms silicon-30.

Nuclear Energy

In the bombardment reactions we have studied so far, only minor changes to the target nuclei occurred. We will now discuss processes that cause major nuclear changes and produce a great deal of energy. Originally, applications of these reactions were used in warfare. More recently, however, the reactions have been applied to generate electrical energy.

The origin of this *nuclear energy* is the conversion of a small amount of matter into radiant energy. If the masses of the reactants and the products of an energy-producing nuclear reaction are measured very precisely, the products are found to have *less* mass than the reactants. This *mass defect* has been converted into an equivalent amount of energy. This conversion is a consequence of Einstein's famous formula ($E = mc^2$), which states that mass and energy can be transformed into one another. The two principal reactions used as sources of energy are nuclear *fission* and nuclear *fusion*.

Nuclear Fission

In the 1930s, Enrico Fermi, an Italian-American physicist, suggested that neutrons be used in bombardment reactions because they are uncharged particles and therefore would not be repelled by the target nuclei. In 1938, German physicists Otto Hahn and Fritz Strassman bombarded uranium atoms with neutrons and discovered that some of the uranium atoms split into two roughly equal fragments. This reaction, known as **nuclear fission**, also produced more neutrons and a large amount of energy.

There are many types of fission reactions; one type is shown below:

$$^{235}_{92}\text{U} + ^{1}_{0}\text{n} \rightarrow \left[^{236}_{92}\text{U}\right] \rightarrow ^{141}_{56}\text{Ba} + ^{92}_{36}\text{Kr} + 3^{1}_{0}\text{n} + \text{energy}$$

The enormous amount of energy released is due to the fact that the combined mass of the reactants (uranium-235 and the neutron) is considerably greater than the combined mass of the products (barium-141, krypton-92, and the 3 neutrons). In addition, the 3 neutrons that are released can be used to cause other uranium-235 nuclei to fission, producing, in turn, 9 neutrons, then 27 neutrons, ... If the fission process is allowed to continue in this manner, an *uncontrolled chain reaction* will occur and the resulting explosion will release a great deal of *destructive* energy. This is the stuff of which fission bombs are made.

SECTION II—ADDITIONAL MATERIAL

7.1A   THE URANIUM-238 DECAY SERIES

As we noted previously, radioactive decay occurs in order to produce stable nuclei. Sometimes, this process requires more than one step. The unstable nuclide $^{238}_{92}$U decays to the stable nuclide $^{206}_{82}$Pb in a series of steps involving both alpha and beta $(-)$ decay.

The accompanying graph indicates how the decay of uranium-238 occurs. Atomic numbers are plotted along the x-axis, and mass numbers along the y-axis. Each diagonal line represents an alpha decay; each horizontal line, a beta $(-)$ decay. The daughter nucleus at each step (atomic number and mass number) is indicated by a dot (•).

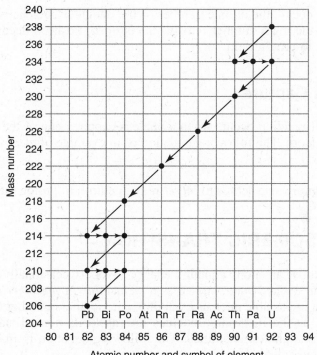

Atomic number and symbol of element

PROBLEM

How does the nuclide $^{214}_{82}$Pb decay?

SOLUTION

We locate the dot corresponding to $^{214}_{82}$Pb on the graph given above and then follow the (horizontal) arrow to the next dot, which corresponds to $^{214}_{83}$Bi. Therefore, $^{214}_{82}$Pb decays to $^{214}_{83}$Bi by beta $(-)$ decay.

TRY IT YOURSELF
How many alpha and beta decays are there in the uranium-238 decay series?

ANSWER
According to the graph, the series has eight alpha decays and six beta ($-$) decays.

Positron (beta plus) Decay

The name *positron* is a combination of the word parts *posit*ive elec*tron*. The positron is the *antiparticle* (i.e., the *antimatter* counterpart) of the electron. The following equation is an example of *positron decay*:

$$^{19}_{10}\text{Ne} \rightarrow {}^{19}_{9}\text{F} + {}^{0}_{+1}\text{e}$$

In positron decay, the atomic number of the daughter nucleus is *decreased* by 1 while the mass number remains *unchanged*. Positron decay *increases* the neutron to proton ratio.

Electron Capture

This type of reaction occurs when a nucleus captures one of the inner electrons of an atom. The following equation is an example of *electron capture:*

$$^{7}_{4}\text{Be} + {}^{0}_{-1}\text{e} \rightarrow {}^{7}_{3}\text{Li}$$

Electron capture also serves to *increase* the neutron to proton ratio.

7.2A   ISOMERIC TRANSITION

The nucleus of an atom contains energy levels. Occasionally, a nucleus will enter an excited state, known as a *metastable* or *mesomeric* state. We symbolize a metastable nucleus by adding m to its mass number (e.g., 99m). Eventually, as the metastable nucleus returns to its normal state, a high-energy gamma photon is emitted. The following equation is an example of *isomeric transition*:

$$^{99\text{m}}_{43}\text{Tc} \rightarrow {}^{99}_{43}\text{Tc} + \gamma$$

The nuclides ^{99}Tc and $^{99\text{m}}\text{Tc}$ are known as *nuclear isomers*.

Reference Table N in Appendix 1 of this book lists the modes of decay for selected radioactive isotopes.

7.3A   DETECTION AND MEASUREMENT OF RADIOACTIVITY

Charged nuclear particles may be separated by electric and magnetic fields. For example, in an electric field (produced by oppositely charged parallel plates) alpha particles are deflected toward the negative plate, and beta ($-$) particles toward the positive plate. Neutrons and gamma ray photons, being uncharged, are not deflected.

Radioactivity may be detected by a number of devices such as photographic film, scintillation counters, electroscopes, bubble and cloud chambers, and (perhaps the best known) the Geiger counter. The tube in the Geiger counter contains a gas that is ionized (electrically charged) by incoming radioactive particles. Each charged gas particle creates a pulse of electric current that produces an audible "click" and/or is counted electronically. The rate of the recorded counts or the frequency of the audible clicks is a measure of the degree of radioactivity present. The details of the other detection devices are described in standard chemistry texts.

Rate of Radioactive Decay

The radioactivity of a substance may be *measured* by the number of disintegrations taking place per unit of time. A substance has an activity of 1 becquerel (Bq) if it experiences one disintegration per second. An older unit, the curie (Ci), was originally based on the activity of 1 gram of radium. One curie equals 3.7×10^{10} becquerels.

7.4A   SOLVING RADIOACTIVE DECAY PROBLEMS

Radioactive decay problems are easily solved if the half-life of the radioactive material is given. We begin by considering a simple problem.

PROBLEM
Supposed that 256 mg of iodine-131 (half-life = 8.0 days) decays for 40 days.
(a) After 40. days, what fraction of the I-131 remains?
(b) What mass of I-131 is present after 40. days of decay?

SOLUTION

(a) We can calculate the number of half-life periods that 40. days represents. if t is the decay period and $t_{0.5}$ is the nuclide's half-life:

$$\text{number of half-life periods} = \frac{t}{t_{0.5}}$$

$$\frac{t}{t_{0.5}} = \frac{40.\,\text{d}}{8.0\,\text{d}} = 5.0$$

For each half-life period, the amount of I-131 is reduced by one-half. Therefore, in 5.0 half-life periods (for 40. days), the fraction of I-131 that remains is as follows:

$$1 \xrightarrow{\ 1\ } \frac{1}{2} \xrightarrow{\ 2\ } \frac{1}{4} \xrightarrow{\ 3\ } \frac{1}{8} \xrightarrow{\ 4\ } \frac{1}{16} \xrightarrow{\ 5\ } \frac{1}{32}\,(0.03125)$$

After 5.0 half-life periods, only $\dfrac{1}{32}$ or 0.03125 of the original amount of I-131 remains.

(b) We can calculate the amount remaining by multiplying the original amount of I-131 by the fraction remaining after 5.0 half-life periods:

$$\text{amount remaining} = 256\,\text{mg} \cdot 0.03125 = 8.0\,\text{g}$$

We can also use the radioactive decay formula. If A_0 = the original amount of a radioactive nuclide, t = the amount of time that the nuclide decays, and $t_{0.5}$ = the nuclide's half-life:

$$\text{fraction remaining} = \frac{1}{2}^{\frac{t}{t_{0.5}}}$$

$$\text{amount remaining} = A_0 \cdot \frac{1}{2}^{\frac{t}{t_{0.5}}}$$

We can easily figure out the answer to the previous problem using a graphing calculator such as the TI-83 or TI-84. To calculate the fraction remaining in part (a), key in the following strokes in order:

$$\boxed{0}\ \boxed{.}\ \boxed{5}\ \boxed{\text{2nd}}\ \boxed{\wedge}\ \boxed{(}\ \boxed{4}\ \boxed{0}\ \boxed{\div}\ \boxed{8}\ \boxed{)}\ \boxed{\text{ENTER}}$$

To calculate the amount remaining in part (b), key in the following strokes in order:

$$\boxed{2}\ \boxed{5}\ \boxed{6}\ \boxed{\times}\ \boxed{\text{2nd}}\ \boxed{[\text{ANS}]}\ \boxed{\text{ENTER}}$$

Using these formulas provides a real bonus—one does not have to use "easy" numbers. For example, consider the following problem.

PROBLEM
The half-life of Kr-85 is 10.73 years. How much of a 412.2-gram sample will remain after 39.22 years of decay?

SOLUTION
Use a calculator:

$$\text{amount remaining} = 412.2 \text{ g} \cdot 0.5^{\left(\frac{39.22 \text{ y}}{10.73 \text{ y}}\right)} = \mathbf{32.72 \text{ g}}$$

TRY IT YOURSELF
Use a calculator:

The half-life of P-32 is 14.28 days. How much of a 0.6768-gram sample will remain after 128.2 days of decay?

ANSWER
0.001343 grams of P-32 remains.

7.5A PARTICLE ACCELERATORS

A *particle accelerator* is a device that uses electric and magnetic fields to provide a (charged) bombarding nuclear particle with sufficient energy to induce a desired nuclear reaction. As an analogy, consider a bullet fired at a wall at a speed of 10 miles per hour. At this slow speed, the energy of the bullet will have hardly any effect on the wall. If the bullet were fired at 1000 miles per hour, however, its effect on the wall would be devastating!

Examples of modern particle accelerators include the *Van de Graaff accelerator,* the *linear accelerator*, the *cyclotron,* the *synchrotron,* and the *large hadron collider* (*LHC*). Detailed descriptions of these devices, which can supply bombarding particles with a range of energies, can be found in standard chemistry and physics texts.

7.6A FISSION REACTORS

A number of physicists recognized that in a *controlled* chain reaction, the fission could provide a continuous source of useful energy. This is the principle behind the *fission reactor*. Fission reactors are used in the production of electrical energy, as research tools, and as a means of producing other radioactive isotopes. The accompanying diagram illustrates a typical fission reactor used to produce electricity.

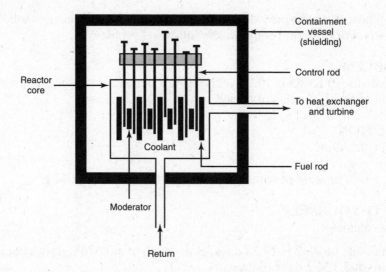

The primary system of a fission reactor has the following components:

- The *containment vessel* (concrete and steel) provides shielding for the reactor.
- The *fuel rods*, located in the *core*, serve as sources of energy. Uranium-233 and uranium-235 are used as fuels.
- The *moderator*, also located in the core, slows the neutrons so that they will be absorbed by the fuel nuclei. These slow neutrons have kinetic energies that are close to the energies of air molecules at room temperature. For this reason, they are known as *thermal neutrons*. Moderators usually consist of water (containing either hydrogen-1 or hydrogen-2), graphite, or beryllium.
- The *control rods* in the core regulate the rate of fission by absorbing neutrons. Control rods are usually made of cadmium or boron.
- The *coolant* (water or liquid sodium) removes thermal energy from the core.
- The *heat exchanger* receives the thermal energy and produces steam for the generation of electrical energy by the secondary system (turbine) of the reactor.

The most common isotope of uranium, uranium-238, comprises about 99 percent of the naturally occurring element but does *not* undergo fission. Although the isotope uranium-235 does undergo fission, it represents less than 1 percent of the naturally occurring element. For this reason, it is necessary to *enrich* uranium (usually to 3–4%) in order to use it as a fissionable fuel.

When uranium-238 absorbs a neutron, it is converted into uranium-239, which then decays in two steps to form plutonium-239:

$$^{238}_{92}U + ^{1}_{0}n \rightarrow ^{238}_{92}U$$

$$^{238}_{92}U \rightarrow ^{239}_{93}Np + ^{0}_{-1}e$$

$$^{239}_{93}Np \rightarrow ^{239}_{94}Pu + ^{0}_{-1}e$$

Plutonium-239 *does* fission, and it has been suggested that *breeder reactors*, which would be able to produce their own fuel, be constructed. No breeder reactors have been operated commercially in the United States, however, because of the serious health and environmental hazards attributable to plutonium and also because of the possibility that the plutonium might be used to produce nuclear weapons.

A number of problems are associated with any fission reactor. For example, the heat energy produced by the reactor contributes to *thermal pollution*. There is also the serious problem of *radioactive waste disposal*. Solid and liquid wastes, such as cesium-237 and strontium-90, are placed in corrosion-resistant containers and stored in isolated underground areas. Wastes with low levels of radioactivity may be diluted until they are considered harmless and then released into the environment. Gaseous wastes, such as krypton-85m, nitrogen-16, and radon-222, are stored and allowed to decay until it is considered safe to release them into the atmosphere.

Nuclear reactors are also used to produce radioactive isotopes for a variety of applications. For example, when the stable isotope $^{59}_{27}Co$ is bombarded with neutrons from a reactor, the radioisotope $^{60}_{27}Co$ is produced. As noted in Section 15.5, this radioisotope is used in cancer therapy.

Nuclear Fusion

The joining of light nuclei to form heavier, more stable nuclei is known as **nuclear fusion** and is the process by which stars, including our own Sun, produce their energy. In stars, fusion consists of a series of reactions that depend on the temperature of the particular star. The *net* fusion reaction in our Sun is as follows:

$$4^{1}_{1}H \rightarrow ^{4}_{2}He + 2^{0}_{+1}e + 2\gamma + 2\nu + energy$$

where γ represents a gamma photon, and ν represents another subatomic particle called a *neutrino*.

For a fusion reaction to occur, very high temperatures are needed to give the positively charged nuclei the energy they require to overcome their mutual repulsion. For this reason, all fusion devices are referred to as *thermonuclear devices*.

For a given mass of fuel, nuclear fusion yields more energy than nuclear fission and there are fewer pollution problems with the products of fusion reactions. For this reason, work is being done to develop fusion reactors as a means of producing power. Four fusion reactions under investigation are as follows:

$$^2_1H + {}^2_1H \rightarrow {}^3_1H + {}^1_1H + \text{energy}$$

$$^2_1H + {}^2_1H \rightarrow {}^4_2He + \text{energy}$$

$$^2_1H + {}^2_1H \rightarrow {}^3_2He + {}^1_0n + \text{energy}$$

$$^2_1H + {}^3_1H \rightarrow {}^4_2He + {}^1_0n + \text{energy}$$

At this time, however, no successful fusion reactor has been constructed. The problems to be overcome include the production of a high *ignition* temperature, the packing of nuclei into a space small enough to allow a sufficient number of collisions, the control of the fusion reaction once it has begun, and the development of materials that can withstand the high operating temperatures and radiation levels of the reactor.

END-OF-CHAPTER QUESTIONS

Some questions have the symbol "§2" in front of the question number. This symbol means that the question is based on Section II material.

1. Which particle is electrically neutral?
(1) proton   (2) positron   (3) neutron   (4) electron

2. Isotopes of the same element have the same number of
(1) protons, neutrons, and electrons
(2) protons, but different numbers of neutrons
(3) neutrons, but different numbers of protons
(4) electrons, but different numbers of protons

3. Which particle has the same mass as an electron, but a positive electric charge?
(1) alpha particle   (2) gamma photon
(3) proton   (4) positron

4. The total number of protons and neutrons in the nuclide ^{37}Cl is
(1) 54   (2) 37   (3) 20   (4) 17

5. In the reaction

$$^{6}_{3}Li + ^{1}_{0}n \rightarrow ^{4}_{2}He + X$$

the species represented by X is

(1) ^{2}H (2) ^{3}H (3) ^{3}He (4) ^{4}He

6. In the reaction

$$^{9}_{4}Be + ^{1}_{1}H \rightarrow ^{4}_{2}He + X$$

which species is represented by X?

(1) ^{8}Li (2) ^{6}Li (3) ^{8}B (4) ^{10}B

7. Given the reaction

$$^{27}_{13}Al + ^{4}_{2}He \rightarrow X + ^{1}_{0}n$$

When the equation is correctly balanced, the nucleus represented by X is

(1) ^{30}Al (2) ^{30}Si (3) ^{30}P (4) ^{30}S

8. In the reaction

$$^{99m}Tc \rightarrow ^{99}Tc + X$$

X represents

(1) an alpha particle   (2) a beta particle
(3) a positron   (4) gamma radiation

9. In the equation

$$^{39}_{19}K \rightarrow ^{39}_{20}Ca + X$$

which particle is represented by X?

(1) a proton   (2) an electron   (3) a positron   (4) a deuteron

10. The nuclear reaction

$$^{14}_{6}C \rightarrow ^{14}_{7}N + ^{0}_{-1}e$$

is an example of

(1) nuclear fusion   (2) nuclear fission
(3) natural transmutation   (4) artificial transmutation

11. Which element has no known stable isotope?

(1) C   (2) K   (3) Po   (4) P

12. When uranium-238 undergoes alpha decay, which daughter nucleus is formed?

(1) ^{234}Th (2) ^{238}Np (3) ^{206}Pb (4) ^{233}U

13. The diagram below represents radiation passing through an electric field.

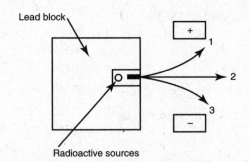

The arrow marked 2 most likely represents
(1) an alpha particle   (2) a beta ($-$) particle
(3) gamma radiation   (4) a proton

14. When a stream of radioactive particles is passed through a pair of oppositely charged parallel plates, which particle would be deflected toward the *negative* plate?
(1) alpha particle   (2) beta ($-$) particle
(3) gamma photon   (4) neutron

15. After 62.0 hours, 1.0 gram remains unchanged from a sample of ^{42}K. How much ^{42}K was in the original sample?
(1) 8.0 g   (2) 16 g   (3) 32 g   (4) 64 g

16. What mass of a 16-gram sample of cobalt-60 will remain unchanged after 15.78 years?
(1) 1.0 g   (2) 2.0 g   (3) 8.0 g   (4) 4.0 g

17. How many days are required for phosphorus-32 to undergo three half-life periods?
(1) 4.77   (2) 14.3   (3) 28.6   (4) 42.9

18. In 6.20 hours, a 100.-gram sample of an isotope decays to 25.0 grams. What is the half-life, in hours, of this isotope?
(1) 1.60   (2) 3.10   (3) 6.20   (4) 12.4

19. How much of an 8-gram sample of radium-226 will remain unchanged at the end of two half-life periods?
(1) 1 g   (2) 2 g   (3) 3 g   (4) 4 g

20. At the end of 12 days, 1/8 of an original sample of a radioactive element remains. What is the half-life of the element?
(1) 36 days   (2) 48 days   (3) 3 days   (4) 4 days

Base your answers to questions 21–23 on the following graph, which represents the decay curve of a certain radioisotope.

Activity versus Time

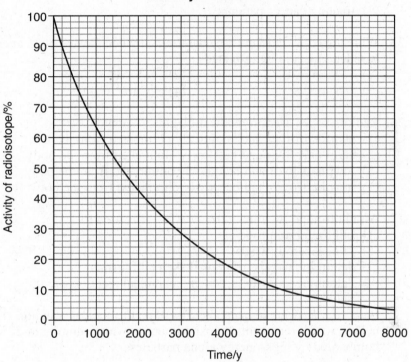

21. After approximately how many years will 70 percent of the radioisotope's activity remain?
(1) 500   (2) 800   (3) 2,800   (4) 8,000

22. On the basis of the graph and Reference Table N, this radioisotope is most likely
(1) ^{14}C   (2) ^{137}Cs   (3) ^{226}Ra   (4) ^{232}Th

23. If the temperature at which the radioisotope decayed had been higher, the half-life of the isotope would have been
(1) less   (2) greater   (3) the same

24. For which application is the ratio of uranium-238 to lead-206 used?
(1) diagnosing thyroid disorders
(2) dating geologic formations
(3) detecting brain tumors
(4) treating cancer patients

25. Which isotopic ratio must be determined when the age of an ancient wooden object is investigated?
(1) ^{235}U to ^{238}U (2) ^{2}H to ^{3}H
(3) ^{16}N to ^{14}N (4) ^{14}C to ^{12}C

26. Radioisotopes used in medical diagnosis should have
(1) short half-lives and be quickly eliminated from the body
(2) short half-lives and be slowly eliminated from the body
(3) long half-lives and be quickly eliminated from the body
(4) long half-lives and be slowly eliminated from the body

27. A radioisotope is called a *tracer* when it is used to
(1) kill bacteria in food
(2) kill cancerous tissue
(3) determine the age of animal skeletal remains
(4) determine the way in which a chemical reaction occurs

Base your answers to questions 28–30 on the nuclear equation given below, which represents an experiment in which aluminum-27 was bombarded by high-energy alpha particles.

$$^{27}_{13}Al + ^{4}_{2}He \rightarrow ^{1}_{0}X + Y$$

28. Particle Y is an isotope of
(1) Al   (2) C   (3) P   (4) Si

29. Particle X is
(1) a proton   (2) an electron   (3) a positron   (4) a neutron

30. This nuclear reaction may be classified as
(1) artificial transmutation
(2) nuclear fission
(3) nuclear fusion
(4) natural radioactive decay

31. Which particle *cannot* be accelerated by the electric or magnetic fields in a particle accelerator?
(1) neutron   (2) proton
(3) alpha particle   (4) beta (−) particle

32. An isotope of which element may be used as a fuel in a fission reaction?
(1) H   (2) C   (3) Li   (4) U

§2 **33.** Which substance is used in the control rods of a nuclear reactor?
(1) B   (2) He   (3) CO_2 (4) Be

§2 **34.** The main purpose of a moderator in a fission reactor is to
(1) emit neutrons   (2) absorb neutrons
(3) slow down neutrons   (4) speed up neutrons

§2 **35.** The atoms of some elements can be made radioactive by
(1) placing them in a magnetic field
(2) bombarding them with high-energy particles
(3) separating them into their isotopes
(4) heating them to a very high temperature

§2 **36.** Particle accelerators are used primarily to
(1) detect radioactive particles
(2) identify radioactive particles
(3) increase a particle's kinetic energy
(4) increase a particle's potential energy

§2 **37.** Which is a gaseous radioactive waste produced during some fission reactions?
(1) nitrogen-16   (2) thorium-232
(3) uranium-235   (4) plutonium-239

38. In a fusion reaction, a major problem in causing the nuclei to fuse into a single nucleus is the
(1) small mass of the nuclei   (2) large mass of the nuclei
(3) attractions of the nuclei   (4) repulsions of the nuclei

39. In the reaction $^{239}_{93}\text{Np} \rightarrow {}^{239}_{94}\text{Pu} + X$, what does X represent?
(1) a neutron   (2) a proton
(3) an alpha particle   (4) a beta particle

40. Which of these types of nuclear radiation has the greatest penetrating power?
   (1) alpha   (2) beta   (3) neutron   (4) gamma

41. What occurs in both fission and fusion reactions?
   (1) Small amounts of energy are converted into large amounts of matter.
   (2) Small amounts of matter are converted into large amounts of energy.
   (3) Heavy nuclei are split into lighter nuclei.
   (4) Light nuclei are combined into heavier nuclei.

42. Given the reaction:

$$_{13}^{27}\text{Al} + {}_{2}^{4}\text{He} \rightarrow X + {}_{0}^{1}\text{n}$$

Which particle is represented by X?

   (1) $_{12}^{28}\text{Mg}$ (2) $_{13}^{28}\text{Al}$ (3) $_{14}^{30}\text{Si}$ (4) $_{15}^{30}\text{P}$

43. A radioactive isotope has a half-life of 2.5 years. Which fraction of the original mass remains unchanged after 10. years?

   (1) $\dfrac{1}{2}$ (2) $\dfrac{1}{4}$ (3) $\dfrac{1}{8}$ (4) $\dfrac{1}{16}$

44. Compared to the half-life and decay mode of the nuclide ^{90}Sr, the nuclide ^{226}Ra has
   (1) a longer half-life and the same decay mode
   (2) a longer half-life and a different decay mode
   (3) a shorter half-life and the same decay mode
   (4) a shorter half-life and a different decay mode

Constructed-Response Questions

1. In a laboratory experiment, a student investigated the activity of a radioisotope. The table below represents the data she took.

Time/min	0	100	200	300	400	500	600	700	800	900	1000
Activity Remaining/%	100	64	45	29	19	12	9	6	5	4	1

    (a) Plot the data on the graph given below, and draw a *best-fit* radioactive decay curve using the data points.

Activity versus Time

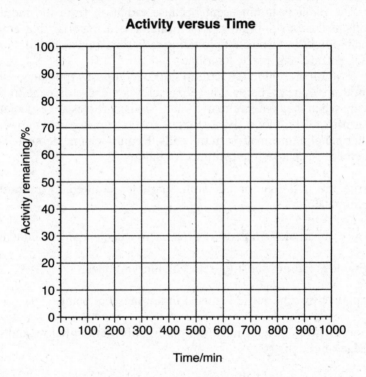

Time/min

    (b) Using the curve you have drawn, determine the half-life of the isotope.

Base your answers to questions 2 through 5 on the reading passage below, the *New York State Regents Reference Tables for Chemistry*, and your knowledge of chemistry.

> In the 1920s, paint used to inscribe the numbers on watch dials was composed of a luminescent (glow-in-the-dark) mixture. The powdered paint base was a mixture of radium salts and zinc sulfide. As the paint was mixed, the powdered base became airborne and drifted throughout the workroom. This caused the contents of the workroom, including the painters' clothes and bodies, to glow in the dark.
>
> The paint is luminescent because radiation from the radium salts strikes a scintillator. A scintillator is a material that emits visible light in response to ionizing radiation. In watch dial paint, zinc sulfide acts as the scintillator.
>
> Radium present in the radium salts decomposes spontaneously, emitting alpha particles. These particles can cause damage to the body when they enter human tissue. Alpha particles are especially harmful to the blood, liver, lungs, and spleen because they can alter genetic information in the cells. Radium can be deposited in the bones because it substitutes for calcium.

2. Write the notation for the alpha particles emitted by radium in the radium salts.

3. How can particles emitted from radioactive nuclei damage human tissue?

4. Why does radium substitute for calcium in bones?

5. Explain why zinc sulfide is used in luminescent paint.

Base your answers to questions 6–8 on the information below and on your knowledge of chemistry.

One fission reaction for U-235 is represented by the balanced nuclear equation below:

$$_{92}^{235}\text{U} + _{0}^{1}\text{n} \rightarrow _{54}^{140}\text{Xe} + _{38}^{94}\text{Sr} + 2_{0}^{1}\text{n}$$

Both radioisotopes produced by this reaction undergo beta decay. The half-life of Xe-140 is 13.6 seconds, and the half-life of Sr-94 is 1.25 minutes.

6. Explain in terms of *both* reactants and products why the reaction represented by the nuclear equation is a fission reaction.

7. In the decay equation of Xe = 140 shown below, what is the missing product?

$$^{140}_{54}Xe \rightarrow \, ^{0}_{-1}e + \underline{}$$

8. Determine the time required for an original 24.0-gram sample of Sr-94 to decay until only 1.5 grams of the sample remains unchanged.

The answers to these questions are found in Appendix 3.

Chapter
Eight

THE ELECTRONIC STRUCTURE OF ATOMS

KEY IDEAS

This chapter continues the development of the atomic model of matter begun in Chapter 2. The Bohr model introduces the idea of quantized electron orbits, and provides an explanation for the emission spectrum of atomic hydrogen. The currently accepted wave-mechanical model extends the placement of electrons within atomic levels, sublevels, and orbitals and is applicable to every element in the Periodic Table.

KEY OBJECTIVES

At the conclusion of this chapter you will be able to:
- Describe the Bohr model and its relationship to atomic spectra.
- Describe the modern wave-mechanical model and its relationship to electron configuration.
- Use the diagonal rule to predict the filling patterns of atoms.
- Define and apply the terms *principal energy level*, *sublevel*, *orbital*, *electron configuration*, *ground state*, *excited state*, *spin state*, *valence electron*, and *Lewis structure* (*electron-dot diagram*).

SECTION I—BASIC (REGENTS-LEVEL) MATERIAL

NYS REGENTS CONCEPTS AND SKILLS

Note: By the time you have finished Section I, you should have mastered the concepts and skills listed below. The Regents chemistry examination will test your knowledge of these items and your ability to apply them.

Concepts are the *basic ideas* that form the body of the Regents chemistry course (what you need to know!).

Skills are the *activities* that demonstrate your mastery of these concepts (how you show that you know them!).

Following each concept or skill is a page reference (given in parentheses) to this chapter.

8.1   Concept:
      Each electron in an atom has its own distinct amount of energy.
      (Page 220)

8.2   Concept:
      When an electron in an atom gains a specific amount of energy, the
      electron is at a higher energy state (excited state). (Page 221)

      Skill:
      Distinguish between ground-state configurations (e.g., 2-8-2) and
      excited-state configurations (e.g., 2-7-3). (Pages 220–221)

8.3   Concept:
      When an electron returns from a higher energy state to a lower
      energy state, a specific amount of energy is emitted. This emitted
      energy can be used to identify the element. (Page 220)

      Skill:
      Identify an element by comparing its bright-line spectrum to given
      spectra. (Page 219)

8.4   Concept:
      In the wave mechanical model (electron cloud model), the electrons
      are in orbitals, which are defined as the regions of most probable
      electron location (ground state). (Page 222)

8.5   Concept:
      The outermost electrons in an atom are called the valence electrons.
      In general, the number of valence electrons affects the chemical
      properties of an element. (Pages 222–223)

      Skills:
      • Draw a Lewis electron-dot structure of an atom. (Page 223)
      • Given an electron configuration, such as 2-8-2, distinguish
        between valence and nonvalence electrons. (Page 223)

8.6   Concept:
      Electron dot diagrams (Lewis structures) can represent the valence
      electron arrangements in elements. (Page 223)

      Skill:
      Draw the Lewis structures for the elements in Periodic Groups: 1,
      2, 13–18. (Page 223)

8.1 INTRODUCTION

In Chapter 2, we examined the development of the atomic models of Dalton, Thomson, and Rutherford. In this chapter, we continue the development of the atomic model into modern times. Our current focus will be on the arrangement of *electrons* within atoms.

8.2 THE BOHR MODEL OF THE ATOM

Rutherford proposed an atomic model in which a positively charged nucleus was surrounded by negative electrons. However, there were serious deficiencies with the Rutherford model's placement of electrons around the nucleus. If the electrons were *stationary*, they should have been drawn into the nucleus, since opposite charges attract one another; if the electrons *orbited* the nucleus (as the planets orbit the Sun), they should have radiated electromagnetic energy continuously. In either case, the atom would collapse! And there was another problem as well. . . .

When an element in the form of a gas at low pressure is excited (by heating or by electricity), it emits visible light. When this light is studied, it is found to be composed of regularly spaced lines. Each element has its own characteristic **visible-line spectrum** by which it can be identified. For example, a portion of the visible-line spectrum of atomic hydrogen is shown in the accompanying diagram. The color of each spectral line is identified by the *wavelength* of the light, which is given in nanometers (nm, 1 nm = 10^{-9} m).

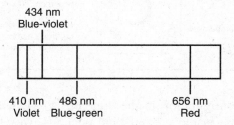

The Rutherford model could not explain why hydrogen (or any other element, for that matter) produced a visible-line spectrum. Did this mean that the Rutherford model had to be abandoned totally? Absolutely not! It meant only that the model needed to be refined.

In 1913, Niels Bohr, a Danish physicist, proposed his own model of the hydrogen atom, a model that answered many of the questions that had confounded Rutherford. In order to do so, however, Bohr had to make a number of assumptions based on the revolutionary ideas first proposed by physicists Max Planck and Albert Einstein.

Bohr proposed a planetary model of the hydrogen atom, as did Rutherford. But in Bohr's model, the electron behaved most unusually:

- The electron could orbit only in certain specified levels, each of which represented a distinct amount of energy. The more energy associated with a level, the farther it was from the nucleus. These **principal energy levels** were said to be *quantized* and were designated by the letters *K, L, M*, and so on. We now designate principal energy levels by numbers—1, 2, 3, 4, and so on.
- The lowest energy level (1 for hydrogen) was called the **ground state**. (Other levels were called **excited states**.) Since the electron could have no lower energy than the ground state, it could not come any closer to the nucleus, and therefore it remained stable.
- In a given orbit, an electron never radiated or absorbed energy. In this case, the electron was said to be in a **stationary state**.
- If an atom absorbed *exactly* the right amount of energy, the electron would rise to a higher energy level. Conversely, if the atom released energy (also in an exact amount), the electron would fall to a lower energy level. The energy that was released appeared as a *photon* of light.

The Periodic Table of the Elements in Appendix 1 lists the ground-state electron configurations of most of the elements, as shown in the accompanying diagram, which is the key to the table:

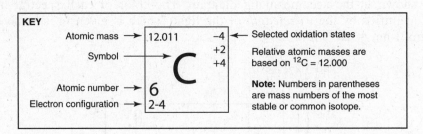

The electron configuration in the diagram is 2-4, which means that an atom of carbon (C) in the ground state contains two electrons in its first principal energy level and four electrons in its second principal energy level.

TRY IT YOURSELF
The element aluminum has an atomic number of 13. Use the Periodic Table to determine the ground-state electron configuration of an atom of aluminum.

ANSWER

The electron configuration of Al is 2-8-3, that is, two electrons in the first energy level, eight electrons in the second energy level, and three electrons in the third energy level, as shown in the accompanying diagram:

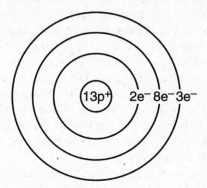

When an atom enters an excited state, one or more electrons are "promoted" to higher levels. For example, the configuration of an excited aluminum atom might be 2-7-4, as shown in the accompanying diagram:

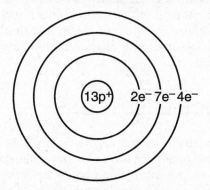

Bohr recognized that *spectral lines are the result of electrons falling from higher energy levels to lower ones.* By applying mathematics to his model, Bohr was able to account precisely for each line in the visible spectrum of hydrogen. This, in itself, provided strong evidence for the correctness of Bohr's model of the hydrogen atom.

Unfortunately, although the Bohr model worked well for hydrogen, it did not work for any other atom—even one with just two electrons (helium). Therefore, once again, the atomic model would have to be modified, hopefully in a way in which the behavior of *all* atoms could be explained. But now, the scientific revolution begun by Planck and Einstein (and continued by Bohr) had attracted a whole new host of players—scientists who would make astounding discoveries about the way nature behaves!

8.3 THE MODERN (WAVE-MECHANICAL) MODEL

The modern model of the atom is based on the premise that an electron (and all other matter) exhibits both particle and wavelike properties, depending on the circumstances. Whereas we can locate a particle with precision, however, we cannot do so in regard to a wave. (For example, we cannot say precisely where a water wave begins or ends.) Therefore, we must find alternative ways of describing an electron. Instead of trying to pinpoint its exact position, we must be content to guess where it is *most probably* located. A very useful analogy is to compare an electron with a cloud. An *electron cloud* is a distribution of probabilities: it is thickest in the regions where we are most likely to find the electron at any given moment and thinnest in the regions where the electron is least likely to be.

This modern model of the atom was developed with a great deal of very high level mathematics well beyond the scope of any high school course. Our goal is to try and understand the model and its applications to the study of chemistry.

In the modern model of the atom, an electron in an atom is described in terms of a concept known as an *atomic orbital*. An **atomic orbital** (**orbital**, for short) may be thought of as the region of space around the nucleus where the probability of locating an electron with a given energy is greatest. Orbitals are associated with characteristic energies, sizes, shapes, and orientations in space.

8.4 VALENCE ELECTRONS

The outermost principal energy level of an atom is important in regard to the way in which simple bonds are formed between atoms. We call this level the *valence level,* and the electrons that occupy the valence level are known as **valence electrons**. If we look through the Periodic Table, or use our diagonal rule, we will discover that valence electrons are found only in the *s* and *p* sublevels. Thus, the number of valence electrons for any element is between one and eight.

PROBLEM
    (a) What is the valence level of sulfur (atomic number 16)?
    (b) How many valence electrons does an atom of sulfur have?

SOLUTIONS
    (a) Referring to the Periodic Table in Appendix 1, we find that the valence level of sulfur is 3.
    (b) An atom of sulfur has six valence electrons.

TRY IT YOURSELF
For an atom of nitrogen (atomic number 7), answer parts (a) and (b) of the problem given on page 222.

ANSWERS
(a) The valence level of nitrogen is 2.
(b) An atom of nitrogen has five valence electrons.

8.5 LEWIS STRUCTURES (ELECTRON-DOT DIAGRAMS)

It would be very useful to be able to see the orbital arrangement of the valence electrons of an element at a glance. Unfortunately, an arrow diagram does not identify the particular element it represents. We can solve this problem by employing a device known as a **Lewis structure** or an **electron-dot diagram**. To draw a Lewis structure, we divide the atom into two parts. The first part, known as the *kernel,* consists of the nucleus plus the inner (non-valence) electrons of the atom. The second part of the atom is its valence electrons.

(Ground-State) Lewis Electron-Dot Diagrams for Elements 1–20

1. H·		11. Na·	
2. He:		12. Mg:	
3. Li·		13. A̤l·	
4. Be:		14. S̤i·	
5. B̈·		15. ·P̈·	
6. C̈·		16. :S̈·	
7. ·N̈·		17. :C̈l·	
8. :Ö·		18. :A̤r:	
9. :F̈·		19. K·	
10. :N̈e:		20. Ca:	

Note: The positions of the electrons around the symbol of the element is entirely arbitrary. However, the number of paired and unpaired electrons must be correct.

223

PROBLEM

Describe an atom of carbon (atomic number 6) in terms of the two-part concept explained on the previous page.

SOLUTION

In a carbon atom, the kernel consists of the carbon nucleus and the two electrons in level 1. The valence electrons are the remaining four electrons in level 2.

TRY IT YOURSELF

Answer the problem given above for an atom of sodium (atomic number 11).

ANSWER

The kernel of a sodium atom consists of the sodium nucleus, two electrons in level 1, and eight electrons in level 2. The valence electron is the single electron in level 3.

SECTION II—ADDITIONAL MATERIAL

8.1A   ATOMIC ORBITALS AND SUBLEVELS

At this point, it is appropriate to discuss atomic orbitals in more detail. Recall that an atomic **orbital** is the region of space where an electron with a specific energy is most likely to be located. There are many different orbitals, and each orbital is associated with a specific letter: $s, p, d, f, g, \ldots$. Fortunately, we need to be concerned only with $s, p, d,$ and f orbitals. Every atomic orbital, no matter what its designation, can accommodate a maximum of two electrons. An orbital that contains no electron is said to be *empty*; an orbital that contains one electron, to be *half-filled*; an orbital that contains two electrons, to be *filled*.

There is only *one* type of s orbital, and it is present on every principal energy level $(1, 2, 3, \ldots)$. The designation "$3s$ orbital" refers to the s orbital on principal energy level 3. All s orbitals have spherical shapes, as shown in the accompanying diagram:

The $p, d,$ and f orbitals are more complicated. There are *three* types of p orbitals (designated as $p_x, p_y,$ and p_z), and they are located on every energy level *except* level 1. All p orbitals have characteristic "dumbbell" shapes, as shown in the accompanying diagram:

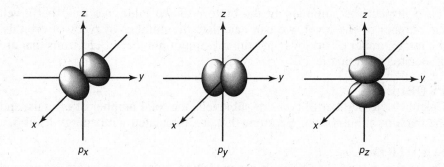

The designation "$4p_y$ orbital" refers to the p_y orbital on energy level 4. There are *five* types of d orbitals, and they are located on every energy level *except* levels 1 and 2. Finally, there are *seven* types of f orbitals, and they are located on every energy level *except* levels 1, 2, and 3. We shall not be concerned with either the labels or the shapes applied to individual d and f orbitals because they are too complex.

The three types of p orbitals, taken together, constitute a p **sublevel** (or **subshell**). Similarly, the five types of d orbitals and the seven types of f orbitals constitute d and f sublevels. Therefore, the designation "$4d$ sublevel" refers to the five d orbitals on energy level 4. The s sublevel is peculiar because it contains only the single s orbital.

The accompanying table indicates how orbitals and sublevels are arranged on individual principal energy levels.

Principal Energy Level	Type(s) of Sublevel	Number of Orbitals	Maximum Number of Electrons
1 Level 1	s 1 type	1 1 orbital	2 2 electrons
2 2 Level 2	s p 2 types	1 3 4 orbitals	2 6 8 electrons
3 3 3 Level 3	s p d 3 types	1 3 5 9 orbitals	2 6 10 18 electrons
4 4 4 4 Level 4	s p d f 4 types	1 3 5 7 16 orbitals	2 6 10 14 32 electrons
SUMMARY			
Level n ($n = 1, 2, 3, \ldots$)	n types (s, p, d, f, others)	n^2 orbitals	$2n^2$ electrons

225

If we inspect the summary at the bottom of the table, we find that given the number of the level, we can calculate the number of *types* of orbitals, the total number of orbitals, and the maximum number of electrons that are associated with that level.

PROBLEM

Calculate the number of *types* of sublevels, the total number of orbitals, and the maximum number of electrons that are associated with energy level 6.

SOLUTION

On level n, n types of sublevels are present; 6 types of sublevels are present on level 6 ($s, p, d, \ldots$).

On level n, there is a total of n^2 orbitals; on level 6, there is a total of 36 orbitals.

Level n can accommodate a *maximum* of $2n^2$ electrons; level 6 can accommodate a *maximum* of 72 electrons.

8.2A ELECTRON CONFIGURATIONS OF ATOMS

We now build up the atoms for the elements in the Periodic Table.

The basic rule for assigning electrons to atoms is that electrons should occupy the lowest energy states possible. Generally, as the principal energy level number increases, the energy of the electron on that level also increases. Therefore, an electron on level 4 ought to have more energy than an electron on level 3. For atoms with more than one electron (i.e., atoms other than hydrogen), however, this is not always the case, as we will see shortly. Sublevels *within a level* are more predictable. The order of increasing energy is as follows: $s < p < d < f$. Therefore, a $5p$ electron has less energy than a $5d$ electron and more energy than a $5s$ electron.

When we have assigned all of the electrons to a particular atom, the next atom will *repeat* that **electron configuration** and then add one new electron. [This is known as the *Aufbau* ("building-up") principle.]

The Diagonal Rule

To determine the relative energies of sublevels, we will make use of a technique, illustrated below, known as the *diagonal rule*.

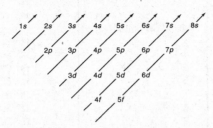

Note in the diagram that each successive line is indented because there are no $1p$, $2d$, or $3f$ sublevels.

The rule is quite simple: We begin at the left and follow each arrow from *tail to head* and then work from *left to right*. The first sublevel we meet is $1s$, and we complete the first arrow. Then we move to the right, where the next sublevel is $2s$. On the third arrow, there are *two* sublevels. Tracing the arrow from tail to head, we meet sublevels $2p$ and $3s$. This series ($1s$, $2s$, $2p$, $3s$) represents the first part of the electron-filling pattern in the Periodic Table.

PROBLEM
Write the electron-filling pattern for the first ten sublevels.

SOLUTION
Using the diagonal rule, we have $1s$, $2s$, $2p$, $3s$, $3p$, $4s$, $3d$, $4p$, $5s$, $4d$.

TRY IT YOURSELF
Complete the electron pattern through the $7s$ sublevel.

ANSWER
$1s$, $2s$, $2p$, $3s$, $3p$, $4s$, $3d$, $4p$, $5s$, $4d$, $5p$, $6s$, $4f$, $5d$, $6p$, $7s$

The answer to this *"Try It Yourself"* exercise provides us with some insights as to how levels and sublevels fill. *Within* a given level, the order of sublevel filling is always $s \rightarrow p \rightarrow d \rightarrow f \rightarrow g$. However, the filling of levels is irregular: we note that level 4 does not *complete* filling ($4f$ sublevel) until level 6 ($6s$ sublevel) has begun!

To apply the diagonal rule, we must remember that the *maximum* numbers of electrons for the sublevels are as follows: $s = 2$; $p = 6$; $d = 10$; $f = 14$; $g = 18$. We simply count electrons until the sublevel is filled, and then we move to the next sublevel in the order given by the diagonal. Each new atom always repeats the pattern of the atom before it.

Let us build up the first 10 elements of the Periodic Table using the diagonal rule. The result is shown in the accompanying table.

Element	Atomic Number	Electron Configuration
H	1	$1s^1$
He	2	$1s^2$
	(Level 1 is complete.)	
Li	3	$1s^2\,2s^1$
Be	4	$1s^2\,2s^2$
B	5	$1s^2\,2s^2\,2p^1$
C	6	$1s^2\,2s^2\,2p^2$
N	7	$1s^2\,2s^2\,2p^3$
O	8	$1s^2\,2s^2\,2p^4$
F	9	$1s^2\,2s^2\,2p^5$
Ne	10	$1s^2\,2s^2\,2p^6$
	(Level 2 is complete.)	

An element that has a total of eight electrons in its *outer* principal energy level, such as neon, is known as a *noble* (or *inert*) gas. Helium is also classified as an inert gas because its only principal energy level is filled (with two electrons). When we reach a noble gas, the *next* element will begin a *new* principal energy level. To conserve space, we usually place the symbol of a noble gas in brackets—for example, [Ne]—rather than write the entire configuration.

PROBLEM
Write the electron configurations of elements 11–20 in the Periodic Table using the diagonal rule and the format given above.

SOLUTION

Element	Atomic Number	Electron Configuration
Na	11	[Ne] $3s^1$
Mg	12	[Ne] $3s^2$
Al	13	[Ne] $3s^2\,3p^1$
Si	14	[Ne] $3s^2\,3p^2$
P	15	[Ne] $3s^2\,3p^3$
S	16	[Ne] $3s^2\,3p^4$
Cl	17	[Ne] $3s^2\,3p^5$
Ar	18	[Ne] $3s^2\,3p^6$
	(Level 3 is *not* complete, but it has 8 valence electrons.)	
K	19	[Ar] $4s^1$
Ca	20	[Ar] $4s^2$

Now suppose we wish to write the electron configuration of element 21 (scandium). Using the diagonal rule, we obtain $1s^2 2s^2 2p^6 3s^2 3p^6 4s^2 3d^1$. The last sublevel of level 3 has begun to fill. We could write this configuration in the "shorthand" notation we used above as [Ar] $4s^2 3d^1$, or we could write it as [Ar] $3d^1 4s^2$ in order to keep the principal energy levels in *numerical order*. Reference Table V in Appendix 2 uses the latter method, but the two notations are entirely equivalent.

TRY IT YOURSELF
Write the electron configuration of bromine (Br), element number 35, (a) according to the diagonal rule and (b) according to our "shorthand" notation.

ANSWERS
(a) $1s^2 2s^2 2p^6 3s^2 3p^6 4s^2 3d^{10} 4p^5$
(b) [Ar] $4s^2 3d^{10} 4p^5$ or [Ar] $3d^{10} 4s^2 4p^5$

The diagonal rule is a *general* rule, and there are exceptions to it. For example, the rule predicts that the electron configuration of an atom of chromium (Cr, atomic number 24) should be [Ar] $3d^4 4s^2$. If we consult Reference Table V in Appendix 2, however, we see that the electron configuration is [Ar] $3d^5 4s^1$. The reason for exceptions such as this one is explored in Chapter 9.

TRY IT YOURSELF
Which element in the Periodic Table lies in the same horizontal row as chromium and also has an irregular filling pattern?

ANSWER
copper (Cu, atomic number 29)

Filling Electrons into Orbitals

At this point, we need to investigate the filling patterns within the *orbitals* of a sublevel. *Any* orbital (regardless of its sublevel) can accommodate zero, one, or two electrons. If the orbital has no electron, we say that it is *empty*; if it has one electron, it is *half-filled*; if it has two electrons, it is *filled*.

Since an s sublevel can accommodate a maximum of two electrons and an orbital can accommodate a maximum of two electrons, it follows that an s sublevel can have only one orbital. Since a p sublevel can accommodate a maximum of six electrons, it follows that a p sublevel has three orbitals. Similarly, a d sublevel has five orbitals and an f sublevel has seven orbitals. It does not matter *where* the sublevel is located; the rule is the same. Thus, the $2p$ sublevel contains three orbitals, as do the $3p$, $4p$, and $5p$ sublevels.

We will indicate the orbitals on a sublevel by using a series of dashes as follows:

___	___ ___ ___	___ ___ ___ ___ ___
s orbital (1) on s sublevel	p orbitals (3) on p sublevel	d orbitals (5) on d sublevel

Electrons possess a property known as **spin**. There are only two spin states for an electron, and we indicate these by vertical arrows: $\uparrow$ and $\downarrow$. These designations are completely arbitrary; we could have used any other pair of matched symbols.

When an orbital is empty, we indicate this by ___.

When an orbital contains only one electron, we indicate this by $\uparrow$.

When an orbital contains two electrons, we indicate this by $\uparrow\downarrow$. This notation shows that the electron "spins" have been "paired," resulting in a more stable condition for the electrons in that orbital.

We are now ready to begin filling the orbitals of each element, using the sublevel chart for elements 1–10 that we developed on page 228.

Electron Configurations

Element	By Sublevels	By Orbitals
H	$1s^1$	$\uparrow$
He	$1s^2$	$\uparrow\downarrow$
Li	$1s^2\,2s^1$	$\uparrow\downarrow$ $\uparrow$
Be	$1s^2\,2s^2$	$\uparrow\downarrow$ $\uparrow\downarrow$
B	$1s^2\,2s^2\,2p^1$	$\uparrow\downarrow$ $\uparrow\downarrow$ $\uparrow$ ___ ___
C	$1s^2\,2s^2\,2p^2$	$\uparrow\downarrow$ $\uparrow\downarrow$ $\uparrow$ $\uparrow$ ___
N	$1s^2\,2s^2\,2p^3$	$\uparrow\downarrow$ $\uparrow\downarrow$ $\uparrow$ $\uparrow$ $\uparrow$
O	$1s^2\,2s^2\,2p^4$	$\uparrow\downarrow$ $\uparrow\downarrow$ $\uparrow\downarrow$ $\uparrow$ $\uparrow$
F	$1s^2\,2s^2\,2p^5$	$\uparrow\downarrow$ $\uparrow\downarrow$ $\uparrow\downarrow$ $\uparrow\downarrow$ $\uparrow$
Ne	$1s^2\,2s^2\,2p^6$	$\uparrow\downarrow$ $\uparrow\downarrow$ $\uparrow\downarrow$ $\uparrow\downarrow$ $\uparrow\downarrow$
		$1s$ $2s$ $2p_x\,2p_y\,\,2p_z$

Notice, in the accompanying table, what happens when we begin to fill carbon (C): the second $2p$ electron does *not* fill the first orbital; it half-fills the second orbital. This pattern is repeated with nitrogen (N). This half-filling occurs because electrons repel each other and will occupy whichever positions are farthest from each other. When oxygen (O) is filled, the last $2p$ electron has no choice but to enter and fill the first orbital. Meanwhile, we have another rule: *Electrons entering a sublevel will half-fill the orbitals in that sublevel before they fill it completely*.

PROBLEM

Using the preceding tables as a guide, list the orbital filling patterns for elements 11–20.

SOLUTION

Electron Configurations

Element	By Sublevels	By Orbitals
Element 10: Ne	$1s^2\,2s^2\,2p^6$ [Ne]	↑↓ ↑↓ ↑↓ ↑↓ ↑↓ 1s 2s $2p_x$ $2p_y$ $2p_z$
Na	[Ne] $3s^1$	[Ne] ↑
Mg	[Ne] $3s^2$	[Ne] ↑↓
Al	[Ne] $3s^2\,3p^1$	[Ne] ↑↓ ↑ _ _
Si	[Ne] $3s^2\,3p^2$	[Ne] ↑↓ ↑ ↑ _
P	[Ne] $3s^2\,3p^3$	[Ne] ↑↓ ↑ ↑ ↑
S	[Ne] $3s^2\,3p^4$	[Ne] ↑↓ ↑↓ ↑ ↑
Cl	[Ne] $3s^2\,3p^5$	[Ne] ↑↓ ↑↓ ↑↓ ↑
Ar	[Ne] $3s^2\,3p^6$ [Ar]	[Ne] ↑↓ ↑↓ ↑↓ ↑↓ 3s $3p_x$ $3p_y$ $3p_z$
K	[Ar] $4s^1$	[Ar] ↑
Ca	[Ar] $4s^2$	[Ar] ↑↓ 4s

In the preceding examples, we listed the electron configurations of the elements that fill their *s* and *p* sublevels and orbitals. In the Periodic Table, these elements are found in Groups 1, 2, and 13–18 (the *s*-block and *p*-block). When we examine these elements in Chapter 9, we will find that their regular properties entitle them to be called *representative elements*.

8.3A LEWIS STRUCTURES AND ATOMIC ORBITALS

In drawing a Lewis structure, we represent the kernel (nucleus plus inner electrons) by the symbol for the element. Thus, C would represent the kernel of an atom of carbon. We then represent the valence electrons as a series of dots surrounding the kernel. However, the dots (electrons) must be placed in a special manner in order to indicate how the valence orbitals are filled. We recall that valence electrons occupy only *s* and *p* sublevels. Since an *s* sublevel has only one orbital and a *p* sublevel has three orbitals (p_x, p_y, and p_z), we have only four positions in which to place dots.

Let us adopt the following scheme for the mythical element X. Imagine that a clock surrounds X. The *s* orbital will be located at 12 o'clock, and the

p_x, p_y, and p_z orbitals at 3, 6, and 9 o'clock, respectively, as indicated in the diagram.

If an orbital contains no electrons, we will leave the position blank; if it has one electron, we will place a single dot (•) at that position; if it has two electrons, we will place a pair of dots (• •) at that position.

We recall that the electron configuration of oxygen is as follows:

$$1s^2 2s^2 2p^4 \qquad \underset{1s}{\uparrow\downarrow} \quad \underset{2s}{\uparrow\downarrow} \quad \underset{2p_x}{\uparrow\downarrow} \quad \underset{2p_y}{\uparrow} \quad \underset{2p_z}{\uparrow}$$

Applying our rules, we can write the Lewis structure for oxygen as follows:

·Ö:

·

PROBLEM
Draw the Lewis structure for carbon (atomic number 6).

SOLUTION
Carbon has four valence electrons ($2s^2 2p^2$). The Lewis structure is

TRY IT YOURSELF
Draw the Lewis structure for calcium (atomic number 20).

ANSWER
Calcium has two valence electrons ($4s^2$). The Lewis structure is

Cä

A final note: The "clock" introduced for drawing Lewis structures is *completely arbitrary*. For example, all of the following are correct for the Lewis structure of Ca:

Cä   Ca: Cạ   :Ca

Some questions have the symbol "§2" in front of the question number. This symbol means that the question is based on Section II material.

1. The characteristic bright-line spectrum of sodium is produced when its electrons
   (1) return to lower energy levels
   (2) jump to higher energy levels
   (3) are lost by the neutral atoms
   (4) are gained by the neutral atoms

2. The following diagram shows the characteristic spectral line patterns of four elements. Also shown are the spectral lines produced by an unknown substance. Which pair of elements is present in the unknown?

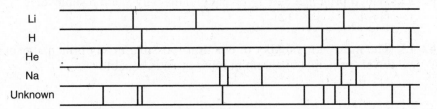

   (1) Li and Na   (2) Na and H   (3) Li and He   (4) He and H

§2 3. The total number of orbitals in an *f* sublevel is
   (1) 1   (2) 5   (3) 3   (4) 7

4. Which principal energy level can hold a maximum of 18 electrons?
   (1) 5   (2) 2   (3) 3   (4) 4

§2 5. The total number of *d* orbitals in the third principal energy level is
   (1) 1   (2) 5   (3) 3   (4) 7

§2 6. The total number of orbitals in the second principal energy level is
   (1) 1   (2) 2   (3) 3   (4) 4

§2 7. The maximum number of electrons that a single orbital of the 3*d* sublevel may contain is
   (1) 5   (2) 2   (3) 3   (4) 4

§2 **8.** What is the maximum number of sublevels in the third principal energy level?
    (1) 1   (2) 2   (3) 3   (4) 4

§2 **9.** Energy level 2 contains a total of
    (1) 8 sublevels   (2) 2 sublevels   (3) 3 sublevels   (4) 4 sublevels

10. As an electron in an atom moves from the ground state to an excited state, the potential energy of the atom
    (1) decreases   (2) increases   (3) remains the same

11. Which principal energy level change by the electron of a hydrogen atom will cause the greatest amount of energy to be absorbed?
    (1) $n = 2$ to $n = 4$ (2) $n = 2$ to $n = 5$
    (3) $n = 4$ to $n = 2$ (4) $n = 5$ to $n = 2$

§2 **12.** Which principal energy level contains only s and p sublevels?
    (1) 1   (2) 2   (3) 3   (4) 4

§2 **13.** The maximum number of electrons possible in any principal energy level (principal quantum number $= n$) is equal to
    (1) n (2) $2n$ (3) n^2 (4) $2n^2$

14. If 18 electrons completely fill a principal energy level, what is the number of the principal energy level?
    (1) 1   (2) 2   (3) 3   (4) 4

§2 **15.** What is the number of sublevels in an energy level whose number of principal energy levels is n?
    (1) n (2) $2n$ (3) n^2 (4) $2n^2$

§2 **16.** Which principal energy level has no f sublevel?
    (1) 5   (2) 6   (3) 3   (4) 4

17. Which configuration represents an excited atom?
    (1) 2-1   (2) 2-8-1   (3) 2-8-2   (4) 2-7-2

18. An excited atom with the configuration 2-8-6-3 is an atom of?
    (1) Na   (2) K   (3) S   (4) Br

§2 **19.** What is the maximum number of electrons that can occupy a $5f$ sublevel?
    (1) 5   (2) 7   (3) 14   (4) 50

20. An atom contains a total of 25 electrons. When the atom is in the ground state, how many different principal energy levels will contain electrons?
(1) 1   (2) 2   (3) 3   (4) 4

§2 **21.** Energy is released when an electron changes from a sublevel of
(1) $1s$ to $2p$ (2) $2s$ to $3s$ (3) $3s$ to $2s$ (4) $3p$ to $5s$

§2 **22.** What is the total number of occupied sublevels in an atom of chlorine in the ground state?
(1) 1   (2) 5   (3) 3   (4) 9

§2 **23.** An atom in the ground state has 7 valence electrons. Which electron configuration could represent the outermost principal energy level of this atom in the ground state?
(1) $3s^13p^6$ (2) $3s^23p^5$ (3) $3s^13p^43d^2$ (4) $3s^23p^43d^1$

§2 **24.** Which electron configuration represents an atom in an excited state?
(1) $1s^22s^22p^63p^1$ (2) $1s^22s^22p^63s^23p^1$
(3) $1s^22s^22p^63s^23p^2$ (4) $1s^22s^22p^63s^2$

25. What is the total number of principal energy levels that are completely filled in an atom of magnesium in the ground state?
(1) 1   (2) 2   (3) 3   (4) 4

§2 **26.** The total number of completely filled orbitals in an atom of nitrogen in the ground state is
(1) 1   (2) 2   (3) 3   (4) 4

§2 **27.** The neutral atom of an element has an electron configuration of 2-8-1. What is the total number of p electrons in this atom?
(1) 6   (2) 2   (3) 10   (4) 11

28. What is the total number of electrons in the second principal energy level of a calcium atom in the ground state?
(1) 6   (2) 2   (3) 8   (4) 18

§2 **29.** Which is the electron configuration of a noble-gas atom in the excited state?
(1) $1s^1$ (2) $1s^12s^1$ (3) $1s^22s^2$ (4) $1s^22s^22p^2$

§2 **30.** Which is an electron configuration of a fluorine atom in the excited state?
(1) $1s^2 2s^2 2p^4$ (2) $1s^2 2s^2 2p^5$
(3) $1s^2 2s^2 2p^4 3s^1$ (4) $1s^2 2s^2 2p^5 3s^1$

§2 **31.** Which orbital notation represents the outermost principal energy level of a phosphorus atom in the ground state?

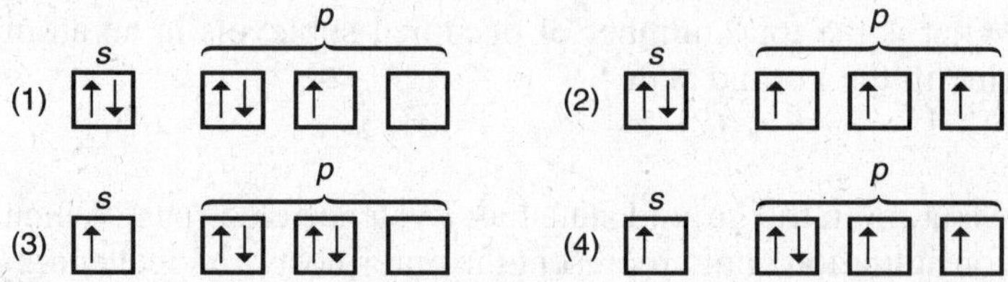

§2 **32.** Which atom in the ground state contains only one orbital that is partially occupied?
(1) Si   (2) Ne   (3) Ca   (4) Na

§2 **33.** An electron in an atom will emit energy when it moves from energy level
(1) $2s$ to $3p$ (2) $2s$ to $2p$ (3) $2p$ to $3s$ (4) $2p$ to $1s$

§2 **34.** The electron configuration of an atom in the ground state is $1s^2 2s^2 2p^3$. The total number of occupied atomic orbitals in this atom is
(1) 6   (2) 5   (3) 3   (4) 4

§2 **35.** An atom contains a total of 25 electrons. When the atom is in the ground state, how many different sublevels will contain electrons?
(1) 7   (2) 2   (3) 3   (4) 5

36. Which is the correct Lewis structure of an atom of sulfur in the ground state?
(1) S: (2) •S: (3) •S: (4) :S:

37. If X is the symbol of a noble-gas atom in the ground state, the Lewis structure of the atom could be
(1) X•   (2) X: (3) •X•   (4) :X:

38. Which is the Lewis structure of an atom of boron in the ground state?
(1) •B: (2) B•   (3) •B: (4) B:

39. Which is the correct Lewis structure of an aluminum atom in the ground state?

(1) Al: (2) Al̇:   (3) •Al̇:   (4) •Al:

40. Usually the term *kernel* includes all parts of the atom except
(1) neutrons   (2) protons
(3) valence electrons   (4) orbital electrons

41. What is the total number of valence electrons of an atom of sulfur in the ground state?
(1) 6   (2) 8   (3) 3   (4) 4

42. Which is an electron configuration of chlorine in the excited state?
(1) 2-8-7   (2) 2-8-8
(3) 2-8-6-1   (4) 2-8-7-1

43. The strength of an atom's attraction for electrons in a chemical bond is the atom's
(1) electronegativity
(2) ionization energy
(3) heat of reaction
(4) heat of formation

44. Which ion has the same electron configuration as an atom of He?
(1) H^- (2) O^{2-} (3) Na^+ (4) Ca^{2+}

45. In the ground state, an atom of which element has two valence electrons?
(1) Cr   (2) Cu   (3) Ni   (4) Se

Constructed-Response Questions

1. Use Reference Table S and the Periodic Table in Appendix 1 to complete the accompanying table.

Element	Symbol	Electron Configuration	Number of Valence Electrons	Lewis Structure
Argon				
Barium				
Carbon				
Chlorine				
Helium				
Nitrogen	N	2-5	5	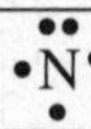 •N•
Oxygen				
Potassium				
Sulfur				
Strontium				

Base your answers to questions 2 and 3 on the electron configuration table shown below.

Element	Electron Configuration
X	2-8-8-2
Y	2-8-7-3
Z	2-8-8-5

2. What is the total number of valence electrons in an atom of electron configuration X?

3. Which electron configuration represents the excited state of a calcium atom?

CHEMICAL PERIODICITY

KEY IDEAS

This chapter focuses on the periodic properties of the elements. After certain general properties are developed, the elements are studied according to the groups in the Periodic Table in which they are located.

KEY OBJECTIVES
At the conclusion of this chapter you will be able to:
- State the periodic law of Moseley.
- Describe the general arrangement of the elements in the modern Periodic Table with regard to electron configuration.
- Describe how properties of metallic elements differ from those of nonmetallic elements.
- Define the term *metalloid*, and use the Periodic Table to indicate which elements are metalloids.
- Define the term *atomic radius*.
- Define the term *ionic radius*, and indicate how the size of an ion compares with the size of its parent atom.
- Indicate how the following general properties of elements vary within the Periodic Table: metallic character, atomic radius, ionization energies, electron affinity, ionic radius, and electronegativity.
- Compare and contrast the properties of the elements in the various representative groups of the Periodic Table.
- Compare some of the properties of the transition elements with those of the representative elements.
- Describe how the properties of the elements vary across a period.

SECTION I—BASIC (REGENTS-LEVEL) MATERIAL

NYS REGENTS CONCEPTS AND SKILLS

Note: By the time you have finished Section I, you should have mastered the concepts and skills listed below. The Regents chemistry examination will test your knowledge of these items and your ability to apply them.

Concepts are the *basic ideas* that form the body of the Regents chemistry course (what you need to know!).

Skills are the *activities* that demonstrate your mastery of these concepts (how you show that you know them!).

Following each concept or skill is a page reference (given in parentheses) to this chapter.

9.1 Concept:
The placement or location of elements on the Periodic Table gives an indication of the physical and chemical properties of that element. The elements in the Periodic Table are arranged in order of increasing atomic number. (Page 244)

Skill:
Explain the placement of an unknown element in the Periodic Table based on its properties. (Pages 244–247)

9.2 Concept:
Elements can be classified by their properties and located in the Periodic Table as metals, nonmetals, metalloids (B, Si, Ge, As, Sb, Te), or noble gases. (Pages 244, 256–258)

Skill:
Classify elements as metals, nonmetals, metalloids, or noble gases by their properties. (Pages 244, 256–258)

9.3 Concept:
Elements can be differentiated by their physical properties. Physical properties such as density, conductivity, malleability, solubility, and hardness differ among elements. (Page 245)

Skill:
Describe the states (phases) of elements at STP. (Pages 256–258)

9.4 Concept:
Elements can be differentiated by chemical properties. Chemical properties determine how an element behaves during a chemical reaction. (Pages 256–258)

9.5 Concept:
Some elements exist in two or more forms in the same phase. Since these forms differ in their molecular or crystal structures, they also differ in their properties. (Pages 256–258)

9.6 Concept:
For Groups 1, 2, and 13–18 in the Periodic Table, elements within the same group have the same number of valence electrons (helium is an exception) and, therefore, have similar chemical properties. (Page 244)

Skill:
Given the chemical formula of a compound (such as XCl or XCl_2), determine the group of an element (X). (Pages 256–258)

9.7 Concept:
The succession of elements within the same group demonstrates characteristic trends: differences in atomic radius, ionic radius, electronegativity, first ionization energy, and metallic (or nonmetallic) properties. (Pages 248–255)

Skill:
Compare and contrast the properties of elements within a single group (Groups 1, 2, 13–18) in the Periodic Table. (Pages 248–255)

9.8 Concept:
The succession of elements across the same period demonstrates characteristic trends: differences in atomic radius, ionic radius, electronegativity, first ionization energy, and metallic (or nonmetallic) properties. (Pages 248–255)

Skill:
Compare and contrast the properties of elements across a period (for Groups 1, 2, 13–18) in the Periodic Table. (Pages 249–258)

9.9 Concept:
When an atom gains one or more electrons, its radius increases. When an atom loses one or more electrons, its radius decreases. (Page 247)

9.1 INTRODUCTION

The first question we need to ask is this: *Why* do we have a Periodic Table of the Elements? The term *periodic* means that a quantity repeats itself at regular intervals. For example, the motion of the Moon around Earth is periodic because, after a certain length of time (1 month), the motion repeats itself. As early as the Middle Ages, scientists recognized that elements could be differentiated by their physical and chemical properties. These properties are also periodic—*if we list the elements properly.*

If we examine the modern Periodic Table of the Elements in Appendix 1, we find that it is almost, but not quite, rectangular in shape. At the top of the table, two elements (hydrogen, H, and helium, He) are separated from the body of the table, as are two rows of elements at the bottom. This table provides the answer to the question asked above: it allows us to place elements with *similar properties* in close proximity.

9.2 THE PERIODIC TABLE IN HISTORY

Historically, the creation of the modern Periodic Table is credited to chemists Dmitry Mendeleev (Russia) and Lothar Meyer (Germany), who did their work during the nineteenth century. The two men, working independently, produced a table in which elements with similar properties were located in the same vertical column. Using his table, Mendeleev was able to predict the properties of such then-undiscovered elements as germanium (Ge) (which he called *eka-silicon*) and gallium (Ga) (which he called *eka-alominum*). Both Mendeleev and Meyer based their tables on the atomic *masses* of the elements. This arrangement led to some problems, however, because the atomic masses of the elements do not increase regularly. Compare, for example, the atomic masses of tellurium (Te) and iodine (I).

In the early twentieth century, physicist Henry Moseley (England) discovered what we now call our modern periodic law: *The periodic properties of the elements are functions of their atomic numbers.* This means that the Periodic Table should be arranged in order of increasing atomic number—which it is! However, the *basis* of the Periodic Table is the (ground-state) *electron configuration* of each element.

9.3 THE MODERN PERIODIC TABLE

The modern Periodic Table is described in terms of **periods** (the horizontal rows) and **groups** (the vertical columns). Each period is indicated by a number, beginning with 1 and *currently* ending with 7. Elements in the same period have the same number of occupied principal energy levels. For example, the elements sodium (Na) and phosphorus (P) are in the same period—3. Therefore, sodium and phosphorus have *three* occupied principal energy levels. The period number of an element also tells us the *valence* level of the element: the valence level of sodium and of phosphorus is also 3.

The groups begin at the left side of the table (Group 1) and continue through Group 18 on the right side. Elements within a group share similar properties, and this likeness is directly related to the similarities in the electron configurations of their valence levels. For this reason, we also call a group a *chemical family*. For example, the elements in Group 13 constitute a chemical family: Atoms of Group 13 elements each have three valence electrons. Consequently, the properties of the elements in this group are similar to one another. (We note, however, that the word *similar* does *not* mean "identical"!)

9.4 PROPERTIES ASSOCIATED WITH PERIODICITY

Before we describe periodic properties in detail, we need to discuss a number of important factors associated with periodicity.

Metallic Character

Approximately two-thirds of the elements in the Periodic Table are metals. Metallic elements have certain unique properties:

- *Luster*, the mirrorlike shine that reflects light well.
- *Conductivity*, the ability to transfer heat and electrons well.
- *Malleability*, the ability to be rolled or hammered into thin sheets.
- *Ductility*, the ability to be drawn into wire.

Examples of metals are sodium (Na), iron (Fe), and mercury (Hg, a liquid at room temperature).

In contrast, nonmetals tend to be brittle (in the solid phase), to lack luster, and to have poor conductivities. Examples of nonmetals include sulfur (S), helium (He), and bromine (Br_2, a liquid at room temperature).

The bold zigzag line in the Periodic Table separates the metallic elements from the nonmetallic elements. We will refer to it as the *metal-nonmetal line*.

A number of elements, such as boron (B), silicon (Si), germanium (Ge), arsenic (As), antimony (Sb), and tellerium (Te), have properties somewhat between those of metals and those of nonmetals. These *semiconductors* are also known as **metalloids** or *semimetals*.

Atomic Radius

If we think of an atom (whether combined with other atoms or not) as having a roughly *spherical* shape, then its size is best reported as a *radius* (called the **atomic radius**), as shown in the accompanying diagram:

Atomic radius

The radius of a "free" atom cannot be measured *directly*. One way to measure the atomic radius is to calculate a quantity called the **internuclear distance**, which is defined to be the distance between two adjacent atoms when they are closest together. The accompanying diagram illustrates the internuclear distance between two adjacent atoms. We can see from the diagram that the radius of an atom is equal to *one-half* of the internuclear distance.

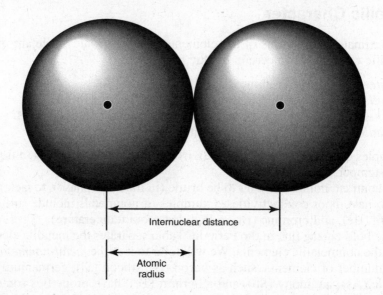

Reference Table S in Appendix 1 lists the atomic radii for most elements. The atomic radius of an atom is reported in *picometers* (pm, 1 pm = 10^{-12} m). For example, according to Table S, the atomic radius of phosphorus (P) is 128 picometers.

First Ionization Energy

An *ion* is an electrically charged atom. Neutral atoms can become charged in either of two ways: they can gain or they can lose electrons. If an atom *loses* one or more electrons, it becomes *positively* charged; if an atom *gains* one or more electrons, it becomes *negatively* charged. The following examples illustrate this principle for an atom of potassium (K) and an atom of sulfur (S):

$$K \rightarrow K^+ + e^-$$
$$S + 2e^- \rightarrow S^{2-}$$

The **first ionization energy** (I_1) measures the ease with which an atom *loses* its *first* electron. To remove an electron from a neutral atom, an amount of energy equal to the first ionization energy must be added:

$$K + I_1 \rightarrow K^+ + e^-$$

The smaller the amount of energy needed, the more easily the electron is lost. Although the first ionization energy may be measured for a single atom, it is generally more useful to measure the amount of energy needed to remove a single electron from each atom in 1 mole of atoms. Therefore, a first ionization energy of 900 kilojoules per mole means that 900 kilojoules is required to remove one electron from each atom in 1 mole of atoms.

First ionization energies of the representative elements range from 393 kilojoules per mole to 2081 kilojoules per mole. Metallic elements generally have low first ionization energies: they tend to lose electrons more easily than nonmetallic elements. This information can be verified by examining Reference Table S in Appendix 1.

PROBLEM
How much energy, in kilojoules, is needed to ionize 0.400 mole of sodium (Na) according to the equation below?

$$Na \rightarrow Na^+ + e^-$$

SOLUTION
According to Reference Table S, the first ionization energy of sodium is 496 kJ/mol, and the amount of energy needed is found by using the factor-label method:

$$0.400 \ \cancel{mol} \cdot 496 \ \frac{kJ}{\cancel{mol}} = 198 \ kJ$$

TRY IT YOURSELF
The first ionization energy of the element boron (B) is 801 kilojoules per mole. How many moles of boron atoms can be ionized to boron ions (B^+) by the addition of 2000. kilojoules of energy?

ANSWER
2.50 mol

Ionic Radius

As we will see in Chapter 10, atoms can combine with other atoms by forming positive and negative *ions*. Therefore, it is frequently useful to measure the sizes of the corresponding ions these atoms produce. This quantity is known as the **ionic radius**. When an atom forms a *positive ion* by losing one or more electrons, the ionic radius is *smaller* than the atomic radius of the parent atom. Conversely, when an atom forms a *negative ion* by gaining one or more electrons, the ionic radius is *larger* than the atomic radius of the parent atom. The following two examples illustrate this principle:

1. The radius of a lithium atom (Li) is 155 picometers; the radius of a lithium ion (Li^+) is 60 picometers.
2. The radius of a chlorine atom (Cl) is 97 picometers; the radius of a chloride ion (Cl^-) is 181 picometers.

Electronegativity

Another way in which atoms can combine is by sharing one or more *pairs* of electrons. The **electronegativity** of an atom measures its ability to attract a pair of electrons when bonded to another atom. The higher the electronegativity, the more the atom attracts the shared pair of electrons.

Electronegativities range from a high of 4.0 for fluorine (F) to a low of 0.7 for francium (Fr). Electronegativity values are given in Reference Table S in Appendix 1 of this book.

9.5 VARIATION OF PERIODIC PROPERTIES AMONG THE ELEMENTS

We now examine how the periodic properties introduced in Section 8.4 vary within the Periodic Table. We will restrict our examination to the *representative* elements (Groups 1, 2, 13–18) because these elements are the simplest to study. In order to draw our conclusions, we will consider how the properties of these elements vary *across a period* (i.e., from left to right across the table) and *down a group* (i.e., from top to bottom down the table). When we require quantitative data, we will focus our attention on Period 3 and Group 16 of the table.

The accompanying table is an abbreviated version of the Periodic Table of the Elements in Appendix 1. It shows the periodic properties of the *representative elements* that we shall study in this section.

Metallic Character

As we proceed down a group, metallic character *increases*. As we proceed across a period, metallic character *decreases*.

Elements display metallic properties when they have *low ionization energies* and *relatively few electrons in their valence levels*. Therefore, we would expect that the greatest metallic character would be displayed by elements located in the *lower left* part of the Periodic Table, and the least metallic character by elements located in the *upper right* part of the table. This is, in fact, correct: the lower elements in Groups 1 and 2 are the most metallic elements; the upper elements in Groups 16 and 17, the most nonmetallic.

SELECTED PERIODIC PROPERTIES OF THE REPRESENTATIVE ELEMENTS

GROUP

Period	1	2	13	14	15	16	17	18
1	H 37 1312 2.1							He 32 2372 (−)
2	Li 155 520 1.0	Be 112 900 1.6	B 98 801 2.0	C 91 1086 2.6	N 92 1402 3.0	O 65 1314 3.5	F 57 1681 4.0	Ne 51 2081 (−)
3	Na 190 496 0.9	Mg 160 736 1.3	Al 143 578 1.6	Si 132 787 1.9	P 128 1012 2.2	S 127 1000 2.6	Cl 97 1251 3.2	Ar 88 1521 (−)
4	K 235 419 0.8	Ca 197 590 1.0	Ga 141 579 1.8	Ge 137 762 2.0	As 139 944 2.2	Se 140 941 2.6	Br 112 1140 3.0	Kr 103 1351 (−)
5	Rb 248 403 0.8	Sr 215 549 1.0	In 166 558 1.8	Sn 162 709 2.0	Sb 159 831 2.1	Te 142 869 2.1	I 132 1008 2.7	Xe 124 1170 2.6
6	Cs 267 376 0.8	Ba 222 503 0.9	Tl 171 589 2.0	Pb 175 716 2.3	Bi 170 703 2.0	Po 167 812 2.0	At 145 (−) 2.2	Rn 134 1037 (−)
7	Fr 270 393 0.7	Ra 233 (−) 0.9	Nh (−) (−) (−)	Fl (−) (−) (−)	Mc (−) (−) (−)	Lv (−) (−) (−)	Ts (−) (−) (−)	Og (−) (−) (−)

metal   nonmetal

metalloid   no data

SYMBOL	→	C
ATOMIC RADIUS / pm	→	91
FIRST IONIZATION ENERGY / (kJ/mol)	→	1086
ELECTRONEGATIVITY	→	2.6

• A dash (−) indicates that data are not available.

These variations are borne out by the heavy metal-nonmetal line in the Periodic Table. Its zigzag shape is shifted further to the right as we move down the table, indicating that metallic properties appear in elements that lie further down in the table. Even the Group 16 element polonium (Po) exhibits metallic character because it is far enough down. We must conclude that metallic character does indeed *increase* down a group.

Atomic Radius

As we proceed down a group, we find that the radii of successive atoms *increase*.

The explanation of this observation lies in the fact that, as the period number increases, *new* energy levels are being occupied. Therefore, the outer electrons of successive elements are *farther* away from the nucleus. In addition, the inner electrons serve to *shield* the outer electrons from the nucleus. These two circumstances lead to an *increase* in radius. The accompanying table and graph illustrate this principle for Group 16 of the Periodic Table.

Elements down Group 16	Atomic Radius / pm
O	65
S	127
Se	140
Te	142

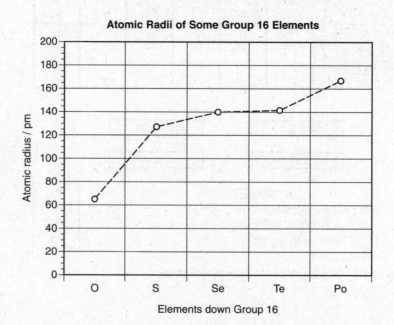

Atomic Radii of Some Group 16 Elements

As we proceed across a period, the radii of successive atoms *decrease*. The explanation of this observation lies in the fact that with each successive element the positive nuclear charge increases, as does the negative electron charge. Since the successive electrons are being added in the *same* principal energy level, the force of attraction between the nucleus and the electrons *increases*, leading to a *decrease* in radius. The accompanying table and graph illustrate this principle for Period 3 of the Periodic Table.

Elements across Period 3	Atomic Radius / pm
Na	190
Mg	160
Al	143
Si	132
P	128
S	127
Cl	97
Ar	88

Atomic Radii of the Period 3 Elements

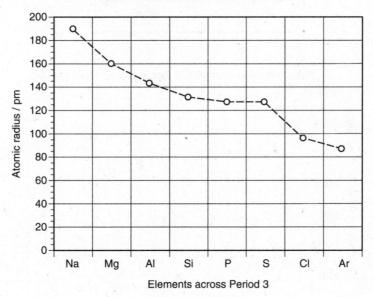

First Ionization Energy

As we proceed down a group, the first ionization energy of each successive element *decreases*.

Recall that the first ionization energy of an atom is the energy needed to remove a single electron from the atom. If the electron is loosely held, the ionization energy is low; if the electron is strongly held, the ionization energy is high. There is an inverse relationship between the radius of an atom and its first ionization energy: the larger the radius, the more weakly its outer electron is held and, consequently, the more easily it is removed.

If we examine the first ionization energies of the elements down Group 16, as shown in the accompanying table and graph, we find a regular decrease with increasing atomic radius, as we predicted.

Elements down Group 16	First Ionization Energy / (kJ/mol)
O	1314
S	1000
Se	941
Te	869
Po	812

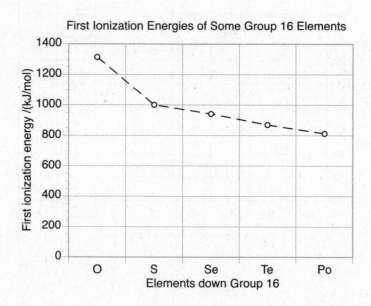

First Ionization Energies of Some Group 16 Elements

As we proceed across a period, the first ionization energy of successive elements *generally increases*. The accompanying table and graph illustrate this trend for Period 3.

We find, however, two surprises: the first ionization energies of aluminum (Al) and sulfur (S) are *lower* than we would expect.

Elements across Period 3	First Ionization Energy / (kJ/mol)
Na	496
Mg	738
Al	578
Si	787
P	1012
S	1000
Cl	1251
Ar	1521

First Ionization Energies of the Period 3 Elements

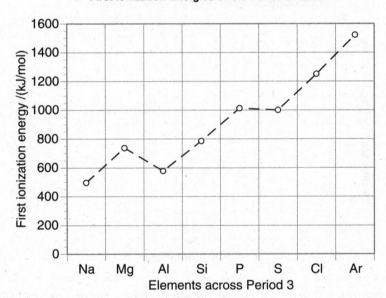

As a result of these two trends in first ionization energies—a *decrease* down a group and an *increase* across a period—we can infer that *elements located at the left side of the Periodic Table and elements located toward the bottom of the table have the greatest tendency to lose electrons.*

Electronegativity

As we proceed down a group, we observe a *decrease* in electronegativity. The variation of electronegativity down a group, as shown in the accompanying table and graph, can be predicted from the variation in the properties discussed above: atomic radius and ionization energy. Elements located successively further down a group have a decreased ability to attract electrons.

Elements down Group 16	Electronegativity
O	3.4
S	2.6
Se	2.6
Te	2.1
Po	2.0

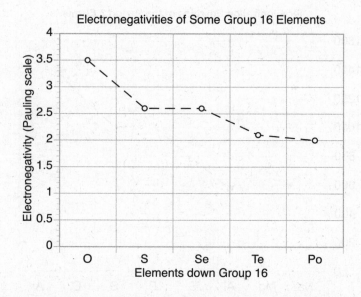

As we proceed across a period, we observe an *increase* in electronegativity. The variation in electronegativity across a period, as shown in the accompanying table and graph, can also be predicted from the variation in the properties discussed above: atomic radius and ionization energy. Elements located successively to the right across a period have an increased ability to attract electrons.

Elements across Period 3	Electronegativity
Na	0.9
Mg	1.3
Al	1.6
Si	1.9
P	2.2
S	2.6
Cl	3.2

Electronegativities of the Period 3 Elements

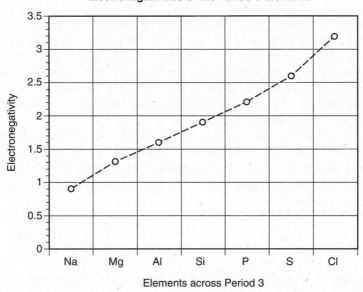

If we examine the Selected Periodic Properties of the Representative Elements on page 249, we can verify these relationships among the representative elements.

The accompanying table provides a general summary of the variation in three periodic properties of the representative elements within the Periodic Table. (Remember: *There are exceptions to each of these generalizations!*)

Periodic Property	Variation down a Group	Variation across a Period
Atomic radius	Increases	Decreases
First ionization energy	Decreases	Increases
Electronegativity	Decreases	Increases

9.6 THE CHEMISTRY OF THE REPRESENTATIVE GROUPS

We now describe the general properties of the representative elements within a single group. Our purpose is *not* to provide an exhaustive description, but rather to point out the similarities and differences among the elements in that group.

Group 1

The elements in Group 1 are called the *alkali metals*. (Each element has one valence electron.) Elements in this group have the most metallic character, have low ionization energies and electronegativities, and have the largest atoms in their period. Group 1 elements form 1+ ions. Because they are highly reactive, they do not occur as free elements in nature and must be prepared by passing an electric current through their molten compounds. We will discuss this technique in Chapter 15.

Hydrogen (H) is *not* considered an alkali metal because its properties are not similar to those of the other elements in Group 1. This exception is due to the small size of the hydrogen atom and the large attraction of the nucleus for the single electron. Hydrogen is associated with Group 1 because it has one valence electron.

Group 2

The elements in Group 2 are known as the *alkaline earth metals*. (Each element has two valence electrons.) An alkaline earth metal has less metallic character than the alkali metal that precedes it. The elements of Group 2 form 2+ ions, although beryllium (Be), which is relatively small, is known to form covalent bonds (see Chapter 10) with elements such as chlorine (Cl, in $BeCl_2$). The alkaline earth metals are reactive and do not occur as free elements in nature.

Group 13

Each element in Group 13 has three valence electrons. The first element of this group, boron (B), is a metalloid. Because of its small size, it does not form a 3+ ion readily, as do the other elements in this group.

Group 14

Each element in Group 14 has four valence electrons. Carbon (C) is a non-metal that occurs in two distinct crystalline forms: graphite and diamond. (These forms are known as *allotropes*.) Silicon (Si) and germanium (Ge) are metalloids that are used widely in the computer industry. Tin (Sn) and lead (Pb) are typical metallic substances.

Group 15

Each element in Group 15 has five valence electrons. Nitrogen (N), the first member of the group, exists in nature as N_2, a diatomic molecule with a triple bond (see Chapter 10). The relative inactivity of nitrogen gas in the atmosphere is due to the fact that considerable energy is needed to break the triple bond. Nitrogen is nonmetallic. Phosphorus (P), also a nonmetal, exists in two allotropic forms: red and white.

Group 16

Each element in Group 16 has six valence electrons. Oxygen (O), a non-metal, exists in nature in two allotropic forms, O_2 and O_3 (ozone). It readily forms a 2− ion. Sulfur (S) is also a nonmetal; it exists in several allotropic forms and can form a 2− ion although it is less reactive than oxygen. Selenium (Se) and tellurium (Te) are classified as metalloids, and polonium (Po) is considered a metal.

Group 17

The elements in Group 17 are known as the *halogens*. Each element has seven valence electrons. All of these elements are nonmetals that form 1− ions. They are the most reactive of the nonmetallic elements, with fluorine (F) being the most reactive element in this group. Fluorine and chlorine (Cl) are gases at room temperature; bromine (Br) is a liquid, and iodine (I) is a solid. Because of their reactivity, the halogens do not occur as free elements in nature.

Group 18

The elements in Group 18 are known as the *noble* (or *inert*) *gases*. Each element other than helium has eight valence electrons. Although helium (He) has only two valence electrons, it is associated with this group because its properties match those of the other elements. The elements of Group 18 are extremely stable and occur as monatomic gases in nature. Although they do not combine readily with other elements, compounds of krypton (Kr) and xenon (Xe) have been prepared.

The Transition Elements

The *transition elements* begin in Period 4 and include Groups 3–11. (Group 12 elements are *not* considered transition elements.) The transition elements exhibit multiple oxidation states because the *two* outermost levels may be involved in chemical combinations. The ions of transition elements are usually colored, both in aqueous solution and as solids.

9.7   THE CHEMISTRY OF A PERIOD

The properties of the elements across a period (left to right) vary much more than the properties of the elements within a single group. Generally, a decrease in atomic radius and an increase in ionization energy and electronegativity occur. The metallic character of successive elements decreases, going from metal to metalloid to nonmetal to noble gas. These changes are accompanied by a gradual change from positive to negative to zero oxidation states.

SECTION II—ADDITIONAL MATERIAL

9.1A   SUBLEVELS AND THE PERIODIC TABLE

Groups 1 and 2 constitute the **s-block** because within these two adjacent groups the *s* sublevels are filled. Groups 13–18 constitute the **p-block** because within these six adjacent groups the *p* sublevels are filled. The *d* sublevels fill in Groups 3–12, and the *f* sublevels fill in the two separated rows at the bottom of the table. These collections of elements are named the **d-block** and the **f-block**, respectively.

The *s*- and *p*-blocks are called the **representative elements**; the *d*-block (through Group 11), the **transition elements**; and the *f*-block, the **lanthanoid** and **actinoid** (or **lanthanide** and **actinide**) *series*.

9.2A SUCCESSIVE IONIZATION ENERGIES

An atom can have as many ionization energies as it has electrons. For example, potassium (K, atomic number 19) has *nineteen* ionization energies. The equations corresponding to the first three ionization energies are as follows:

$$K + I_1 \rightarrow K^+ + e^-$$
$$K^+ + I_2 \rightarrow K^{2+} + e^-$$
$$K^{2+} + I_3 \rightarrow K^{3+} + e^+$$

Successive ionization energies can become very large because of the considerable difficulty in separating negative electrons from increasingly positive ions.

9.3A ELECTRON AFFINITY

Electron affinity measures the energy change that occurs when an atom *gains* a single electron and forms a negative ion. The more energy *released* by the atom, the more stable it becomes when it gains an electron. In this case, the electron affinity is reported as a *negative* number. In some atoms, however, the tendency to gain an electron is negligible, and the electron affinity is reported as zero. In general, nonmetals gain electrons more easily than metals, as the accompanying table shows.

Element	Classification	Electron Affinity (kJ/mol)
Li	Metal	−60
Na	Metal	−53
Mg	Metal	0
C	Nonmetal	−122
S	Nonmetal	−200
Cl	Nonmetal	−349

9.4A ADDITIONAL ASPECTS OF FIRST IONIZATION ENERGY

Explaining the Irregularities of Aluminum and Sulfur

The explanation for the irregularity of *aluminum* lies in the fact that the outer electron is in the 3*p* sublevel. As a result this electron has more energy, and therefore *less* energy is needed to remove it from the atom. The irregularity of *sulfur* is explained by the fact that its outer electron is the

259

fourth electron in the $3p$ sublevel. Removing this electron will leave a half-filled $3p$ sublevel that has more stability. This removal requires *less* energy than removing a single electron in phosphorus, which already has a half-filled $3p$ sublevel. Consequently, the first ionization energy for sulfur is *less* than that for phosphorus.

An Important Principle of First Ionization Energy

These observations allow us to state an important principle: *Half-filled and fully filled sublevels provide an atom with additional stability.* In Chapter 7 (page 229), we saw that two elements, chromium (Cr) and copper (Cu), have irregular electron-filling patterns. This principle explains the observed patterns: The $3d^54s^1$ configuration of chromium allows the atom to have two half-filled sublevels; the $3d^{10}4s^1$ configuration of copper allows the atom to have one fully filled sublevel and one half-filled sublevel.

9.5A VARIATION OF SUCCESSIVE IONIZATION ENERGIES

When we examine the successive ionization energies of the first seven elements across Period 3, as shown in the accompanying table, we see a progressive increase for each element (i.e., $I_1 < I_2 < I_3 < \ldots$). The increases are not smooth, however. For each of the first six elements, there is a very large increase in ionization energy (indicated in the *shaded* boxes in the table). For aluminum (Al), this increase occurs at I_4.

Elements across Period 3	Electron Configuration	Successive Ionization Energies /(kJ/mol)							
		I_1	I_2	I_3	I_4	I_5	I_6	I_7	I_8
Na	[Ne]$3s^1$	496	4562						
Mg	[Ne]$3s^2$	738	1451	7733					
Al	[Ne]$3s^23p^1$	578	1817	2745	11,577				
Si	[Ne]$3s^23p^2$	787	1577	3232	4356	16,091			
P	[Ne]$3s^23p^3$	1012	1907	2914	4964	6274	21,267		
S	[Ne]$3s^23p^4$	1000	2252	3357	4556	7004	8496	27,107	
Cl	[Ne]$3s^23p^5$	1251	2298	3822	5159	6542	9362	11,018	33,604

In general, when an atom has had all of its *valence* electrons removed, the removal of the *next* electron requires the absorption of a very large quantity of energy because this electron lies within the *noble-gas core* of the atom. Such electrons are held very tightly by the nucleus of the atom.

TRY IT YOURSELF

Explain why the large energy increase for phosphorus (P) occurs at I_6.

ANSWER

An atom of phosphorus has *five* valence electrons $(3s^23p^3)$. The sixth electron lies within the noble-gas core of the phosphorus atom.

Because of the great difficulty in removing these inner-core electrons, we can infer that they are unlikely to be removed or (as we shall see in Chapter 9) to participate in any type of chemical combination between atoms. Consequently, we can expect that, according to the table, sodium (Na), magnesium (Mg), and aluminum (Al) will be able to lose one, two, and three electrons, respectively, without too much difficulty.

TRY IT YOURSELF

According to the table, what is the maximum number of electrons you would expect phosphorus (P) and sulfur (S) to be able to lose?

ANSWER

Five and six, respectively.

In reality, phosphorus and sulfur do *not* lose electrons; they tend to *gain* three and two electrons, respectively.

PROBLEM

What happens when an atom of sulfur gains two electrons?

SOLUTION

The sulfur atom becomes an ion with a charge of $2-$ (S^{2-}). The electron configuration of this ion is $[Ne]3s^23p^6$, which is the configuration of argon [Ar].

9.6A SYNTHETIC ELEMENTS

A number of elements do *not* occur in nature. These *synthetic elements* include technetium (Tc, atomic number 43) and all of the elements with atomic numbers larger than 92 (the *transuranium* elements), all of which are radioactive. Technetium was the first element to be produced artificially by bombardment; it is also found as a by-product of fission reactors. Transuranium elements are also produced by bombardment reactions, the largest element discovered to date is Oganesson (Og), with an atomic number of 118.

The names and symbols of elements 104–118 are given in the accompanying table.

Atomic Number	Name	Symbol
104	Rutherfordium	Rf
105	Dubnium	Db
106	Seaborgium	Sg
107	Bohrium	Bh
108	Hassium	Hs
109	Neitnerium	Mt
110	Darmstadtium	Ds
111	Roentgenium	Rg
112	Copernicium	Cn
113	Nihonium	Nh
114	Flerovium	Fl
115	Moscovium	Mc
116	Livermorium	Lv
117	Tennessine	Ts
118	Organesson	Og

Naming Newly Discovered Elements

When a new element is discovered, it is given a temporary (Latin) name and a three-letter symbol that is related to its atomic number. For example, when flerovium (Fl, 114) was discovered, it was assigned the temporary name *ununquadium* (symbol, Uuq). Each letter of the symbol corresponds to a digit in the element's atomic number: (U = 1, u = 1, q = 4). Each letter has a Latin root and these roots are joined to form the element's temporary name (*un*, *un*, *quad*, [*ium*]). The accompanying table lists Latin roots and the symbols corresponding to each digit from 0 through 9. As usual the first letter in the three-letter symbol is capitalized.

Digit	Root	Symbol
0	nil	n
1	un	u
2	bi	b
3	tri	t
4	quad	q
5	pent	p
6	hex	h
7	sept	s
8	oct	o
9	enn	e

TRY IT YOURSELF
In the far, far future element 258 may be discovered. Provide its temporary name and symbol by referring to the table given immediately above.

ANSWER
bipentoctium, Bpo

END-OF-CHAPTER QUESTIONS

Some questions have the symbol "§2" in front of the question number. This symbol means that the question is based on Section II material.

  1. The pair of elements with the most similar chemical properties are
  (1) Mg and S   (2) Ca and Br
  (3) Mg and Ca   (4) S and Ar

  2. More than two-thirds of the elements in the Periodic Table of the Elements are
  (1) metalloids   (2) metals
  (3) nonmetals   (4) noble gases

  3. In the Periodic Tables, all the elements in Group 16 have the same number of
  (1) valence electrons (2) energy levels
  (3) protons   (4) neutrons

§2 **4.** In the ground state, how many electrons are in the outermost s sublevel of each element in Group 17?
  (1) 5   (2) 2   (3) 7   (4) 8

§2 **5.** The lanthanide and actinide series of elements are found, respectively, in Periods
  (1) 6 and 7   (2) 2 and 3
  (3) 3 and 4   (4) 4 and 5

  6. The elements in the present Periodic Table are arranged according to their
  (1) atomic numbers   (2) atomic masses
  (3) oxidation states (4) mass numbers

§2 **7.** The element whose electron configuration is $1s^2 2s^2 2p^6 3s^2$ is a
  (1) metalloid   (2) metal
  (3) noble gas   (4) nonmetal

263

8. Which is most likely the first ionization energy, in kilojoules per mole, of a metal?
   (1) 520   (2) 1012   (3) 1314   (4) 2081

9. Which element will form an ion whose ionic radius is larger than its atomic radius?
   (1) K   (2) F   (3) Li   (4) Mg

10. Atoms of metallic elements tend to
    (1) gain electrons and form negative ions
    (2) gain electrons and form positive ions
    (3) lose electrons and form negative ions
    (4) lose electrons and form positive ions

11. Which are two properties of most nonmetals?
    (1) low ionization energy and good electrical conductivity
    (2) high ionization energy and poor electrical conductivity
    (3) low ionization energy and poor electrical conductivity
    (4) high ionization energy and good electrical conductivity

12. Which element is considered malleable?
    (1) Au   (2) H   (3) S   (4) Rn

13. Element M has an electronegativity of less than 1.2 and reacts with bromine to form the compound MBr_2. Element M could be
    (1) Al   (2) Na   (3) Ca   (4) K

14. Which of the following particles has the *smallest* radius?
    (1) Na   (2) K   (3) Na^+ (4) K^+

15. Which element in Period 3 has the least tendency to lose an electron?
    (1) Ar   (2) Na   (3) P   (4) Al

16. Which ion is smaller than its corresponding neutral atom?
    (1) Al^{3+} (2) Br^- (3) O^{2-} (4) N^{3-}

17. Nonmetals in the solid state are poor conductors of heat and tend to
    (1) be brittle
    (2) be malleable
    (3) have luster
    (4) have good electrical conductivity

18. The atomic radius of an oxygen atom is
    (1) 167 pm   (2) 64 pm   (3) 31 pm   (4) 598 pm

19. As the atoms of the elements from atomic number 3 to atomic number 9 are considered in sequence from left to right across the Periodic Table, the atomic radius of each successive atom is
(1) smaller, and the nuclear charge is less
(2) smaller, and the nuclear charge is greater
(3) larger, and the nuclear charge is less
(4) larger, and the nuclear charge is greater

20. As the elements in Group 1 are considered in order of increasing atomic number, the atomic radius of each successive element increases. This is primarily due to an increase in the number of
(1) neutrons in the nucleus
(2) electrons in the outermost shell
(3) unpaired electrons
(4) principal energy levels

21. In Period 2, as the elements are considered from left to right, there is a decrease in
(1) ionization energy   (2) atomic mass
(3) metallic character   (4) nonmetallic character

22. Which ion has the largest radius?
(1) I^- (2) Cl^- (3) Br^- (4) F^-

23. Which element in Group 15 has the most metallic character?
(1) N   (2) P   (3) As   (4) Bi

24. Which ion has the smallest radius?
(1) Na^+ (2) K^+ (3) Mg^{2+} (4) Ca^{2+}

25. Which ion has the largest radius?
(1) Na^+ (2) K^+ (3) Mg^{2+} (4) Ca^{2+}

26. Which aqueous solution has a color?
(1) $BaSO_4(aq)$ (2) $CuSO_4(aq)$
(3) $SrSO_4(aq)$ (4) $MgSO_4(aq)$

27. Which element is so active chemically that it occurs naturally only in compounds?
(1) K   (2) Ag   (3) Cu   (4) S

28. Which element is a solid at room temperature and standard pressure?
(1) Br   (2) I   (3) Hg   (4) Ne

29. Ozone is an allotropic form of the element
(1) O   (2) P   (3) S   (4) C

30. In which group do the elements usually form chlorides that have the general formula MCl_2?
(1) 1   (2) 2   (3) 17   (4) 18

31. Which of the following elements is most likely to form a compound with radon?
(1) I   (2) F   (3) Na   (4) Ca

32. An element whose atoms have the electron configuration 2-8-18-1 is
(1) a transition element   (2) a noble gas
(3) an alkali metal   (4) an alkaline earth

33. Which is the most active nonmetal in the Periodic Table of the Elements?
(1) Na   (2) F   (3) I   (4) Cl

34. Element X is in Group 1, and element Y is in Group 16. A compound formed from these two elements is most likely to have the formula
(1) X_2Y (2) XY_2 (3) X_2Y_7 (4) X_7Y_2

35. As the elements are considered from the top to the bottom of Group 15, which sequence in properties occurs?
(1) metal $\rightarrow$ metalloid $\rightarrow$ nonmetal
(2) metal $\rightarrow$ nonmetal $\rightarrow$ metalloid
(3) metalloid $\rightarrow$ metal $\rightarrow$ nonmetal
(4) nonmetal $\rightarrow$ metalloid $\rightarrow$ metal

36. A chloride dissolves in water to form a colored solution. The chloride could be
(1) HCl   (2) KCl   (3) $CaCl_2$ (4) $NiCl_2$

37. Which metal atoms can form ionic bonds by losing electrons from both the outermost and next-to-outermost principal energy levels?
(1) Fe   (2) Pb   (3) Mg   (4) Ca

38. Which of the following periods contains the greatest number of metallic elements?
(1) 1   (2) 2   (3) 3   (4) 4

39. Which sequence of properties occurs as the atomic numbers increase across Period 3?
(1) metal → metalloid → nonmetal → noble gas
(2) metal → nonmetal → noble gas → metalloid
(3) nonmetal → metalloid → metal → noble gas
(4) nonmetal → metal → noble gas → metalloid

§2 **40.** A newly discovered element has been given the provisional symbol Ubb. According to the chart on page 262, the atomic number of this element is
(1) 112   (2) 122   (3) 211   (4) 221

41. Which list of elements consists of a metal, a metalloid, and a noble gas?
(1) aluminum, sulfur, argon
(2) magnesium, selenium, sulfur
(3) sodium, silicon, argon
(4) silicon, phosphorus, chlorine

42. Element X reacts with copper to form the compounds CuX and CuX_2. In which group of the Periodic Table is element X found?
(1) Group 1   (2) Group 2   (3) Group 13   (4) Group 17

43. All elements on the modern Periodic Table are arranged in order of increasing
(1) atomic mass
(2) nuclear mass
(3) number of neutrons per atom
(4) number of protons per atom

44. The valence electron of which atom in the ground state has the greatest amount of energy?
(1) cesium   (2) lithium   (3) rubidium   (4) sodium

Constructed-Response Questions

Base your answers to questions 1 and 2 on the information below.

You are given samples of Na, Ar, As, and Rb.

1. Which two of the given elements have the most similar chemical properties?

2. Explain your answer in terms of the Periodic Table of the Elements.

Base your answers to questions 3–6 on the information below.

Potassium ions are essential to human health. The movement of dissolved potassium ions, $K^+(aq)$, into and out of a nerve cell allows that cell to transmit an electrical impulse.

3. What is the total number of electrons in a potassium ion?

4. Explain, in terms of atomic structure, why a potassium ion is smaller than a potassium atom.

5. What property of potassium ions allows them to transmit an electrical impulse?

6. Based on the Periodic Table, explain why the elements Na and K have similar chemical properties.

Base your answers to questions 7–9 on the information below and on your knowledge of chemistry.

There are six elements in Group 14 of the Periodic Table. One of these elements had the temporary name ununquadium (symbol Uuq). It is now known as flerovium (symbol Fl).

7. Identify an element in Group 14 that is a metalloid.

8. Explain in terms of electron shells why each successive element in Group 14 has a larger atomic radius as you consider the elements in order of increasing atomic number.

9. State the expected number of valence electrons in a ground-state atom of flerovium.

Base your answers to questions 10–12 on the information below and on your knowledge of chemistry.

10. Compare the radius of a fluoride ion with the radius of a fluorine atom.

11. Explain in terms of subatomic particles why the radius of a lithium ion is smaller than the radius of a lithium atom.

12. Describe the general trend in atomic radius as each element in Period 2 is considered in order from left to right.

The answers to these questions are found in Appendix 3.

Chapter Ten

CHEMICAL BONDING AND MOLECULAR SHAPE

KEY IDEAS

This chapter focuses on the ways in which atoms, ions, and molecules link together. The properties of substances are discussed in terms of their chemical bonds.

KEY OBJECTIVES

At the conclusion of this chapter you will be able to:

- Explain how energy and stability are related to chemical bond formation.
- Describe how ionic bonding occurs.
- Draw Lewis structures for simple ions.
- Describe how covalent bonding occurs.
- Predict whether a bond between atoms is ionic or covalent.
- Distinguish between polar and nonpolar bonds.
- Draw Lewis structures for covalently bonded molecules.
- Describe coordinate covalent bonding.
- Describe network and metallic bonding.
- Define the terms *polar*, *nonpolar*, and *dipole* as they apply to molecules.
- Describe how the polarity of a molecule is related to its symmetry.
- Define the term *intermolecular attraction* (*force*).
- Define and apply the following terms as they apply to inter-molecular attractions: *London dispersion forces, dipole-dipole attraction, hydrogen bonding*, and *van der Waals forces*.
- Describe how molecule-ion attractions occur.
- Relate chemical bond types to the properties of substances.
- Define the term *resonance structure*, and explain how this alternative feature plays a part in covalent bonding.
- Describe how the octet rule is *not* obeyed in certain molecules.
- Use the VSEPR model to predict the shape of a covalently bonded molecule or ion.
- Describe how orbital overlap leads to the formation of a covalent bond.
- Define the terms *sigma bond* and *pi bond*, and indicate how each type formed from the overlapping of atomic orbitals.

- Define the terms *bond energy* and *bond length*, and describe how these quantities are related.
- Define the term *hybrid orbital*, and indicate how sp, sp^2, and sp^3 orbitals are formed.

SECTION I—BASIC (REGENTS-LEVEL) MATERIAL

NYS REGENTS CONCEPTS AND SKILLS

Note: By the time you have finished Section I, you should have mastered the concepts and skills listed below. The Regents chemistry examination will test your knowledge of these items and your ability to apply them.

Concepts are the *basic ideas* that form the body of the Regents chemistry course (what you need to know!).

Skills are the *activities* that demonstrate your mastery of these concepts (how you show that you know them!).

Following each concept or skill is a page reference (given in parentheses) to this chapter.

10.1 Concept:
When a bond is broken, energy is absorbed. When a bond is formed, energy is released. (Pages 272–273)

10.2 Concepts:
- Atoms attain a stable valence electron configuration by bonding with other atoms. (Page 274)
- Noble gases have stable valence configurations and tend not to bond. (Page 272)

Skill:
Determine the noble-gas configuration an atom will have as a result of bonding. (Pages 273–274)

10.3 Concept:
Two major categories of compounds are ionic and molecular (covalent) compounds. (Pages 273–276)

10.4 Concept:
Chemical bonds are formed when valence electrons are:
(a) transferred from one atom to another (ionic) (Page 273);
(b) shared between atoms (covalent) (Page 275);
(c) mobile within a metal (metallic) (Pages 280–281).

Skill:
. Demonstrate bonding using Lewis (electron-dot) structures representing:
(a) the transferring of valence electrons in ionic bonding
   (Page 274);
(b) the sharing of valence electrons in covalent bonding
   (Page 275);
(c) the formation of a stable octet as a result of bonding
   (Page 275).

10.5   Concept:
Electron-dot diagrams (Lewis structures) can represent the valence electron arrangement in compounds and ions. (Pages 273–276)

10.6   Concept:
In a multiple covalent bond, more than one pair of electrons are shared between two atoms. (Page 275)

10.7   Concept:
Electronegativity indicates how strongly an atom of an element attracts electrons in a chemical bond. Electronegativity values are assigned according to arbitrary scales. (Pages 276–277)

10.8   Concept:
The electronegativity difference between two bonded atoms is used to determine the degree of polarity in the bond. (Pages 276–277)

Skill:
Distinguish between nonpolar covalent and polar covalent bonds. (Pages 276–277)

10.9   Concepts:
• Metals tend to react with nonmetals to form ionic compounds. (Page 273)
• Nonmetals tend to react with other nonmetals to form molecular (covalent) compounds. (Page 275)
• Ionic compounds containing polyatomic ions have both ionic and covalent bonding. (Page 276)

10.10   Concept
The polarity of a molecule can be determined by the shape of the molecule and the distribution of charge around it. Symmetrical (nonpolar) molecules include CO_2, CH_4, and diatomic elements. Asymmetrical (polar) molecules include HCl, NH_3, and H_2O. (Page 282)

10.11   Concepts:
• Intermolecular forces created by an unequal distribution of charge result in varying degrees of attraction between molecules. (Pages 283–287)

- Hydrogen bonding is an example of a strong intermolecular force. (Pages 285–286)

Skill:
Explain each of the following in terms of intermolecular forces:
(a) vapor pressure;
(b) evaporation rate;
(c) phase changes (Pages 283–287).

10.12 Concept:
Compounds can be differentiated by their chemical and physical properties. (Page 289)

10.13 Concept:
Physical properties of substances can be explained in terms of chemical bonds and intermolecular forces. (These properties include conductivity, malleability, solubility, hardness, melting point, and boiling point.) (Page 289)

Skill:
Compare the physical properties of substances based on chemical bonds and intermolecular forces. (Page 289)

10.1 BONDING AND STABILITY

Now that we have studied the structure of atoms, let us learn how atoms and molecules bond to one another. The first question we need to ask is this: Why do chemical bonds exist at all?

An atom becomes more stable when it acquires the electron configuration of a noble gas. If the valence level of a "stabilized" atom is the first principal energy level, it will contain *two* electrons; otherwise, it will contain *eight* electrons.

Normally, however, the valence levels of atoms are not isoelectronic with the valence levels of noble gases. In order to achieve this state, atoms *bond* with one another in a variety of ways. In most cases, bonded atoms acquire eight electrons in their valence levels. In this case, we say that the bonding atoms have obeyed the **octet rule**.

As any system becomes more stable, its energy *decreases*. This decrease occurs when atoms form chemical bonds with one another. For example, if the energy of two isolated hydrogen (H) atoms is compared with the energy of an H_2 molecule (in which two atoms of hydrogen are *bonded* together), we find that the energy of the molecule is less than the energy of the two isolated atoms. Therefore, the H_2 molecule is more stable than the two isolated hydrogen atoms, as shown in the accompanying diagram.

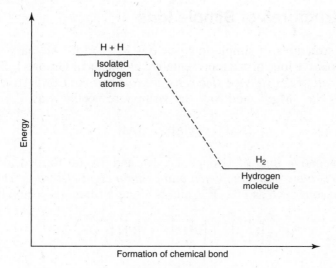

Energy

H + H

Isolated
hydrogen
atoms

H_2

Hydrogen
molecule

Formation of chemical bond

10.2   IONIC BONDING

Electron Transfer

To achieve noble-gas configurations, a pair of atoms may fill their valence levels by *transferring electrons* between them to form ions with opposite charges. The following examples illustrate how an electron is lost by an atom of potassium and is transferred to an atom of fluorine:

$$K \rightarrow K^+ + e^-$$

$$F + e^- \rightarrow F^-$$

When the pair of ions have been created, their opposite charges cause them to attract strongly. We call this attraction an **ionic bond**. When an ionic bond is formed in a compound such as potassium fluoride (KF), the attraction is not limited to *pairs* of ions. Oppositely charged ions arrange themselves in a regular array (known as a crystal lattice), as shown in the diagram on page 274.

From the diagram it can be seen that KF does *not* exist as individual units or *molecules*. This is true of ionic compounds in general. The formula of an ionic compound simply tells us the *ratio* of positive ions to negative ions. We call this ratio a **formula unit**. In the case of KF, the ratio is 1 to 1.

For electron transfer to occur, one atom must be able to lose electrons readily (i.e., have a low ionization energy) and the other atom must be able to gain electrons readily (i.e., have a sufficiently negative electron affinity). In Section 10.4, we will learn to predict whether two atoms will be able to transfer electrons between them and to form an ionic bond.

Lewis Structures of Simple Ions

The Lewis structures of simple ions such as K^+ and F^- are easy to draw. When the positive ions of the representative elements of Groups 1, 2, and 13 are formed, *all of the valence electrons are lost* and the Lewis structures of ions such as Na^+, Mg^{2+}, and Al^{3+} are written, respectively as:

$$[Na]^+ \quad [Mg]^{2+} \quad [Al]^{3+}$$

When the negative ions of Groups 17, 16, and 15 are formed, *all of the valence levels have been completed and contain eight electrons*. The Lewis structures of ions such as F^-, O^{2-}, and N^{3-} are written, respectively, as:

$$\left[:\ddot{F}: \right]^- \quad \left[:\ddot{O}: \right]^{2-} \quad \left[:\ddot{N}: \right]^{3-}$$

SECTION OF POTASSIUM FLUORIDE (KF) CRYSTAL LATTICE
(TWO-DIMENSIONAL)

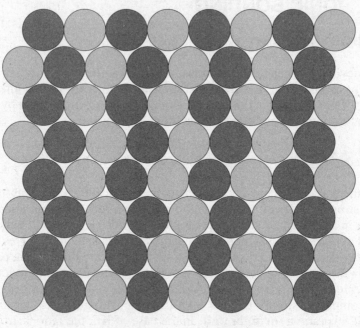

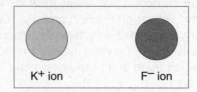

K⁺ ion   F⁻ ion

10.3 COVALENT BONDING

Certain pairs of atoms do not have a strong tendency to transfer electrons between them. Then how do they complete their valence levels? These atoms *share* electrons between them. Electron sharing always occurs in *pairs,* and two atoms may normally share one, two, or three pairs of electrons. The atoms bond together because the shared pair or pairs belong to *both* atoms. This type of bonding is called **covalent bonding**. The following examples show how it is accomplished:

$$\ddot{:}F\underset{xx}{\overset{xx}{:}}F\underset{x}{\overset{x}{x}} \qquad H \underset{xx}{\overset{xx}{:}}Cl\overset{x}{x}$$

In each molecule a single pair of electrons is shared between the atoms. This type of arrangement is called a *single covalent* bond. Using chemical shorthand, we can let a line represent a pair of electrons. Then, the molecules can be represented as F—F and H—Cl.

However, the sharing of the electron pair is not the same in both molecules. In the F_2 molecule, the pair of electrons is shared *equally* by both atoms. This type of bonding is known as **nonpolar covalent** bonding. In the HCl molecule, however, the electron pair is not shared equally; the chlorine atom has a greater attraction for the shared electrons than the hydrogen atom. This type of bonding is known as **polar covalent** bonding.

Multiple Covalent Bonds

Let us construct a molecule of O_2. An oxygen atom has six valence electrons; and in order to place eight electrons in the valence level of each oxygen atom, it is necessary that the atoms share *two* pairs of electrons, as shown below:

$$:\ddot{O}::\ddot{O}:$$

This is known as a *double* covalent bond and may be represented as $:\ddot{O} = \ddot{O}:$. (If you take more advanced courses in chemistry, you will learn why this diagram does not adequately represent the structure of O_2. For the present, however, we will accept it as written.)

PROBLEM
How is a molecule of N_2 constructed?

SOLUTION
A nitrogen atom has five valence electrons. Therefore, it is necessary that the two atoms share *three* pairs of electrons (known as a triple bond). This is represented as follows:

$$:N:::N: \quad \text{or} \quad :N\equiv N:$$

Coordinate Covalent Bonding

We can draw the Lewis structure for the ammonium (NH_4^+) ion by adding an H^+ ion to the NH_3 molecule:

$$H \overset{\times}{\cdot} \overset{\cdot\cdot}{N} \overset{\times}{\cdot} H \quad \xleftarrow{\quad H^+ \quad} \qquad \longrightarrow \qquad \left[\begin{array}{c} H \\ H \overset{\times}{\cdot} \overset{\cdot\cdot}{N} \overset{\times}{\cdot} H \\ H \end{array} \right]^+$$

Note that in the NH_3 molecule the hydrogen atoms are bonded to the nitrogen atom with covalent bonds in which each atom contributes one electron to the shared pair. The hydrogen *ion,* however, lacks electrons; and when it bonds, the nitrogen atom must contribute *both* electrons to the shared pair. This type of bond, known as a *coordinate covalent* bond, is found in many molecules and polyatomic ions.

Once the coordinate covalent bond is formed, *it is exactly the same as any other covalent bond.* If we could "view" the nitrogen-hydrogen bonds in an ammonium ion, for example, each bond would be seen to be entirely equivalent to the others.

PROBLEM
Draw the Lewis structure for the H_3O^+ ion, starting with H_2O and H^+.

SOLUTION

$$H \overset{\times}{\cdot} \overset{\cdot\cdot}{O} : \quad \xleftarrow{\quad H^+ \quad} \qquad \longrightarrow \qquad \left[\begin{array}{c} H \\ H \overset{\times}{\cdot} \overset{\cdot\cdot}{O} : \\ H \end{array} \right]^+$$

This polyatomic ion, known as the *hydronium* ion, plays a key role in acid–base chemistry.

10.4 ELECTRONEGATIVITY AND BONDING

How can we tell whether two atoms form an ionic, polar covalent, or nonpolar covalent bond? We compare the *electronegativities* of the atoms. In Chapter 4 we learned that *electronegativity* measures the relative attraction of an atom for shared electrons. The higher the electronegativity, the more the atom attracts the shared pair of electrons. Electronegativities (given in Reference Table S in Appendix 1 of this book) range from a high of 3.98 for fluorine (F) to a low of 0.70 for cesium (Cs).

It is the electronegativity *difference* (Δ) that enables us to predict the type of bond between two atoms. We calculate Δ by subtracting the smaller electronegativity value from the larger one.

First, however, we must recognize that no bond is *completely* ionic or covalent: it is a mixture of both types. We can relate the nature of the bond to the difference in the electronegativities of the bonded atoms. If the difference in the electronegativities of the bonded atoms (Δ) is:

- 0.0–0.4, we classify the bond as *nonpolar covalent*.
- 0.5–1.6, we classify the bond as *polar covalent*.
- 1.7 or above, we classify the bond as *ionic*.

This information is summarized in the accompanying table.

Range of Δ*	Bond Type	Example
0.0–0.4	Nonpolar covalent	BrCl (Δ = 0.2)
0.5–1.6	Polar covalent	HBr (Δ = 0.9)
≥1.7	Ionic	KF (Δ = 3.2)

*Other books may use different ranges to determine bond types. At what precise value of Δ a bond is nonpolar covalent, polar covalent, or ionic is really a judgment call.

TRY IT YOURSELF
Using the electronegativity values in Reference Table S in Appendix 1, classify the type of bond that each of the following pairs of atoms is likely to produce:
(a) Na, Br   (b) H, O   (c) N, N

ANSWERS
(a) ionic (Δ = 2.1) (b) polar covalent (Δ = 1.3) (c) nonpolar covalent (Δ = 0)

10.5 DRAWING THE LEWIS STRUCTURES OF COVALENT MOLECULES AND POLYATOMIC IONS

It is one thing to know which atoms are present in a covalent molecule or polyatomic ion; it is quite another to be able to say *how* the atoms are joined. Fortunately, there is a set of five rules that we can use in order to "build" molecules and polyatomic ions. When we follow these rules, we end up with a representation of a molecule known as a **Lewis structure**, a term familiar to us from Chapter 8 and illustrated for simple ions on page 274 of this chapter. There are (as always!) exceptions to these rules, but we will not worry about them for the moment.

Let us apply the set of rules in order to "build" a molecule of water (H_2O):

1. Count the total number of valence electrons in the molecule or ion.

   In this example, the water molecule has eight valence electrons: six from the oxygen atom and one from each hydrogen atom.

2. If the particle is an ion, add one electron for each negative charge present; subtract an electron for each positive charge present.

   In this example, the water molecule is neutral. No electrons are added or subtracted.

3. Draw a skeleton in which all of the atoms are connected by single bonds. It is always advisable to try a *symmetrical* structure first. Remember: A single bond represents a pair of electrons that belongs to both of the atoms it connects.

   Our skeleton looks like this:

$$H-O$$
$$|$$
$$H$$

   (In Section 10.9 we will learn why the skeletons of H_2O and certain other molecules are written at an angle, rather than in a straight line.)

4. Complete the valence level of each atom in the molecule by adding *unshared pairs* of electrons as necessary. The valence level of a hydrogen atom is complete with two electrons; assume that all other atoms are complete with eight electrons.

   In this example, the valence levels of both hydrogen atoms are already complete. The oxygen atom needs two unshared pairs of electrons:

$$H-\overset{\cdot\cdot}{\underset{|}{O}}:$$
$$H$$

5. Compare the total number of electrons in the skeleton with the number you arrived at in steps 1 and 2. If they agree, you are finished.

   In this example, the numbers of electrons agree and our molecule is "built."

PROBLEM

Draw the Lewis structure of the ClO_2^- ion.

SOLUTION

Each numbered step corresponds to the same one in the rules given above.

1. The number of valence electrons is 19 (6 from each oxygen and 7 from chlorine).
2. We *add* 1 electron since the ion has a charge of $1-$. This brings the total to 20.
3. The proposed (symmetrical) skeleton is:

$$O-Cl$$
$$|$$
$$O$$

4. We complete the octet of each atom by adding pairs of *unshared* electrons:

$$\left[\; \ddot{\underset{\cdot\cdot}{O}}-\ddot{\underset{|}{Cl}}: \atop :\ddot{O}: \;\right]^{-}$$

5. The number of electrons in our skeleton (20) agrees with our count from steps 1 and 2, and we propose the result of step 4 as our Lewis structure.

TRY IT YOURSELF

Draw the Lewis structure of the NH_4^+ ion.

ANSWER

$$\left[\begin{array}{c} H \\ | \\ H-N-H \\ | \\ H \end{array}\right]^{+}$$

What happens if the number of electrons in the skeleton does *not* agree with the number of electrons supplied by the atoms? The next example illustrates how we handle this situation.

Suppose we wish to draw the Lewis structure of the C_2H_4 molecule. Using our set of rules, we arrive at the skeleton:

$$\begin{array}{cc} H & H \\ | & | \\ H-C-C-H \\ \cdot\cdot & \cdot\cdot \end{array}$$

When we count the electrons in the skeleton, we find that we have 14. We should have 12 (four from each carbon and one from each hydrogen)! We can correct this situation by introducing *double* or *triple* bonds.

When we form a double bond, we remove *two* unshared pairs of electrons from adjacent atoms and replace them with the double bond. This has the effect of *lowering* the electron count by 2.

Triple bonds are formed by removing *four* unshared pairs of electrons and replacing them with the triple bond. This has the effect of lowering the electron count by 4.

Using this technique, we arrive at the Lewis structure:

$$H-\underset{\underset{H}{|}}{C}=\underset{\underset{H}{|}}{C}-H$$

TRY IT YOURSELF
Draw the Lewis structures of the following molecules and ion:
(a) C_2H_2 (b) CO_2 (c) NO_2^-

ANSWERS

(a) $H-C\equiv C-H$ (b) $\ddot{O}=C=\ddot{O}$ (c) $\left[\ddot{O} \overset{\ddot{N}}{\diagup} \diagdown \ddot{O} \right]^-$

10.6 NETWORK SOLIDS

Certain atoms, such as carbon, are able to form covalent bonds with other like atoms in a three-dimensional arrangement called a **network solid**. The three-dimensional network solid that consists of carbon is known as the familiar substance *diamond*. Another allotrope of carbon, *graphite,* also exists as a modified network solid, but it consists of a series of loosely associated *two-dimensional sheets*.

Other network solids include silicon carbide (SiC) and silicon dioxide (SiO_2). Network solids, by their very nature, do *not* form individual molecules. As we shall see at the end of Section 10.11, network solids have unique properties because of their extended covalent bonding.

10.7 METALLIC SUBSTANCES

As we learned in Chapter 8, metals have four specific properties (luster, conductivity, malleability, ductility) that arise from the way in which metal atoms bond to one another. Atoms that form metallic substances have low ionization energies and relatively unoccupied valence levels. These conditions allow the valence electrons the freedom and the room in which to move. In a metallic substance, the valence orbitals of the atoms merge with one another, and the valence electrons are *delocalized* over the entire metallic crystal. As a result the valence electrons do not belong to a single atom; they belong to the entire crystal. The remaining part of each atom, which consists of the *kernel* (a positive ion), forms the backbone of the crystal. The accompanying diagram represents a section of a metallic crystal.

Section of Metallic Crystal

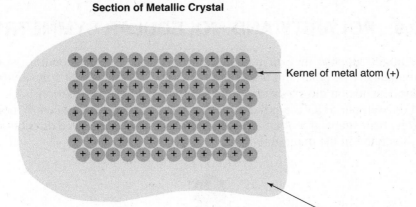

Kernel of metal atom (+)

Mobile valence electrons (−)

10.8   DIPOLES AND POLAR MOLECULES

When a bond is classified as polar covalent (such as the H-O bond), the atom with the higher electronegativity (O in this example) has a greater attraction for the shared electrons than the atom with the lower electronegativity (H in this example). As a result, the electron pair is drawn closer to the high-electronegativity atom and the charge around the two atoms becomes unbalanced. The "negative" end of the bond is centered around the atom with the higher electronegativity, and the "positive" end around the atom with the lower electronegativity.

We call this charge imbalance a **dipole.** The positive and negative ends of the dipole are represented by the symbols $\delta+$ and $\delta-$, respectively. The H-O bond may then be represented as follows:

$$\delta+ \ \text{H} - \text{O} \ \delta-$$

The positive and negative ends of a dipole are not real charges (such as positive and negative *ions*) because no electrons have actually been transferred between the atoms. The dipole represents only an *unbalanced* charge distribution along the bond.

In certain instances, the polar covalent bonds in a molecule will cause the *entire molecule* to have an unbalanced charge distribution. Such molecules are affected by electric fields and are also known as dipoles or as **polar** *molecules*. The HCl molecule has an *unbalanced* charge distribution and is a *polar* molecule. In contrast, the Ne (monatomic) and F_2 (diatomic) molecules have *balanced* charge distributions and are **nonpolar** molecules.

10.9   POLARITY AND MOLECULAR SYMMETRY

We might suppose that *all* molecules with polar covalent bonding would automatically be polar, but this is *not* the case. Whether or not a *molecule* is polar depends on the *symmetry* of its (electron) charge distribution.

For example, H_2O is a polar molecule because its polar covalent bonding *and* its bent shape give it an *asymmetrical* (unbalanced) charge distribution, as illustrated in the diagram below.

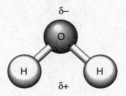

H_2O: A Polar Molecule

In contrast, CO_2 is *not* a polar molecule. While both C=O bonds are polar covalent, the linear shape of the molecule gives it a *symmetrical* (balanced) charge distribution. As a result any dipole arising from one of the C=O bonds is canceled by the other, as shown below.

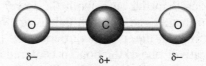

CO_2: A Nonpolar Molecule

In summary:

- Molecules containing only nonpolar covalent bonds and all monatomic molecules are nonpolar.
  Examples include He, Ar, N_2, P_4, and C (diamond).
- Molecules containing polar covalent bonds with symmetrical charge distributions are nonpolar.
  Examples include CO_2, BF_3, CCl_4, and $BeCl_2$.
- Molecules containing polar covalent bonds with asymmetrical charge distributions are polar.
  Examples include HF, H_2O, NH_3, SO_2, and N_2O.

10.10 INTERMOLECULAR FORCES

As its name implies, **intermolecular forces** are those attractions that exist between molecules (or between molecules and ions). We know that molecules do attract one another because gaseous substances condense to form liquids and these, in turn, condense to form solids. Intermolecular forces are also responsible for the solubilities of substances. For example, molecular substances such as table sugar (sucrose) and table salt (sodium chloride) dissolve in water because of the association of water molecules with the molecules of sucrose and the ions present in sodium chloride. All of these attractions are known collectively as **van der Waals forces**. These intermolecular forces include **dipole forces**, **London dispersion forces**, **hydrogen bonding**, and **ion-dipole forces**.

One important way to compare the strengths of intermolecular forces is to examine the boiling points of various substances. Substances with stronger intermolecular attractions have *higher* boiling points than those with weaker intermolecular attractions.

Dipole Forces

Dipole forces exist among polar molecules such as HCl. Each of the molecules of this substance has both a positive and a negative end. Seeing how dipole attractions can occur is easy. The positive end of the dipole is located near the hydrogen, and the negative end is near the chlorine. Since opposite charges attract, the molecules line up so oppositely charged dipoles are close to one another. As the temperature of HCl gas is reduced, these dipole attractions are responsible for the formation of the liquid phase. This is illustrated below:

$$\delta+ \text{ H} \text{—} \text{Cl } \delta- \cdots \delta+ \text{ H} \text{—} \text{Cl } \delta- \cdots \delta+ \text{ H} \text{—} \text{Cl } \delta- \cdots \delta+ \text{ H} \text{—} \text{Cl } \delta-$$

(The solid lines represent *covalent bonds*. The dotted lines represent *dipole forces*.)

Hydrogen chloride dissolves readily in water because of the dipole forces that exist between HCl and H_2O molecules. The diagram below is a simplified representation of this association.

$$\cdots\text{H}\text{—}\text{Cl}\cdots \begin{matrix} \text{H} \\ \diagdown \\ \text{O} \\ \diagup \\ \text{H} \end{matrix} \cdots\text{H}\text{—}\text{Cl}\cdots \begin{matrix} \text{H} \\ \diagdown \\ \text{O} \\ \diagup \\ \text{H} \end{matrix} \cdots$$

Generally, dipole forces depend on the *polarities* of the molecules involved. The more polar the molecule is, the stronger the dipole forces and the higher the boiling point. In order to compare the strengths of dipole forces, we must examine substances that have molar masses ($\mathcal{M}$) close in value. The table below compares three pairs of substances and their respective boiling points. Each pair consists of nonpolar and polar substances with similar molar masses.

Boiling Points of Nonpolar and Polar Substances

Nonpolar Substance	$\mathcal{M}$ (g/mol)	BP (°C)	Polar Substance	$\mathcal{M}$ (g/mol)	BP (°C)
SiH_4	32	−112	PH_3	34	−87
GeH_4	77	−90	AsH_3	78	−55
Br_2	160	+59	ICl	162	+97

In each pair, the polar substance has a higher boiling point than the corresponding nonpolar substance with a similar molar mass due to the presence of dipole forces in each polar substance.

London Dispersion Forces

Wait! There's more! Let us examine the preceding table once again. If we compare the nonpolar substances (SiH_4, GeH_4, and Br_2) and the polar substances (PH_3, AsH_3, and ICl) separately, we observe another trend. As the molar masses of the substances increase, the corresponding boiling points also increase. This trend is shown in the accompanying graph.

Molar Mass and Boiling Point

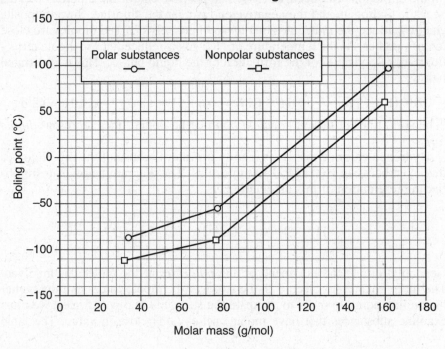

London dispersion forces arise because all molecules, both polar and non-polar, can create temporary dipoles for short periods of time. The resulting forces increase with increasing molecular mass and decrease with increasing distance between adjacent molecules.

Hydrogen Bonding

According to the previous table, the molar mass has a direct influence on the attractions between molecules, as reflected in the boiling point. Nevertheless, some substances seem to defy this general rule. Let's examine the boiling points of the hydrogen compounds of the elements found in Groups 15, 16, and 17. The following table provides these data.

Boiling Points of Various Hydrogen Compounds

Periodic Group	Period	Substance	$\mathcal{M}$ (g/mol)	Boiling Point (°C)
15	2	NH_3	17	−33
	3	PH_3	34	−87
	4	AsH_3	78	−55
	5	SbH_3	125	−17
16	2	H_2O	18	+100
	3	H_2S	34	−60
	4	H_2Se	81	−41
	5	H_2Te	130	−2
17	2	HF	20	+19
	3	HCl	36	−85
	4	HBr	81	−67
	5	HI	128	−35

When we graph these data, we observe some unusual results. We can readily see that each compound formed from a Period 2 element (nitrogen, oxygen, or fluorine) has an unusually high boiling point when compared with those of the other compounds within its group. How can we explain this observation?

First, we note that all of the compounds listed contain *hydrogen*. Second, the unusual behavior occurs only when hydrogen is bonded to a *small, highly electronegative element*, namely nitrogen, oxygen, or fluorine. We attribute this unusual behavior to a special case of dipole attraction called **hydrogen bonding**. This combination of elements allows the hydrogen atom of one molecule to approach the nitrogen, oxygen, or fluorine atom of a *neighboring* molecule very closely, creating an unusually strong attraction between the two polar molecules.

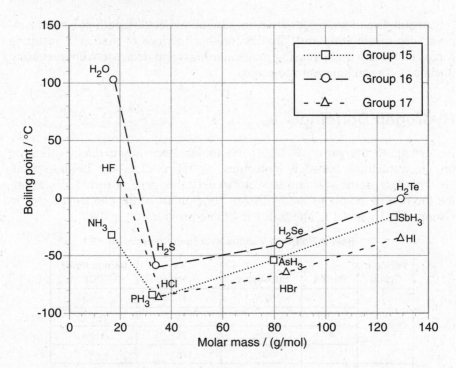

It must be emphasized that hydrogen bonding is *not* covalent bonding. It is a stronger type of dipole force and occurs *between neighboring molecules*. The following diagram provides examples of various types of hydrogen bonds.

The solid lines represent covalent bonds; the dotted lines, hydrogen bonds.

Ion-Dipole Forces

It is well known that ionic compounds such as NaCl (table salt) dissolve readily in water. This solubility is due to the attraction of the Na^+ and Cl^- ions for the negative and positive ends of polar water molecules. As NaCl is added to water, water molecules surround and separate the ions, destroying the crystal structure in a process known as *hydration*. The accompanying diagram illustrates how molecules and ions attract.

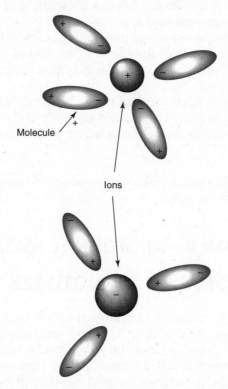

Molecule

Ions

10.11 PHYSICAL AND CHEMICAL PROPERTIES OF BONDED SUBSTANCES: A SUMMARY

The properties of a substance depend on the type of bonding present within the substance. The accompanying chart summarizes the properties of bonded substances.

- *Ionic substances* have high melting and boiling points because of the strong electrical attraction of the ions for one another. These substances usually dissolve in a polar solvent such as water, and their crystals are hard and brittle.

Ionic substances conduct electricity in solution and in the liquid phase, but not in the solid phase because of the rigidity of the ionic crystal.

- *Polar covalent substances* have intermediate melting and boiling points. They also dissolve in polar solvents, but they do not conduct electricity well unless they ionize in solution. Polar covalent substances do not conduct well in either the liquid or the solid phase.
- *Nonpolar substances* have the lowest melting and boiling points because of the weak attraction due to dispersion forces. Nonpolar substances dissolve in nonpolar solvents and do not conduct electricity well in any phase or in solution. Nonpolar crystals tend to be soft and waxy.
- *Metallic substances* have melting and boiling points that vary from low to high, as do the hardnesses of metallic crystals. Metals do not dissolve well in either polar or nonpolar solvents. Metals conduct electricity in the solid and liquid phases.
- *Network substances* have extraordinarily high melting points and form very hard, brittle crystals. They do not generally dissolve in polar or nonpolar solvents, and they are poor conductors of electricity in the solid and liquid phases.

The chart on page 289 summarizes the properties of substances that are consequences of their bond types.

SECTION II—ADDITIONAL MATERIAL

10.1A   RESONANCE STRUCTURES

In the structure below, *either* oxygen atom can be associated with the double bond. Therefore, we could write *two* Lewis structures for the NO_2^- ion. When two or more plausible Lewis structures can be written for a molecule or an ion, each structure is called a **contributing resonance structure**. The *actual* structure of the molecule or ion is a *composite* of its contributing resonance structures and is called a **resonance hybrid**. In the case of the NO_2^- ion, the resonance hybrid is viewed as a structure in which the double bond is actually extended across *both* oxygen atoms, as shown below:

Crystal Type	Particles in Crystal	Principal Attractive Forces Between Particles	Melting Point	Electrical Conductivity of Liquid	Characteristics of Crystal	Conditions for Formation	Examples
Ionic crystals	Positive and negative ions	Electrostatic attractions between ions Very strong	High	High They also conduct electricity in solution.	Hard, brittle. Most dissolve in polar solvents.	Formed between atoms with a large difference (≥ 1.70) in electronegativity.	All salts All metal hydrides KCl CaF_2 $CsBr$ MgO $BaCl_2$ $NaCl$
Covalent network crystals	Atoms	Covalent bonds Very strong	Very high	Poor	Very hard. Insoluble in most ordinary liquids. Covalent bonds extend from one atom to another in a continuous pattern.	Most formed by two elements of Group 14 or by elements whose average periodic group number is 14.	Diamond SiC AlN BeO $CuCl_2$ Mg_2Si Mg_2Sn
Metallic crystals	Positive metal atom kernels and mobile valence electrons	Metallic bonds Strong	Most are high.	Very high	Most are hard, malleable, ductile. High thermal conductivity. Generally insoluble in liquids. Usually soluble in molten metals.	Formed by elements with low first ionization energies and vacant valence levels.	Cu Bi Zn Fe Li Pt Ca Pb La V Cr $CuZn_3$
Molecular crystals (a) Polar	Polar molecules	Dipole forces Intermediate strength Can be strengthened by hydrogen bonds.	Intermediate	Very low	More fragile than ionic crystals. Most are soluble in polar solvents. Usually liquids or solids at room temperature.	Formed from molecules with asymmetrical charge distributions. Polar covalent bonds are formed between atoms having a moderate difference (0.41–1.69) in electronegativity.	All acids Many organic compounds PF_3 H_2O NH_3 SO_3 $CHCl_3$
(b) Nonpolar	Atoms or nonpolar molecules	Dispersion forces Weak	Low	Extremely low	Very soft. Most are soluble in nonpolar or slightly polar solvents. Usually gases at room temperature.	Formed from atoms or from molecules containing symmetrical charge distributions. Nonpolar covalent bonds are formed between like atoms or atoms having a small difference (≤ 0.40) in electronegativity.	All diatomic molecules Ar Cl_2 H_2 S_8 CH_4 I_2 CCl_4 Ne N_2 CO_2 SF_6 BF_3

10.2A ADDITIONAL TOPICS IN BONDING

Exceptions to the Octet Rule

All of the molecules and ions we have constructed so far have been based on the assumption that all atoms (other than hydrogen) need to have *eight* electrons in their valence levels as a result of bond formation. In reality, this is not always the case. Certain atoms may have fewer—or more—than eight electrons in their valence levels when they bond. The accompanying table lists five molecules that are exceptions to the octet rule.

Molecule	Lewis Structure	Condition of Central Atom's Valence Level
$BeCl_2$	:C̈l — Be — C̈l:	Be has *4* electrons in its valence level.
BF_3	:F̈ \ / F̈: B | :F̈:	B has *6* electrons in its valence level.
NO_2	:O = N — Ö:	N has *7* electrons in its valence level. One of these electrons is *unpaired.*
PCl_5	:C̈l: | C̈l: :C̈l — P / | \ C̈l: :C̈l:	P has *10* electrons in its valence level.
SF_6	:F̈: :F̈ \ | / F̈: S :F̈ / | \ F̈: :F̈:	S has *12* electrons in its valence level.

The NO_2 molecule is unusual because one of its valence electrons is *unpaired*. Molecules and atoms with unpaired electrons are known as *free radicals* and are very reactive. In living tissues, free radicals are known to be destructive agents and are believed to be associated with certain serious health problems.

Molecular Shape and VSEPR

Now that we have learned to construct molecules and ions, we need to know how to predict the three-dimensional shapes they assume. Molecules and ions with just two atoms are easy: they are linear. We can predict the shapes of other molecules and ions by using the **Valence Shell Electron Pair Repulsion** model (**VSEPR**).

VSEPR is based on four premises:

1. Shared and unshared pairs of electrons repel each other.
2. An unshared pair of electrons repels more strongly than a shared pair.
3. *For the purposes of this model,* a single, a double, and a triple bond are all considered equivalent.
4. The shape of a molecule or ion is the result of the shared and unshared pairs of electrons being placed as far from each other as possible.

To apply VSEPR, we look at the *central atom* of the molecule or ion and count the number of shared and unshared pairs of electrons that are associated with it. (Remember: A double or triple bond counts as *one* shared pair of electrons.)

The simplest molecules and ions have *no* unshared pairs of electrons. For example, let us consider the shape of the CO_2 molecule. The carbon atom has two double bonds but no unshared pairs of electrons. The *two* shared pairs of electrons are farthest apart when the angle between them is 180°. Therefore, the CO_2 molecule is *linear:* $O{=}C{=}O$.

The accompanying tables and diagrams summarize the shapes of molecules and ions with no unshared pairs of electrons on the central atom. We designate the central atom as A and each atom bonded to it as B.

Number of Shared Pairs	Formula	Shape of Molecule	Bond Angle(s)	Example(s)
2	AB_2	Linear	180°	CO_2, $BeCl_2$
3	AB_3	Trigonal planar	120°	BF_3, NO_3^-
4	AB_4	Tetrahedral	109.5°	CH_4, NH_4^+
5	AB_5	Trigonal bipyramidal	120°, 90°	PCl_5
6	AB_6	Octahedral	90°	SF_6

Arrangement of Electron Pairs about a Central Atom (A) in a Molecule, and Geometry of Some Simple Molecules and Ions in Which the Central Atom Has No Lone Pairs

Number of Electron Pairs	Arrangement of Electron Pairs*	Molecular Geometry*	Examples
2	180° :—A—: Linear	B—A—B Linear	$BeCl_2$, $HgCl_2$
3	120° Trigonal planar	Trigonal planar	BF_3
4	109.5° Tetrahedral	Tetrahedral	CH_4, NH_4^+
5	90° 120° Trigonal bipyramidal	Trigonal bipyramidal	PCl_5
6	90° 90° Octahedral	Octahedral	SF_6

*Only solid lines represent bonds; the dashed lines show the overall shapes.

TRY IT YOURSELF

Draw the Lewis structure of the SO_3 molecule, and use the VSEPR model to determine its shape.

ANSWER

The SO_3 molecule is a resonance hybrid with the following structure:

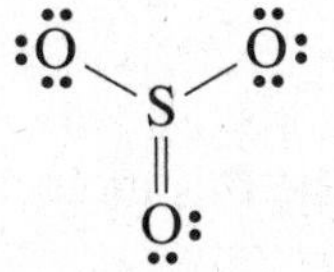

$$:\ddot{O}\diagdown_{\textstyle S}\diagup \ddot{O}:$$
$$\underset{\displaystyle \ddot{O}:}{\big\|}$$

Since it has "three" shared and no unshared pairs of electrons, its shape is *trigonal planar*.

Now let us consider molecules and ions whose central atoms have unshared pairs of electrons in addition to their shared pairs. In the H_2O molecule, the oxygen atom has two single bonds and two unshared pairs of electrons, for a total of *four* pairs of electrons. If *all* of these pairs were shared, we would expect, from the tables and diagrams above, that the shape would be *tetrahedral*.

Here is the surprise: The four electron pairs *do* form a tetrahedron! However, *the overall shape of the molecule is determined only by the atoms that are present,* not by the unshared pairs of electrons. For this reason, the H_2O molecule is *bent* (or *V-shaped*) as shown below:

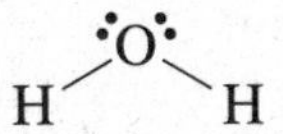

$$\underset{H \diagup \quad \diagdown H}{:\ddot{O}:}$$

Since the unshared pairs of electrons repel more strongly than the bonded pairs, the bonded pairs are forced closer together and the angle between the oxygen and the two hydrogen atoms is less than the expected tetrahedral angle of $109.5°$. (The angle is $104.5°$.)

The accompanying table and diagrams summarize some of the shapes of molecules and ions with both shared and unshared pairs of electrons on the central atom. The symbol A represents the *central* atom. The symbol B represents the *peripheral* atom(s). The symbol E represents an *unshared* pair or pairs of electrons.

Number of Shared Pairs	Number of Unshared Pairs	Formula	Shape	Example(s)
2	1	AB_2E	Bent	NO_2^-
2	2	AB_2E_2	Bent	H_2O
3	1	AB_3E	Trigonal pyramidal	NH_3, H_3O^+

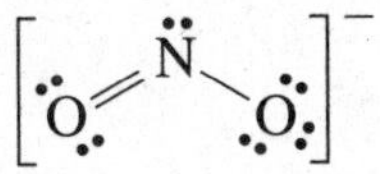

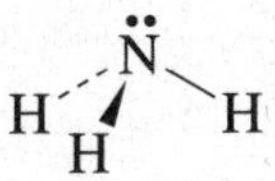

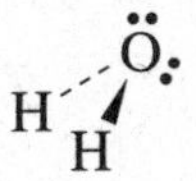

293

The "wedge" (◢) in the second and third diagrams indicates that the molecule is *three-dimensional*. The bond with the wedge projects *out of the paper toward you*.

TRY IT YOURSELF

Draw the Lewis structure of the PF_3 molecule, and use the VSEPR model to determine its shape.

ANSWER

The PF_3 molecule has one unshared and three shared pairs of electrons. Its shape is trigonal pyramidal, as shown below.

$$F \overset{\ddot{P}}{\underset{F}{\diagup}} F$$

How Covalent Bonding Occurs: Orbital Overlap

We already know that electrons and electron pairs reside in atomic orbitals. How, then, are electrons shared by two atoms? The answer is that the individual orbitals *overlap* to form the covalent bond. The following examples show how *orbital overlap* occurs.

When a hydrogen molecule is formed from two isolated hydrogen atoms, the $1s$ orbitals, each containing one electron, overlap to produce the single bond, as illustrated in the accompanying diagram:

Formation of a Hydrogen Molecule

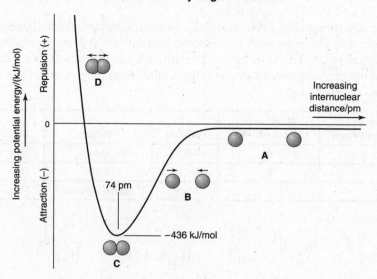

The energy diagram is read from *right to left*.

- At point **A**, the two hydrogen atoms are *isolated*, that is, they do not attract or repel each other. By convention, the potential energy of a system consisting of isolated atoms is set (very close) to 0 kilojoule per mole.
- At point **B**, the hydrogen atoms have moved closer together and a force of attraction exists between them. This configuration is more stable than the one in **A** because the potential energy of the system has been *lowered*. By convention, a force of attraction is represented by a *negative* potential energy value.
- At point **C**, the 1*s* orbitals of the hydrogen atoms have overlapped sufficiently to produce the *maximum* force of attraction and the *minimum* potential energy (−436 kilojoule per mole). At this point, the internuclear distance between the atoms is 74 picometers, and a *stable* H_2 molecule has formed.

At internuclear distances less than 74 picometers, the potential energy of the system rises rapidly and the force between the atoms changes from one of attraction to one of repulsion. Point **D** represents one such *unstable* configuration.

Sigma and Pi Bonds

When a chlorine molecule is formed from two chlorine atoms, the two 3*p* orbitals, each containing one electron, overlap to produce the single bond, as illustrated below.

When a hydrogen chloride molecule is formed from a hydrogen and a chlorine atom, the 1*s* orbital of the hydrogen and the 3*p* orbital of the chlorine, each containing one electron, overlap to produce the single bond, as illustrated below.

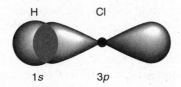

295

Each of these single bonds lies along a line that joins the nuclei of the two bonding atoms. Such bonds are called **sigma (σ) bonds**. *Almost all single bonds are sigma bonds.*

There is another way in which *p* orbitals can overlap. Since they have a "dumbbell" shape, they are able to overlap sideways (i.e., *perpendicular* to the line joining the nucleii of the bonding atoms). Such bonds are called **pi (π) bonds** and are involved in the formation of double and triple bonds. The formation of a pi bond is illustrated below.

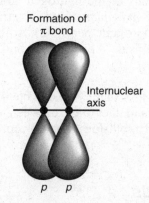

Formation of
π bond

Internuclear axis

p   p

A double bond consists of one sigma bond and one pi bond. *A triple bond* consists of one sigma bond and two pi bonds.

Therefore, a double or triple bond is more complicated than the simple "sharing" of two or three pairs of electrons. In fact, pi bonds are more reactive than sigma bonds, as we will learn in Chapter 11: Organic Chemistry.

Bond Energy and Bond Length

How can we measure the "strength" of a chemical bond? One way is to measure the energy needed to break the bond. This quantity is known as the **bond dissociation energy** or, simply, the **bond energy**. In practice, bond energies are measured for a large variety of compounds and average values are compiled for the bond energy between two atoms. The accompanying table lists a number of average bond energies.

The larger the average bond energy, the stronger the bond. If we examine carbon-to-carbon bonds, for example, we see that the C≡C bond is stronger than the C=C bond, which is, in turn, stronger than the C—C bond. As the table shows, this relationship holds in general. The order of bond energies between two bonded atoms is as follows: triple bond > double bond > single bond.

Bond	Bond Energy / (kJ/mol)	Bond	Bond Energy / (kJ/mol)	Bond	Bond Energy / (kJ/mol)
H−H	436	H−F	565	H−Cl	427
O−H	467	N−H	391	C−H	413
C−C	347	C=C	614	C≡C	839
C−O	358	C=O	745*	C≡O	1072
C−N	305	C=N	615	C≡N	891
N−N	160	N=N	418	N≡N	941
O−O	146	O=O	495		

*The average bond energy for C=O in CO_2 is 799 kJ/mol.

Another quantity directly related to energy is the **bond length**, which is the distance between the nuclei of two bonded atoms. The bond length is measured in picometers ($1 \, pm = 10^{-12}$ m). In general, the shorter the bond length, the larger the bond energy. The accompanying table lists the bond lengths for carbon-to-carbon and carbon-to-nitrogen bonds.

Bond	Bond Length / pm
C−C	154
C=C	134
C≡C	120
C−N	143
C=N	138
C≡N	116

As we can see from the table, the order of bond lengths is as follows: triple bond < double bond < single bond.

Calculation of Heats of Reaction from Bond Energies

We can test the idea that a chemical reaction involves both the making and breaking of chemical bonds by examining the energies needed to make and break the bonds. The table at the top of the page lists the energies needed to *break* selected bonds. Since bond-breaking is an endothermic process, the values of bond energies are always reported as *positive* numbers. However, this same table also represents bond *formation* if we change the values to *negative* numbers, since bond formation is exothermic.

Let us examine this reaction:

$$CH_4(g) + 2O_2(g) \rightarrow CO_2(g) + 2H_2O(g)$$

| 1 mole | 2 moles | 1 mole | 2 moles |

from the point of view that specific bonds are formed and broken as the reaction occurs. We begin by rewriting the reaction to show the bonds present in each substance:

$$\begin{array}{c} H \\ | \\ H-C-H \\ | \\ H \end{array} + 2O{=}O \rightarrow O{=}C{=}O + 2\ \begin{array}{c} O \\ \diagup\ \diagdown \\ H\quad H \end{array}$$

| 1 mole | 2 moles | 1 mole | 2 moles |

When we inspect the diagram, we see that the reaction involves the *breaking* of *4* moles of C—H bonds and *2* moles of O=O bonds. We also see that *2* moles of C=O bonds and *4* moles of O—H bonds have been *formed*. Using the values given in the table on page 287, we can calculate the net energy change over the course of this reaction:

4 mol of C—H bonds broken	=	$4 \text{ mol} \cdot (+413 \text{ kJ/mol})$	= +1652 kJ
2 mol of O=O bonds broken	=	$2 \text{ mol} \cdot (+495 \text{ kJ/mol})$	= + 990 kJ
2 mol of C=O bonds formed	=	$2 \text{ mol} \cdot (-799 \text{ kJ/mol})$	= −1598 kJ
4 mol of O—H bonds formed	=	$4 \text{ mol} \cdot (-467 \text{ kJ/mol})$	= −1868 kJ
		Net energy change	= − 824 kJ

The value −824 kilojoule represents an estimate of the *heat of reaction* (ΔH) *as calculated from bond energies*. It lies within 3 percent of the value of this heat of reaction (−803 kJ) calculated from the data given in Reference Table W-4, Standard Thermodynamic Data, in Appendix 2. The close agreement between the estimated and the measured values of ΔH provides additional evidence that a chemical reaction involves the making and breaking of bonds.

Hybridized Atomic Orbitals

The VSEPR model that we studied on pages 291–294 is a valuable tool for predicting molecular shape. However, it does not really explain *why* molecules such as $BeCl_2$, BF_3, and CH_4 assume their particular shapes (linear, trigonal planar, and tetrahedral, respectively). In particular, the ground-state electron configurations of Be, B, and C seem to suggest that they should *not* form molecules with these shapes.

One explanation for the shapes of $BeCl_2$, BF_3, and CH_4 is provided by the concept of **orbital hybridization**, in which s and p atomic orbitals are "blended" together in various ways.

$BeCl_2$: Before the hybridization process, the ground-state configuration of the Be atom looks like this:

$$\underset{1s}{\underline{\uparrow\downarrow}} \quad \underset{2s}{\underline{\uparrow\downarrow}} \qquad \underset{2p}{\underline{}\,\underline{}\,\underline{}}$$

The first step in the hybridization process involves the "promotion" of one of the $2s$ electrons into an empty $2p$ orbital, as shown below:

$$\underset{1s}{\underline{\uparrow\downarrow}} \quad \underset{2s}{\underline{\uparrow}} \qquad \underset{2p}{\underline{\uparrow}\,\underline{}\,\underline{}}$$

The final step involves the creation of *two hybrid orbitals*, known as *sp* orbitals, as shown below:

$$\underset{1s}{\underline{\uparrow\downarrow}} \qquad \underset{2(sp)}{\underline{\uparrow}\,\underline{\uparrow}} \qquad \underset{2p}{\underline{}\,\underline{}}$$

Each of the orbitals lies 180° apart and allows $BeCl_2$ to be linear.

BF_3: Before the hybridization process, the ground-state configuration of the B atom looks like this:

$$\underset{1s}{\underline{\uparrow\downarrow}} \quad \underset{2s}{\underline{\uparrow\downarrow}} \qquad \underset{2p}{\underline{\uparrow}\,\underline{}\,\underline{}}$$

The first step in the hybridization process involves the "promotion" of one of the $2s$ electrons into an empty $2p$ orbital, as shown below:

$$\underset{1s}{\underline{\uparrow\downarrow}} \quad \underset{2s}{\underline{\uparrow}} \qquad \underset{2p}{\underline{\uparrow}\,\underline{\uparrow}\,\underline{}}$$

The final step involves the creation of *three hybrid orbitals*, known as sp^2 orbitals, as shown below:

$$\underset{1s}{\underline{\uparrow\downarrow}} \qquad \underset{2(sp^2)}{\underline{\uparrow}\,\underline{\uparrow}\,\underline{\uparrow}} \qquad \underset{2p}{\underline{}}$$

Each of the orbitals lies 120° apart and allows BF_3 to be trigonal planar.

CH_4: Before the hybridization process, the ground-state configuration of the C atom looks like this:

$$\underset{1s}{\underline{\uparrow\downarrow}} \quad \underset{2s}{\underline{\uparrow\downarrow}} \qquad \underset{2p}{\underline{\uparrow}\,\underline{\uparrow}\,\underline{}}$$

The first step in the hybridization process involves the "promotion" of one of the 2s electrons into an empty 2p orbital, as shown below:

$$\underline{\uparrow\downarrow} \quad \underline{\uparrow} \qquad \underline{\uparrow} \;\; \underline{\uparrow} \;\; \underline{\uparrow}$$

$$1s \quad\; 2s \qquad\qquad 2p$$

The final step involves the creation of *four hybrid orbitals*, known as sp^3 orbitals, as shown below:

$$\underline{\uparrow\downarrow} \qquad \underline{\uparrow} \;\; \underline{\uparrow} \;\; \underline{\uparrow} \;\; \underline{\uparrow}$$

$$1s \qquad\qquad 2(sp^3)$$

Each of the orbitals lies 109.5° apart and allows CH_4 to be tetrahedral.

END-OF-CHAPTER QUESTIONS

Some questions have the symbol "§2" in front of the question number. This symbol means that the question is based on Section II material.

1. Energy is released when the atoms of two elements bond to form a compound. Compared to the total potential energy of the atoms before bonding, the total potential energy of the atoms after bonding is
   (1) lower, and the compound formed is stable
   (2) lower, and the compound formed is unstable
   (3) higher, and the compound formed is stable
   (4) higher, and the compound formed is unstable

2. Which statement is true concerning the following reaction?

   $$N(g) + N(g) \rightarrow N_2(g) + energy$$

   (1) A bond is broken and energy is absorbed
   (2) A bond is broken and energy is released
   (3) A bond is formed and energy is absorbed
   (4) A bond is formed and energy is released

3. As a chemical bond forms between hydrogen and chlorine atoms, the potential energy of the atoms
   (1) decreases   (2) increases   (3) remains the same

4. A Ca^{2+} ion differs from a Ca atom in that the Ca^{2+} ion has
   (1) more protons   (2) fewer protons
   (3) more electrons   (4) fewer electrons

5. In which pair do the members have identical electron configurations?
   (1) S^{2-} and Cl^-
   (2) S and Ar
   (3) K and Na^+
   (4) Cl^- and K

6. Which compound is ionic?
   (1) HCl
   (2) $CaCl_2$
   (3) SO_2
   (4) N_2O

7. What is the energy-level electron configuration for a Be^{2+} ion?
   (1) 1
   (2) 2
   (3) 2-1
   (4) 2-2

8. Which formula represents a molecular solid?
   (1) NaCl(s)
   (2) $C_6H_{12}O_6$(s)
   (3) Cu(s)
   (4) CsF(s)

9. What kind of bond is formed in the reaction below?

$$H:\overset{..}{\underset{..}{O}}:H + H^+ \longrightarrow \left[H:\overset{..}{\underset{..}{O}}:H \right]^+$$

   (1) metallic
   (2) hydrogen
   (3) network
   (4) coordinate covalent

10. Which element at STP exists as monatomic molecules?
    (1) N
    (2) O
    (3) Cl
    (4) Ne

11. Two atoms of element *A* unite to form a molecule with the formula A_2. The bond between the atoms in the molecule is
    (1) electrovalent
    (2) ionic
    (3) nonpolar covalent
    (4) polar covalent

12. Which formula represents a molecular solid?
    (1) NaCl(s)
    (2) $C_6H_{12}O_6$(s)
    (3) Cu(s)
    (4) KF(s)

13. Which compound contains both covalent and ionic bonds?
    (1) KCl
    (2) NH_4Cl
    (3) $MgCl_2$
    (4) CCl_4

14. In which compound does the bond between the atoms have the *least* ionic character?
    (1) HF
    (2) HCl
    (3) HBr
    (4) HI

15. When a reaction occurs between atoms with ground-state electron configurations 2-1 and 2-7, the predominant type of bond formed is
    (1) polar covalent
    (2) nonpolar covalent
    (3) ionic
    (4) metallic

§2 **16.** Which of the following molecules has a central atom with fewer than eight electrons in its valence level?
(1) NH_3 (2) BF_3 (3) PCl_5 (4) H_2O

§2 **17.** Which molecule has an atom with an unpaired electron in its valence level?
(1) H_2O (2) NO_2 (3) BF_3 (4) Cl_2

§2 **18.** The shape of a CCl_4 molecule is
(1) linear   (2) trigonal planar
(3) tetrahedral   (4) bent (V-shaped)

§2 **19.** A double bond, such as N=O, consists of
(1) one sigma (σ) bond, only
(2) one pi (π) bond, only
(3) one sigma bond and one pi bond
(4) two pi bonds, only

§2 **20.** Compared to the length of a C—C bond, the length of a C=C bond is
(1) less   (2) greater   (3) equal

21. Which substance is classified as a network solid?
(1) SiO_2 (2) Li_2O (3) H_2O (4) CO_2

22. What type of bonding is present within a network solid?
(1) hydrogen   (2) covalent   (3) ionic   (4) metallic

23. Which element has mobile valence electrons?
(1) S   (2) N   (3) Ca   (4) Cl

24. Which molecule is nonpolar and has a symmetrical shape?
(1) HCl   (2) CH_4 (3) H_2O (4) NH_3

25. Which formula represents a polar molecule?
(1) CH_4 (2) Cl_2 (3) NH_3 (4) N_2

26. Which structural formula represents a nonpolar symmetrical molecule?

(1)
$$
H-O-H
$$
(2)
$$
H-\overset{\overset{\displaystyle H}{|}}{\underset{\underset{\displaystyle H}{|}}{C}}-H
$$
(3) H—F
(4)
$$
H-N-H\ (\text{with H below})
$$

27. Which substance is made up of molecules that are dipoles?
    (1) N_2 (2) H_2O (3) CH_4 (4) CO_2

28. The weakest intermolecular forces exist between molecules of
    (1) $C_2H_6(\ell)$ (2) $C_3H_8(\ell)$ (3) $C_4H_{10}(\ell)$ (4) $C_5H_{12}(\ell)$

29. The forces of attraction that exist between *nonpolar* molecules are called
    (1) London dispersion   (2) ionic
    (3) covalent   (4) electrovalent

30. The table below shows boiling points for the elements listed.

Element	Normal Boiling Point (°C)
Fluorine	−188.1
Chlorine	−34.6
Bromine	+58.8
Iodine	+184.4

    Which statement best explains the pattern of boiling points relative to molecular size?
    (1) Stronger London dispersion forces occur in larger molecules.
    (2) Weaker London dispersion forces occur in larger molecules.
    (3) Stronger hydrogen bonds occur in larger molecules.
    (4) Weaker hydrogen bonds occur in larger molecules.

31. London dispersion forces of attraction between molecules always decrease with
    (1) increasing molecular size and increasing distance between the molecules
    (2) increasing molecular size and decreasing distance between the molecules
    (3) decreasing molecular size and increasing distance between the molecules
    (4) decreasing molecular size and decreasing distance between the molecules

32. Which of the following liquids has the *weakest* London dispersion forces of attraction between its molecules?
    (1) $Xe(\ell)$ (2) $Kr(\ell)$ (3) $Ne(\ell)$ (4) $He(\ell)$

33. In which of the following liquids are hydrogen bonds between molecules the strongest?
    (1) $HI(\ell)$ (2) $HBr(\ell)$ (3) $HCl(\ell)$ (4) $HF(\ell)$

34. Hydrogen bonds are formed between molecules when hydrogen is covalently bonded to an element that has a
    (1) small atomic radius and low electronegativity
    (2) large atomic radius and low electronegativity
    (3) small atomic radius and high electronegativity
    (4) large atomic radius and high electronegativity

35. The kind of attraction that results in the dissolving of sodium chloride in water is
    (1) ion-ion   (2) molecule-ion
    (3) atom-atom   (4) molecule-atom

36. The attraction between water molecules and an Na^+ ion or a Cl^- ion occurs because water molecules are
    (1) linear
    (2) symmetrical
    (3) polar
    (4) nonpolar

37. A characteristic of ionic solids is that they
    (1) have high melting points
    (2) have low boiling points
    (3) conduct electricity
    (4) are noncrystalline

38. A pure substance melts at 38°C and does not conduct electricity in either the solid or liquid phase. The substance is classified as
    (1) ionic
    (2) metallic
    (3) electrovalent
    (4) molecular

39. Which is a property of most nonmetallic solids?
    (1) high thermal conductivity
    (2) high electrical conductivity
    (3) brittleness
    (4) malleability

40. Which of these compounds contains the most polar bond?
    (1) H–Br   (2) H–Cl   (3) H–F   (4) H–I

41. Which property is defined as the ability of a substance to be hammered into thin sheets?
    (1) conductivity (2) malleability (3) melting point (4) solubility

42. At STP, graphite and diamond are two solid forms of carbon. Which statement explains why these two forms of carbon differ in hardness?
(1) Graphite and diamond have different ionic radii.
(2) Graphite and diamond have different molecular structures.
(3) Graphite is a metal, but diamond is a nonmetal.
(4) Graphite is a good conductor of electricity, but diamond is a poor conductor of electricity.

43. Based on Reference Table S in Appendix 1, an atom of which of the following elements has the *weakest* attraction for electrons in a chemical bond?
(1) polonium   (2) sulfur   (3) selenium   (4) tellurium

44. Which substance in the table below has the strongest intermolecular forces?

Substance	Molar Mass (g/mol)	Boiling Point (kelvins)
HF	20.01	293
HCl	36.46	188
HBr	80.91	207
HI	127.91	237

(1) HF   (2) HCl   (3) HBr   (4) HI

45. Which compound contains both ionic and covalent bonds?
(1) KI   (2) $CaCl_2$ (3) CH_2Br_2 (4) NaCN

46. Which statement explains why a CO_2 molecule is nonpolar?
(1) Carbon and oxygen are both nonmetals.
(2) Carbon and oxygen have different electronegativities.
(3) The molecule has a symmetrical distribution of charge.
(4) The molecule has an asymmetrical distribution of charge.

Constructed-Response Questions

1. Label each of the following compounds as ionic or covalent:
(a) MgO   (b) SO_2 (c) H_2S
(d) BaF_2 (e) CCl_4

2. Write the Lewis structure for the hydrogen peroxide molecule (H_2O_2).

305

Base your answers to questions 3–6 on the information below. Each molecule listed is formed by sharing electrons between atoms when the atoms within the molecule are bonded together.

Molecule A: Cl_2

Molecule B: CCl_4

Molecule C: NH_3

3. Draw the electron-dot (Lewis) structure for the NH_3 molecule.

4. Explain why CCl_4 is classified as a nonpolar molecule.

5. Explain why NH_3 has stronger intermolecular forces of attraction than Cl_2.

6. Explain how the bonding in KCl is different from the bonding in molecules A, B, and C.

Base your answers to questions 7–11 on the information below and on your knowledge of chemistry.

The Lewis electron-dot diagrams for three substances are shown below.

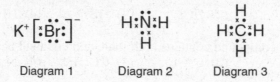

Diagram 1   Diagram 2   Diagram 3

7. Describe in terms of valence electrons how the chemical bonds form in the substance represented in Diagram 1.

8. Determine the total number of electrons in the bonds between the nitrogen atom and the three hydrogen atoms represented in Diagram 2.

9. Explain in terms of distribution of charge why a molecule of the substance represented in Diagram 3 is nonpolar.

10. Draw a Lewis electron-dot diagram for a molecule of Br_2.

11. Identify the noble gas with the same electron configuration as the positive ion represented in Diagram 1.

The answers to these questions are found in Appendix 3.

Chapter
Eleven

ORGANIC CHEMISTRY

KEY IDEAS

This chapter focuses on the structure and reactions of the element carbon in organic compounds. The main series of the hydrocarbons and organic compounds containing oxygen and nitrogen are studied in detail.

KEY OBJECTIVES

At the conclusion of this chapter you will be able to:

- Define the term *organic chemistry*, and list the sources of organic materials.
- Compare general properties of organic compounds with those of inorganic compounds.
- Describe the bonding of carbon in simple organic compounds, and write structural formulas for such compounds.
- Define the term *hydrocarbon*, and describe how hydrocarbons are obtained from petroleum.
- Define the term *homologous series*.
- Describe the alkane, alkene, alkyne, and benzene series of hydrocarbons in terms of their general formulas, structural formulas, isomers, and IUPAC names.
- Define the term *functional group*.
- Describe, in terms of their structures and IUPAC names, the following organic oxygen compounds: alcohols, aldehydes, ketones, ethers, and organic acids.
- Describe the following organic nitrogen compounds: amines, amino acids, and amides.
- Define the term *polymer*.
- Describe the following organic reactions: substitution, addition, fermentation, esterification, saponification, combustion, condensation polymerization, and addition polymerization, and provide examples of these reactions.

SECTION I—BASIC
(REGENTS-LEVEL) MATERIAL

NYS REGENTS CONCEPTS AND SKILLS

Note: By the time you have finished Section I, you should have mastered the concepts and skills listed below. The Regents chemistry examination will test your knowledge of these items and your ability to apply them.

Concepts are the *basic ideas* that form the body of the Regents chemistry course (what you need to know!).

Skills are the *activities* that demonstrate your mastery of these concepts (how you show that you know them!).

Following each concept or skill is a page reference (given in parentheses) to this chapter.

11.1 Concept:
  • Organic compounds contain carbon atoms that bond to one another in chains, rings, and networks to form a variety of structures. (Pages 309–310)
  • Organic compounds can be named using the IUPAC system. (Pages 314–316)

11.2 Concepts:
  • Hydrocarbons are compounds that contain only carbon and hydrogen. (Page 311, alkanes)
  • Saturated hydrocarbons contain only single carbon-carbon bonds. (Pages 311–312, alkenes and alkynes)
  • Unsaturated hydrocarbons contain at least one multiple (double or triple) carbon-carbon bond. (Pages 316–318)

  Skill:
  Draw structural formulas for
      (a) alkanes,
      (b) alkenes,
      (c) alkynes,
  containing a maximum of ten carbon atoms. (Pages 312–318)

11.3 Concepts:
  • Various categories of organic compounds include:
      (a) organic acids (Page 321);
      (b) alcohols (Pages 319–320);
      (c) esters (Pages 321–322);
      (d) aldehydes (Page 320);
      (e) ketones (Pages 320–321);

        (f)   ethers (Page 321);

        (g) halides (Pages 323–324);

        (h) amines (Page 322);

        (i)   amides (Page 323);

        (j)   amino acids (Page 322).

- Each of these categories contains molecules that differ in their structures. (Page 318)
- Functional groups impart distinctive physical and chemical properties to organic compounds. (Page 318)

Skills:
- Classify an organic compound based on its structural or condensed structural formula. (Pages 318–323)
- Given the IUPAC name for an organic compound, draw a structural formula for that compound that includes the appropriate functional group or groups (on a straight-chain hydrocarbon backbone). (Page 319)

11.4 Concept:

Isomers of organic compounds have the same molecular formula but different structures and properties. (Pages 313–314)

11.5 Concept:

Types of organic reactions include:

        (a)   addition (Page 325);

        (b) substitution (Page 324);

        (c)   polymerization (Page 327);

        (d) esterification (Pages 325–326);

        (e)   fermentation (Page 325);

        (f)   saponification (Page 326);

        (g) combustion (Page 326).

Skills:
- Identify types of organic reactions. (Pages 324–327)
- Given a balanced equation for an organic reaction, determine a missing reactant or product. (Pages 324–327)

11.1 ORGANIC CHEMISTRY IS . . . ?

Organic chemistry is essentially the study of carbon and its compounds. Why do we devote an entire chapter to just one element? The answer lies in the fact that the properties of organic compounds and the reactions that they undergo are generally quite different from those of their inorganic counterparts. Organic chemistry is responsible for many of the modern materials we use, and it is the basis of the chemistry of life.

In organic chemistry, the *structure* of a molecule is important in determining its properties and reactivity. While the number of inorganic compounds is in the order of tens of thousands, the number of organic compounds is in the order of millions. The main reason for the existence of so many organic compounds is the ability of carbon atoms to bond with *other carbon atoms* in an almost endless variety of chains, rings, and networks. The raw materials for making organic compounds come from petroleum, plant, and animal sources.

11.2 COMPARISON OF ORGANIC AND INORGANIC COMPOUNDS

To make the difference between organic and inorganic compounds more apparent, let us compare the general properties of each class.

Solubility

Inorganic compounds are generally soluble in polar solvents such as water, while organic compounds are generally soluble in nonpolar solvents. This fact explains why the oil in a salad dressing (organic) does not "mix" with the vinegar (a polar solvent).

Electrolytic Behavior in Solution

When dissolved in solution, many inorganic compounds conduct an electric current because of ionization or dissociation of the solute. This is generally *not* the case with organic compounds.

Melting and Boiling Points

Generally, inorganic compounds have higher melting and boiling points than organic compounds. While it is not unusual for some inorganic compounds to melt at temperatures above 500°C, this is rarely true of organic compounds.

Rate of Reaction

In general, inorganic compounds react more rapidly than organic compounds because the bond rearrangements in inorganic reactions are usually simpler and fewer in number than the bond rearrangements in organic reactions.

Bonding

The properties detailed above support the view that the existence of ions or polar covalent bonds is commonly associated with inorganic compounds. In organic compounds, however, the bonding is usually covalent with little or no polarity. Carbon has four valence electrons, and in organic compounds it forms a total of four covalent bonds. We will see that carbon atoms may form single, double, or triple bonds by sharing one, two, or three pairs of electrons.

11.3 HYDROCARBONS AND HOMOLOGOUS SERIES

The simplest organic compounds contain only carbon and hydrogen and are called **hydrocarbons**. Hydrocarbons are obtained from the refining of petroleum, a complex mixture of natural hydrocarbons. Petroleum is the starting point for the production of plastics, textiles, rubber, and detergents. The petroleum is separated into components with different *boiling points* by the process of **fractional distillation**. Another process, called **cracking**, is used to break the larger hydrocarbon molecules into smaller ones. This procedure increases the yield of the more important fractions with lower boiling points (such as gasoline). The simplest hydrocarbon is *methane* (CH_4), which is the chief component of natural gas.

Hydrocarbons with related structures and properties are usually separated into "families" known as **homologous series**. Each member of a series differs from the next member by a single carbon atom. With each successive member, the properties of the compounds, such as melting and boiling points, change in a regular way. This classification into series greatly simplifies the study of organic chemistry.

The Alkane Series

Alkanes are hydrocarbons that contain only single bonds between carbon atoms. The first member of the alkane series is *methane* (CH_4). In Chapter 9, we discovered that methane is a *tetrahedral* molecule because carbon forms four sp^3 hybrid orbitals—as it does in *every* alkane molecule. The accompanying table lists a number of alkanes together with their molecular formulas and phases at STP.

From the table we see that the name of each member of the alkane series begins with a prefix [e.g., *prop-* ("three") or *pent-* ("five")] that indicates the number of *carbon* atoms in the formula. All alkane names end with the suffix *-ane*. Reference Table P in Appendix 1 also summarizes the names of the organic prefixes for 1–10 carbon atoms.

311

Name of Alkane	Molecular Formula	Phase at STP
*meth*ane	CH_4	Gas
*eth*ane	C_2H_6	Gas
*prop*ane	C_3H_8	Gas
*but*ane	C_4H_{10}	Gas
*pent*ane	C_5H_{12}	Liquid
*hex*ane	C_6H_{14}	Liquid
*hept*ane	C_7H_{16}	Liquid
*oct*ane	C_8H_{18}	Liquid
*non*ane	C_9H_{20}	Liquid
*dec*ane	$C_{10}H_{22}$	Liquid
...	...	...
*octadec*ane	$C_{18}H_{38}$	Solid

In the table we note also that each alkane contains one more carbon and two more hydrogens than the preceding alkane. The *general* formula for an alkane with n carbon atoms is $C_n H_{2n+2}$. We also note that the phase of the substance at STP changes from gas to liquid to solid as the masses of successive molecules increase. This phase change is due to the increased strength of *London dispersion forces,* which we studied in Section 10.10, in more massive molecules.

Chemists write the formulas of organic compounds in several ways:

- A **molecular formula**, such as C_3H_8, indicates how many atoms of each element are present, but it fails to provide information about how the atoms are *connected*.
- A **dashed structural formula** uses straight lines to represent chemical bonds. For example, propane (C_3H_8) is written as follows:

$$
\begin{array}{ccccccc}
 & H & & H & & H & \\
 & | & & | & & | & \\
H-\!\!\!&C&\!\!\!-\!\!\!&C&\!\!\!-\!\!\!&C&\!\!\!-H \\
 & | & & | & & | & \\
 & H & & H & & H &
\end{array}
$$

For convenience, the angles between bonds are usually drawn as either 90° or 180°. The advantage of a dashed structural formula lies in its ability to display the connections among the atoms. However, more complex formulas can be cumbersome to write in this style.

- A **condensed structural formula** represents a compromise between the simplicity of a molecular formula and the information provided by a dashed structural formula. In this style, atoms are usually written *without* dashes, and hydrogen atoms are written to the *right* of the atom to which they are connected. For example, the condensed structural formula of propane is

$$CH_3CH_2CH_3.$$

In this book, dashes occasionally *will* be used with more complex formulas, when the dashes make the structure of the molecule more apparent. For example, the molecule whose dashed structural formula is

$$
\begin{array}{c}
\quad\quad H \\
\quad\quad | \\
H-C-H \\
\quad\quad | \\
H\quad\; |\quad\; H \\
|\quad\; |\quad\; | \\
H-C-C-C-H \\
|\quad\; |\quad\; | \\
H\quad H\quad H
\end{array}
$$

has its condensed structural formula written as follows:

$$
\begin{array}{c}
CH_3 \\
| \\
CH_3CHCH_3
\end{array}
$$

There are other ways of writing formulas for organic compounds, but they are beyond the scope of this book.

Isomers

The condensed structural formula of butane is of particular interest because *two* molecules have the molecular formula C_4H_{10}. Their structures and common names are as follows:

$$
CH_3CH_2CH_2CH_3 \qquad\qquad
\begin{array}{c}
CH_3 \\
| \\
CH_3CHCH_3
\end{array}
$$

butane **isobutane**

Substances that have the same molecular formula but are *different* compounds are called **isomers**. Isomers differ from one another in their physical and chemical properties.

Two or more compounds are **constitutional isomers** if they have the same molecular formula but their atoms are connected in a different order. The two butane molecules just shown are constitutional isomers.

PROBLEM
Pentane (C_5H_{12}) consists of three constitutional isomers.
Write the condensed structural formula for each isomer.

SOLUTION

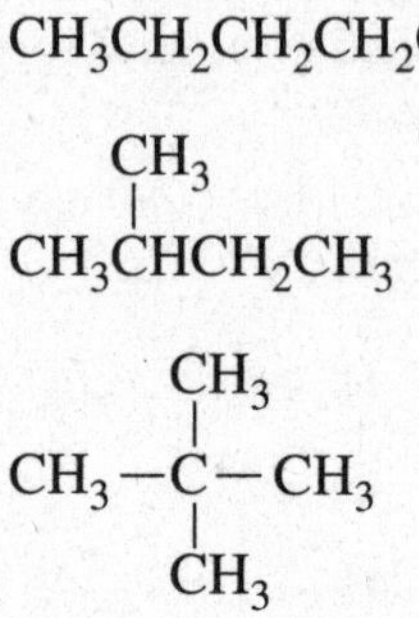

$$CH_3CH_2CH_2CH_2CH_3 \qquad \text{[pentane]}$$

$$\begin{array}{c} CH_3 \\ | \\ CH_3CHCH_2CH_3 \end{array} \qquad \text{[isopentane]}$$

$$\begin{array}{c} CH_3 \\ | \\ CH_3 - C - CH_3 \\ | \\ CH_3 \end{array} \qquad \text{[neopentane]}$$

TRY IT YOURSELF

Hexane consists of five constitutional structural isomers. Draw the dashed structural formulas for any *three* of them.

ANSWER

The five dashed structural formulas shown below are presented as hydrocarbon *skeletons*; that is, the hydrogen atoms have been omitted.

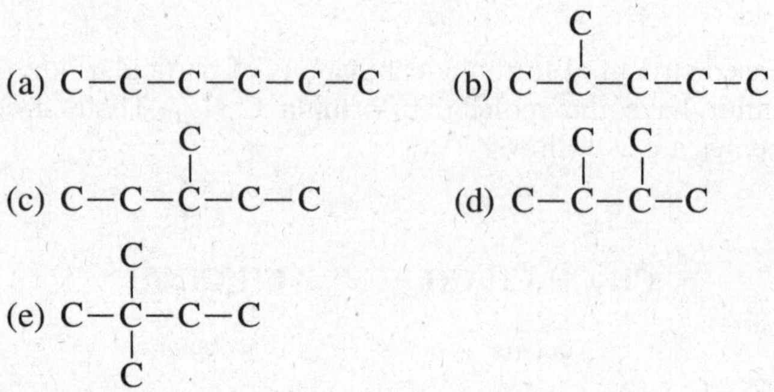

(a) C—C—C—C—C—C

$$\text{(b)} \quad \begin{array}{c} C \\ | \\ C - C - C - C - C \end{array}$$

$$\text{(c)} \quad \begin{array}{c} C \\ | \\ C - C - C - C - C \end{array}$$

$$\text{(d)} \quad \begin{array}{c} C \;\; C \\ | \;\; | \\ C - C - C - C \end{array}$$

$$\text{(e)} \quad \begin{array}{c} C \\ | \\ C - C - C - C \\ | \\ C \end{array}$$

The IUPAC Naming System

As the number of carbon atoms in an organic compound increases, the number of constitutional isomers rises dramatically. Providing a different name for each and every isomer is an impossible task unless a simple and logical way to name organic compounds is available. The *International Union of Pure and Applied Chemistry* (IUPAC) has developed such a system, in which organic compounds are named in much the way people are: each has a *family* name and one or more *given* names. The procedure for naming an organic compound is as follows:

• If a hydrocarbon has *no* branches and has only single carbon-carbon bonds, it is named according to the table given on page 312. For example,

314

$CH_3CH_2CH_2CH_3$ is named *butane*. (Some texts place the symbol *n-* before the name, as in *n*-butane.)

- First the *parent structure* is determined. The parent structure of an organic compound is found by counting the number of atoms in the *longest, unbroken chain of carbon atoms*. For example, in the skeleton shown below:

$$C-C-C-C-C-C$$
$$\quad\ |\qquad\qquad\quad |$$
$$\quad\ C\qquad\qquad\quad C$$

there are *six* carbon atoms in the longest, unbroken chain. Since only single bonds are present, the name of the parent structure is *hexane*.

In this structure:

$$C-C-C-C-C-C-C-C$$
$$\quad\ |\qquad\qquad\qquad\qquad |$$
$$\quad\ C\qquad\qquad\qquad\qquad C$$
$$\quad\ |$$
$$\quad\ C$$

the longest, unbroken chain contains *nine* carbon atoms, again all with single bonds. No one has said the atoms have to be in a straight line—just an *unbroken* line! The arrows show how the longest chain is formed:

$$C-C\to C\to C\to C\to C\to C\to C$$
$$\quad\ \uparrow\qquad\qquad\qquad\qquad |$$
$$\quad\ C\qquad\qquad\qquad\qquad C$$
$$\quad\ \uparrow$$
$$\quad\ C$$

The parent structure of this compound is named *nonane*.

- The full name of an organic compound is determined by modifying the name of the parent structure according to the *functional groups* that are attached to the parent.
- If *no* groups are attached, the parent name is the name of the hydrocarbon. For example, the hydrocarbon shown below is named *butane*.

$$\begin{array}{ccccc} & H & H & H & H \\ & | & | & | & | \\ H- & C- & C- & C- & C-H \\ & | & | & | & | \\ & H & H & H & H \end{array} \qquad \text{or} \qquad CH_3CH_2CH_2CH_3$$

butane

- When a carbon group is attached to the parent, the group is named by attaching the suffix *-yl* to the prefix that indicates how many carbon atoms

315

the group has. The three simplest carbon groups (written in *condensed* form) are these:

$$CH_3- \qquad\qquad methyl$$
$$CH_3CH_2- \text{ or } C_2H_5- \qquad ethyl$$
$$CH_3CH_2CH_2- \text{ or } C_3H_7- \qquad propyl$$

The hydrocarbon shown below is named 2-*methylpropane* (written as one word) because the parent (propane) has three carbon atoms and a one-carbon group is attached to it. The number 2 is added to show precisely *where* the group is located on the parent. The carbon atoms of the parent are numbered in order so that the group is assigned the *lowest* number possible—2 in this example.

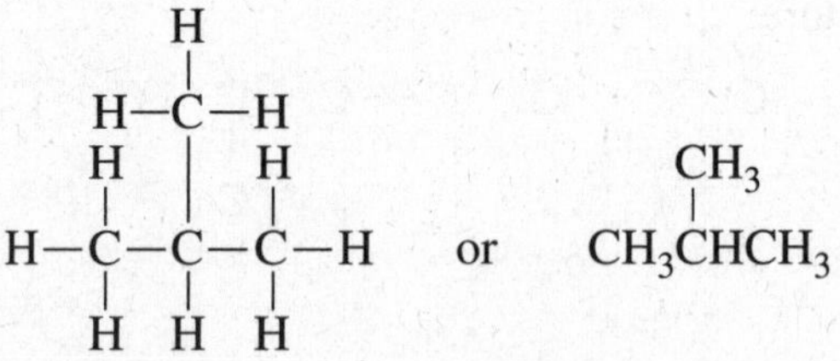

2-methylpropane

PROBLEM
Name, according to the IUPAC system, the three constitutional isomers of pentane whose condensed structural formulas are shown on page 314.

SOLUTION
The first isomer is named *pentane,* as shown. The second isomer (isopentane) is named 2-*methylbutane.* The third isomer (neopentane) is named 2,2-*dimethylpropane* because the parent structure has three carbon atoms, and each of two *one-carbon groups* is located on the middle carbon of the parent. Note that each group is entitled to its own number.

TRY IT YOURSELF
Name the five constitutional isomers of hexane illustrated on page 315.

ANSWERS
(a) hexane　　　(b) 2-methylpentane　　　(c) 3-methylpentane
(d) 2,3-dimethylbutane　　(e) 2,2-dimethylbutane

The Alkene Series

Alkenes are hydrocarbons that contain one carbon-carbon *double* bond. The first member of the alkene series, shown below, is *ethene* (C_2H_4, commonly called *ethylene*).

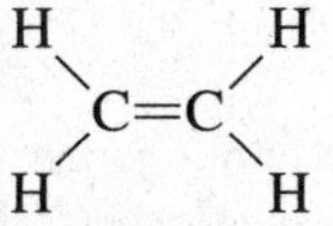

$$H_2C=CH_2$$

ethene

The name of each member of the alkene series begins with a prefix that indicates the number of *carbon* atoms in the formula. *All* alkene names end with the suffix *-ene*. The *general* formula for an alkene with n carbon atoms is C_nH_{2n}.

PROBLEM
What is the formula of *propene*?

SOLUTION
Propene is an alkene with three carbon atoms. According to the general formula for alkenes (C_nH_{2n}), it also has six hydrogen atoms. Its molecular formula is C_3H_6, and its structural formula is as follows:

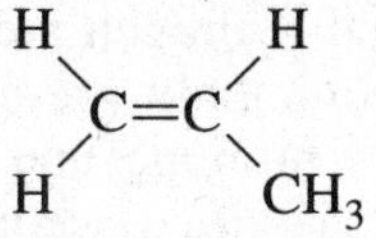

propene

Note that each carbon atom has a total of four bonds attached to it.

TRY IT YOURSELF
What are the formulas of *butene*?

ANSWER
There are three constitutional isomers of butene. All have the molecular formula C_4H_8. Two of the constitutional isomers differ in the position of the double bond, as shown below.

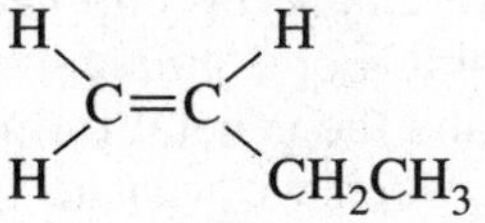

1-butene 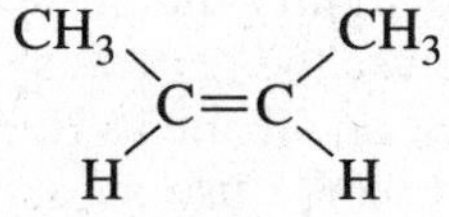   **2-butene**

The third isomer has the following structure:

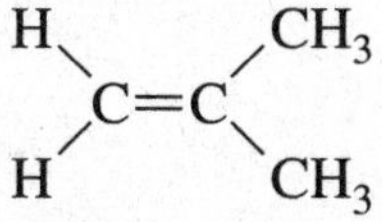

2-methylpropene

317

The Alkyne Series

Alkynes are hydrocarbons that contain one carbon-carbon *triple* bond. The first member of the alkyne series is *ethyne* (C_2H_2, commonly called *acetylene*):

$$H-C\equiv C-H$$

ethyne

The name of each member of the alkyne series begins with a prefix that indicates the number of *carbon* atoms in the formula. *All* alkyne names end with the suffix -*yne*. The *general* formula for an alkyne with n carbon atoms is C_nH_{2n-2}.

Reference Table Q in Appendix 1 contains a summary of the various types of hydrocarbons.

11.4 FUNCTIONAL GROUPS

Elements *other* than carbon and hydrogen are found in most organic compounds. The elements most commonly present are oxygen, nitrogen, and the halogens (fluorine, chlorine, bromine, and iodine). An organic molecule containing another element is classified according to the manner in which the atoms of this element are *arranged* within the molecule. Such an arrangement of atoms is known as a **functional group**. The presence of functional groups imparts distinctive chemical and physical properties to organic compounds.

Alkyl Groups

The hydrocarbon groups methyl (CH_3-), ethyl (CH_3CH_2-), propyl ($CH_3CH_2CH_2-$), and so on, are known collectively as **alkyl groups**. An alkyl group is formed when a hydrogen atom is removed from an *alkane*. When we want to signify the presence of an alkyl group *without identifying it specifically*, we use the general symbol R. For example, we could write the general formula of an alkane as $R-H$, and the general formula of an alkyne as $R_1-C\equiv C-R_2$. (We use *subscripts* when we want to indicate the possibility that the alkyl groups may be different.) Alkyl groups are particularly useful in writing the general formulas of compounds containing functional groups.

The accompanying table describes the functional groups that we will study in this chapter.

Functional Group	Name of Group	General Formula	Name of Compound
$-OH$	Hydroxyl	$R-OH$	Alcohol
$-O-$	–	R_1-O-R_2	Ether
$-\overset{\overset{\displaystyle O}{\|\|}}{C}-$	Carbonyl	$R-\overset{\overset{\displaystyle O}{\|\|}}{C}-H$	Aldehyde
		$R_1-\overset{\overset{\displaystyle O}{\|\|}}{C}-R_2$	Ketone
$-\overset{\overset{\displaystyle O}{\|\|}}{C}-OH$	Carboxyl	$R-\overset{\overset{\displaystyle O}{\|\|}}{C}-OH$	Organic acid
$-\overset{\overset{\displaystyle O}{\|\|}}{C}-O-$	–	$R_1-\overset{\overset{\displaystyle O}{\|\|}}{C}-O-R_2$	Ester
$-NH_2$	Amino	$R-NH_2$	Amine
$-\overset{\overset{\displaystyle O}{\|\|}}{C}-NH_2$	Amido	$R-\overset{\overset{\displaystyle O}{\|\|}}{C}-NH_2$	Amide
$-X^*$	–	$R-X$	Haloalkane (Halide)

*X is a halogen atom (F, Br, Cl, or I).

This information is also summarized in Reference Table R in Appendix 1.

The Hydroxyl Group: Alcohols

All *alcohols* contain the *hydroxyl* functional group ($-OH$). Alcohols can be classified in two ways: (1) according to the number of hydroxyl groups present in a molecule, and (2) according to the way in which the carbon atoms are arranged within the alcohol molecule.

An alcohol is named according to the IUPAC system by adding the suffix *-ol* to the name of the parent hydrocarbon. The simplest *monohydroxy* alcohol contains one carbon atom and is named *methanol* (commonly known as *methyl alcohol* or *wood alcohol*). Methanol is a popular solvent. The next member of this series, *ethanol* (commonly known as *ethyl alcohol* or *grain alcohol*), contains two carbon atoms. Ethanol is used as an antiseptic and is also a component of alcoholic beverages. The structural formulas of methanol and ethanol are as follows:

$$\begin{array}{c} H \\ | \\ H-C-O-H \\ | \\ H \end{array} \qquad \begin{array}{c} H \quad\; H \\ | \quad\; | \\ H-C-C-O-H \\ | \quad\; | \\ H \quad\; H \end{array}$$

$$CH_3OH \qquad\qquad CH_3CH_2OH$$
$$\textbf{methanol} \qquad\qquad \textbf{ethanol}$$

Below the dashed structural formulas for methanol and ethanol are the *condensed structural formulas* for these compounds.

The Carbonyl Group: Aldehydes and Ketones

One of the most important functional groups in organic chemistry is the **carbonyl** group, which consists of a carbon atom connected to an oxygen atom by a double bond:

$$\searrow C = O$$

Aldehydes

In **aldehydes**, the carbon of the carbonyl group is bonded to at least *one* hydrogen atom. An aldehyde is named according to the IUPAC system by adding the suffix *-al* to the name of the parent hydrocarbon. Here are the general formula and two examples (*methanal* and *ethanal*):

$$\begin{array}{c} R \\ \diagdown \\ \quad C=O \\ \diagup \\ H \end{array} \qquad \begin{array}{c} H \\ \diagdown \\ \quad C=O \\ \diagup \\ H \end{array} \qquad \begin{array}{c} CH_3 \\ \diagdown \\ \quad C=O \\ \diagup \\ H \end{array}$$

$$\textbf{aldehyde} \qquad \textbf{methanal} \qquad \textbf{ethanal}$$

Methanal (commonly known as *formaldehyde*) is used as a biological preservative.

Ketones

A **ketone** has *no* hydrogen atoms directly attached to the carbonyl group. According to the IUPAC system, a ketone is named by attaching the suffix *-one* to the name of the parent hydrocarbon. The general formula and an example (*propanone*) are given below.

$$\begin{array}{c} R_1 \\ \diagdown \\ \quad C=O \\ \diagup \\ R_2 \end{array} \qquad \begin{array}{c} CH_3 \\ \diagdown \\ \quad C=O \\ \diagup \\ CH_3 \end{array}$$

$$\textbf{ketone} \qquad\qquad \textbf{propanone}$$

320

Propanone (commonly known as *acetone*) is the principal solvent in nail polish remover.

Ethers

Ethers have the *general* formula R_1-O-R_2. The ether, $C_2H_5-O-C_2H_5$ has the IUPAC name *ethoxyethane*, but it is more commonly known as *diethyl ether*. It was once a popular surgical anesthetic. Ethers may be prepared by dehydrating primary alcohols in the presence of H_2SO_4:

$$2CH_3CH_2OH \xrightarrow{H_2SO_4} C_2H_5OC_2H_5 + H_2O$$

The Carboxyl Group: Organic Acids

The name **carboxyl** is a contraction of *carbonyl* and *hydroxyl*. The carboxyl functional group has this formula:

$$\overset{\displaystyle O}{\underset{\displaystyle -C-OH}{\|}}$$

which is usually abbreviated as COOH. The presence of the carboxyl group confers acidic properties on the organic compound. According to the IUPAC system, an organic acid is named by adding the suffix *-oic* and the word *acid* to the name of the parent hydrocarbon.

The first member of the series is *methanoic acid* (commonly called *formic acid*), whose formula is HCOOH. The second member is *ethanoic acid* (commonly called *acetic acid*). A 5–6 percent aqueous solution of ethanoic acid is popularly known as *vinegar*. The formula for ethanoic acid is CH_3COOH.

Esters

Esters have this general formula:

$$\overset{\displaystyle O}{\underset{\displaystyle R_1-C-O-R_2}{\|}}$$

They are prepared by the reaction between an organic acid and an alcohol (see Section 14.5). An ester is named as a derivative of the acid and alcohol that formed it. For example, the ester

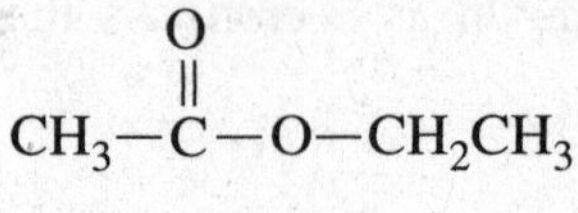

$$CH_3-\overset{\overset{\displaystyle O}{\parallel}}{C}-O-CH_2CH_3$$

is formed by the reaction between *ethanoic acid* and *ethanol*. Its IUPAC name, *ethyl ethanoate*, is derived from the combination of the *alkyl group* found in the alcohol (*ethyl*) and the name of the acid, with the suffix *-ate* replacing *-ic*.

Esters usually have a pleasant odor. The distinctive aromas of many fruits (e.g., bananas) and plants (e.g., wintergreen) are due to esters. The type of organic acid and the type of alcohol present in the ester determine its aroma.

Organic Nitrogen Compounds

Amines

The **amino group** consists of a nitrogen atom bonded to two hydrogen atoms: $-NH_2$.

A hydrocarbon in which one or more hydrogen atoms is replaced by an amino group is known as an *amine*. The simplest amine is *aminomethane*; its structural formula is shown below.

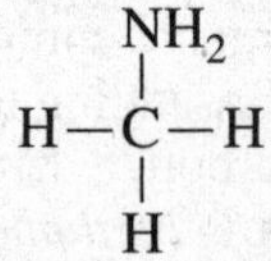

$$H-\overset{\overset{\displaystyle NH_2}{|}}{\underset{\underset{\displaystyle H}{|}}{C}}-H$$

aminomethane

Amino Acids

An **amino acid** is an organic acid that contains one or more amino groups. The simplest amino acids are *aminoethanoic acid* and *aminopropanoic acid,* known, respectively, as *glycine* and *alanine*. Their structures are as follows:

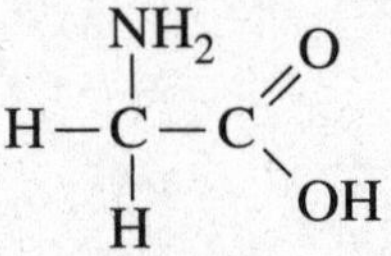

glycine 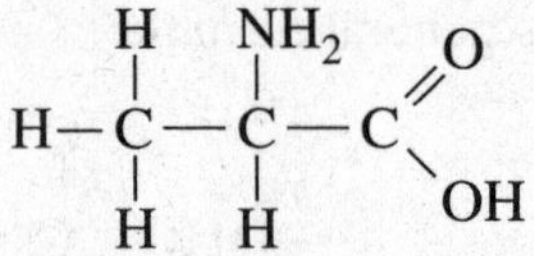   **alanine**

Amino acids are the "building blocks" of proteins and also have other functions in living organisms.

Amides

An **amide** is a derivative of an organic acid in which an amino group ($-NH_2$) replaces the $-OH$ on the carboxyl group. Here is the general formula for an amide:

$$\begin{array}{c} O \\ \parallel \\ R-C-NH_2 \end{array}$$

In the IUPAC system, an amide is named by dropping *-oic* from the name of the acid, and adding *-amide* in its place. For example, the amide whose formula is

$$\begin{array}{c} O \\ \parallel \\ CH_3-C-NH_2 \end{array}$$

is derived from *ethanoic acid*. Its IUPAC name is *ethanamide*.

While a detailed study of amides is well beyond the scope of this book, we will see in Section 11.5 that compounds called *polyamides* or, as they are commonly known, *peptides* can be produced by a reaction between amino acids. Polyamides are of great importance in biology.

Haloalkanes (Halides)

A **haloalkane** is a compound in which one or more *hydrogen* atoms in an alkane is replaced by a corresponding number of halogen atoms. The general formula of a haloalkane is $R-X$, in which X represents a halogen atom.

According to the IUPAC system, a haloalkane is named by placing the combining form of the halogen name (*fluoro-, chloro-, bromo-, iodo-*) before the name of the alkane. If necessary, a prefix (*di-, tri-*, etc.) and numbers are assigned to indicate the number of halogen atoms present and their locations.

The formulas and IUPAC names of two haloalkanes are given below.

$$\begin{array}{c} H \\ | \\ H-C-Cl \\ | \\ H \end{array} \qquad \begin{array}{cc} Br & Br \\ | & | \\ H-C-C-H \\ | & | \\ H & H \end{array}$$

chloromethane   1,2-dibromoethane

PROBLEM

Name the following haloalkanes according to the IUPAC system:

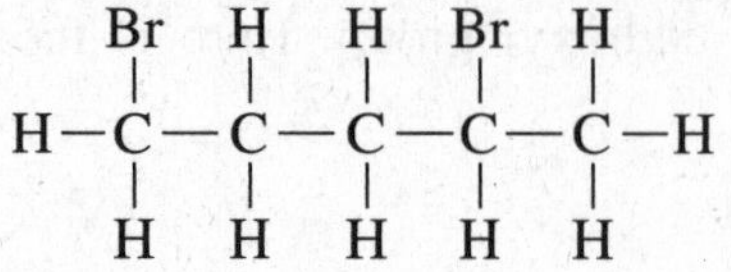

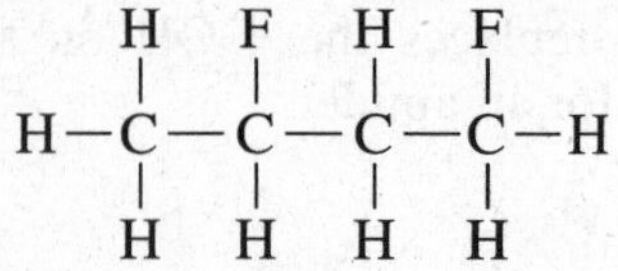

$$
\begin{array}{ccccc}
\text{Br} & \text{H} & \text{H} & \text{Br} & \text{H} \\
| & | & | & | & | \\
\text{H}-\text{C}-\text{C}-\text{C}-\text{C}-\text{C}-\text{H} \\
| & | & | & | & | \\
\text{H} & \text{H} & \text{H} & \text{H} & \text{H}
\end{array}
\qquad
\begin{array}{cccc}
\text{H} & \text{F} & \text{H} & \text{F} \\
| & | & | & | \\
\text{H}-\text{C}-\text{C}-\text{C}-\text{C}-\text{H} \\
| & | & | & | \\
\text{H} & \text{H} & \text{H} & \text{H}
\end{array}
$$

SOLUTION

The haloalkane on the left has a parent hydrocarbon with five carbon atoms; its name is pentane. Counting from the left, bromine atoms are located on the first and fourth carbon atoms. The haloalkane's name is 1,4-dibromopentane.

The haloalkane on the right has a parent hydrocarbon with four carbon atoms; its name is butane. Since the name must have the *smallest* numbers possible, we count the carbons from right to left; the fluorine atoms are located on the first and third carbon atoms. The haloalkane's name is 1,3-difluorobutane.

11.5   ORGANIC REACTIONS

Organic reactions generally occur at a slower rate than inorganic reactions, and they usually involve the functional groups of the reactants.

Substitution

In a **substitution** reaction one type of atom or group is replaced with another type. Halogens such as chlorine and bromine may be substituted for the hydrogen atoms of alkanes. Halogen substitution is a stepwise process, as shown below.

$$
\begin{aligned}
CH_4 + Cl_2 &\rightarrow HCl + CH_3Cl & &(\text{chloromethane}) \\
CH_3Cl + Cl_2 &\rightarrow HCl + CH_2Cl_2 & &(\textit{di}\text{chloromethane}) \\
CH_2Cl_2 + Cl_2 &\rightarrow HCl + CHCl_3 & &(\textit{tri}\text{chloromethane}) \\
CHCl_3 + Cl_2 &\rightarrow HCl + CCl_4 & &(\textit{tetra}\text{chloromethane})
\end{aligned}
$$

In each step, one hydrogen atom is replaced with a chlorine atom from Cl_2. The remaining chlorine atom combines with the substituted hydrogen to form hydrogen chloride (HCl).

Addition

Addition is a reaction in which atoms are "added across" a double or triple bond. As a result of addition, a double bond is converted into a single bond, and a triple bond into a double bond.

The addition of bromine to ethene is shown below.

$$\begin{array}{c}H\\ \diagdown\\ H\diagup\end{array}C=C\begin{array}{c}H\\ \diagup\\ \diagdown H\end{array} + Br_2 \rightarrow H-\underset{\underset{H}{|}}{\overset{\overset{Br}{|}}{C}}-\underset{\underset{H}{|}}{\overset{\overset{Br}{|}}{C}}-H$$

ethene   1.2-dibromoethane

Addition reactions take place more rapidly than substitution reactions because the pi (π) bonds in alkenes and alkynes are weaker and more easily broken than the sigma (σ) bonds. Addition is characteristic of unsaturated compounds. The addition of halogen atoms, such as chlorine and bromine, is a common test for unsaturation in hydrocarbons. The addition of hydrogen to an unsaturated compound is usually called *hydrogenation*; it is used commercially to solidify liquid vegetable oils.

Fermentation

Fermentation is a biological process in which oxidation occurs in the absence of oxygen. *Enzymes* produced by living organisms act as the catalysts for this process. For example, yeast enzymes will ferment glucose (a simple sugar) to produce ethanol and carbon dioxide.

$$C_6H_{12}O_6 \xrightarrow{\text{yeast enzymes}} 2C_2H_5OH + 2CO_2$$

glucose   ethanol   carbon dioxide

Esterification

Esterification is the reaction between an acid and an alcohol. The products of an esterification reaction are an *ester* and water:

Esterification: Acid + Alcohol → Ester + Water

Esterification occurs slowly and must be catalyzed by acids or bases. It is an example of a *dehydration* or *condensation* reaction. The reaction is reversible and is then called *hydrolysis* because the ester is broken down with water. Illustrations of esterification and hydrolysis follow.

Chapter Eleven ORGANIC CHEMISTRY

When long-chain organic acids (called *fatty acids*) are esterified with glycerol, the ester is known as a *lipid* or *fat*. If the fatty acids contain a large number of double bonds within the parent hydrocarbons, the fat is said to be unsaturated and is usually liquid. Such is the case with vegetable oils. Animal fats, which are solid, are known as saturated fats because the parent hydrocarbons contain relatively few double bonds.

Saponification

Saponification is the hydrolysis of a fat with a base such as sodium hydroxide (NaOH). The products of saponification are glycerol and salts of the fatty acids, known collectively as *soap:*

$$\text{Saponification: Fat + NaOH} \rightarrow \text{Glycerol + Soap}$$

Soap is a long molecule that is nonpolar at one end and ionic at the other. The nonpolar end can form attractions with other nonpolar substances such as oil, and the ionic end can attract polar solvents such as water. Therefore, soap can bring about the emulsion of oil and water. This property is known as *detergency*.

Combustion

Saturated hydrocarbons such as methane react with oxygen at suitably high temperatures to produce carbon dioxide and water:

$$CH_4 + 2O_2 \rightarrow CO_2 + 2H_2O$$

This property of **combustion** makes alkanes suitable as fuels. In a *limited* supply of oxygen, carbon monoxide or carbon (soot) is produced.

$$2CH_4 + 3O_2 \rightarrow 2CO + 4H_2O$$
$$CH_4 + O_2 \rightarrow C + 2H_2O$$

Polymers and Polymerization

A **polymer** is a large molecule composed of many repeating units called *monomers*. In *polymerization,* the monomer units are joined to produce the polymer:

$$n \text{ monomer} \rightarrow (\text{monomer})_n$$

n units of monomer   **polymer**

Natural polymers include proteins (monomer: amino acids), cellulose and starch (monomer: glucose). *Synthetic polymers* include the plastic polyethylene (monomer: ethene), nylon, and polyester.

SECTION II—ADDITIONAL MATERIAL

11.1A   STEREOISOMERISM

Stereoisomers are isomers whose atoms have different *arrangements in space*. Actually, there are *two* stereoisomers of 2-butene! The way in which carbon-carbon double bonds are formed causes the atoms around each carbon to be held rigidly in position. The two stereoisomers of 2-butene are shown below:

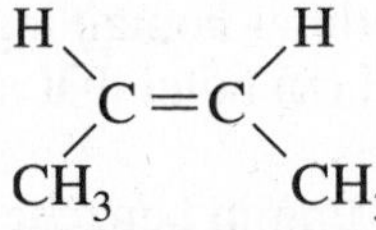

cis-2-butene 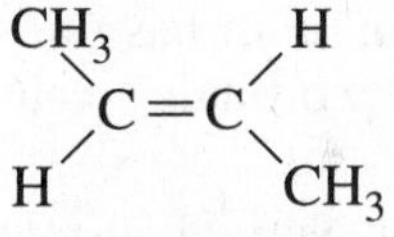   *trans*-2-butene

In the first stereoisomer, known as the *cis* form, the methyl groups are closer together than in the second stereoisomer, the *trans* form. Therefore, we name the isomer on the left *cis*-2-butene and the one on the right *trans*-2-butene.

TRY IT YOURSELF
In the compound 1,2-dichloroethene, one hydrogen atom on each carbon atom is replaced by a *chlorine* atom. Write the formulas of the two stereoisomers of 1,2-dichloroethene.

ANSWER

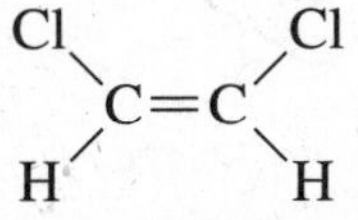

cis-1,2-dichloroethene 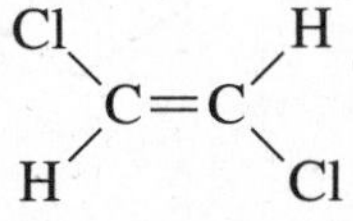   *trans*-1,2-dichloroethene

11.2A   THE BENZENE SERIES

The homologous **benzene** series of hydrocarbons—also known as the **aromatic hydrocarbons**—has the general formula C_nH_{2n-6}. The first member of the series is named *benzene* and has the molecular formula C_6H_6. The structure of benzene (and of its family) is quite unique: its first and last carbons join together, and for this reason it is known as a *ring* compound. Its structure may be represented as follows:

$$
\begin{array}{c}
H \\
\mid \\
H{-}C{=}C{-}H \\
\mid \quad\quad \mid \\
C \quad\quad C \\
\mid \quad\quad \mid \\
H{-}C{=}C{-}H \\
\mid \\
H
\end{array}
$$

Bonding in Aromatic Compounds

Benzene (and its derivatives) are flat, hexagonal molecules, and the bond angles around each carbon atom are 120°. Each of the six carbon atoms in benzene forms sigma (σ) bonds with the two carbon atoms adjacent to it. In addition, each carbon atom has one 2*p* orbital containing a single electron. The overlap of these *p* orbitals creates a pi (π) bond that is *extended over the entire molecule*.

For this reason, the carbon-carbon bonding in benzene consists of neither single nor double bonds; benzene has the properties of *both* a saturated and an unsaturated hydrocarbon. Usually, the benzene ring is drawn as a hexagonal skeleton, with the hydrogen atoms omitted, and the extended pi bond is usually represented by a dashed or a solid circle in the center of the hexagon. In an older, but less accurate, way of depicting the benzene molecule, a hexagon with three alternating double bonds is drawn above. The modern representations of the benzene molecule—hexagons with a dashed circle and a solid circle—are shown in the accompanying diagrams.

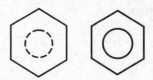

11.3A PRIMARY, SECONDARY, AND TERTIARY ALCOHOLS

Alcohols (as well as certain other types of organic compounds) are classified as *primary*, *secondary*, or *tertiary* because of the differences in the chemical behavior among these types of alcohols. The classification is based on the *number of carbon atoms attached to the carbon atom that contains the hydroxyl group*:

- An alcohol in which *zero* or *one* carbon atom is bonded to the hydroxyl-containing carbon atom is known as a **primary alcohol**. The general formula and an example (*ethanol*) are given below.

$$R-\underset{\underset{\textstyle H}{|}}{\overset{\overset{\textstyle H}{|}}{C}}-OH \qquad CH_3-\underset{\underset{\textstyle H}{|}}{\overset{\overset{\textstyle H}{|}}{C}}-OH$$

primary alcohol   **ethanol**

- An alcohol in which *two* carbon atoms are bonded to the hydroxyl-containing carbon atom is known as a **secondary alcohol**. The general formula and an example (*2-propanol*, commonly called *isopropyl alcohol* or *rubbing alcohol*) are given below.

$$R_1-\underset{\underset{\textstyle R_2}{|}}{\overset{\overset{\textstyle H}{|}}{C}}-OH \qquad CH_3-\underset{\underset{\textstyle CH_3}{|}}{\overset{\overset{\textstyle H}{|}}{C}}-OH$$

secondary alcohol   **2-propanol**

The number 2 in 2-propanol stands for the position of the hydroxyl group.
- An alcohol in which *three* carbon atoms are bonded to the hydroxyl-containing carbon atom is known as a **tertiary alcohol**. The general formula and an example (*2-methyl– 2-propanol*) are given below.

$$R_1-\underset{\underset{\textstyle R_2}{|}}{\overset{\overset{\textstyle R_3}{|}}{C}}-OH \qquad CH_3-\underset{\underset{\textstyle CH_3}{|}}{\overset{\overset{\textstyle CH_3}{|}}{C}}-OH$$

tertiary alcohol   **2-methyl–2-propanol**

11.4A DIHYDROXY AND TRIHYDROXY ALCOHOLS

As their names imply, *dihydroxy* and *trihydroxy alcohols* contain two and three hydroxyl groups, respectively. Dihydroxy alcohols are also known as *glycols*. The most important dihydroxy alcohol is *1,2-ethanediol* (commonly called *ethylene glycol*), which is a component of automobile antifreeze. The most important trihydroxy alcohol is *1,2,3-propanetriol* (commonly called *glycerol*, which plays a part in biological processes). The structures of these compounds are given below.

$$
\begin{array}{cc}
\text{OH} \quad \text{OH} & \text{OH} \quad \text{OH} \quad \text{OH} \\
| \qquad | & | \qquad | \qquad | \\
\text{H}-\text{C}-\text{C}-\text{H} & \text{H}-\text{C}-\text{C}-\text{C}-\text{H} \\
| \qquad | & | \qquad | \qquad | \\
\text{H} \quad \text{H} & \text{H} \quad \text{H} \quad \text{H}
\end{array}
$$

1,2,-ethanediol   1,2,3-propanetriol
(ethylene glycol)   (glycerol)

11.5A TYPES OF POLYMERIZATION

Condensation Polymerization

In **condensation polymerization**, the monomers are joined by a *dehydration* reaction. The first step in this process is illustrated below, as two monomer units (in this case, dihydroxy alcohols) condense to form a *dimer* and water.

$$
\begin{array}{c}
\text{H} \quad \text{H} \qquad\qquad \text{H} \quad \text{H} \qquad\qquad\qquad \text{H} \quad \text{H} \qquad \text{H} \quad \text{H} \\
| \quad | \qquad\qquad | \quad | \qquad\qquad\qquad | \quad | \qquad | \quad | \\
\text{HO}-\text{C}-\text{C}-\text{O}(\text{H}+\text{HO})-\text{C}-\text{C}-\text{OH} \rightarrow \text{HO}-\text{C}-\text{C}-\text{O}-\text{C}-\text{C}-\text{OH}+\text{H}_2\text{O} \\
| \quad | \qquad\qquad | \quad | \qquad\qquad\qquad | \quad | \qquad | \quad | \\
\text{H} \quad \text{H} \qquad\qquad \text{H} \quad \text{H} \qquad\qquad\qquad \text{H} \quad \text{H} \qquad \text{H} \quad \text{H}
\end{array}
$$

monomer   monomer   dimer   water

As additional units are joined to the dimer, the polymer is gradually built up. For condensation polymerization to occur, the monomer must have at least two functional groups that can undergo dehydration.

Examples of synthetic condensation polymers include silicones, nylons, and polyesters.

In living systems, two amino acids can undergo a condensation reaction in which the $-$OH group on the first molecule and the $-$NH$_2$ group on the second molecule interact to eliminate a molecule of H_2O. The two molecules then join to produce a special type of amide known as a *dipeptide*. The bond joining the two amino acids is known as a *peptide link*. The accompanying diagram illustrates how this occurs.

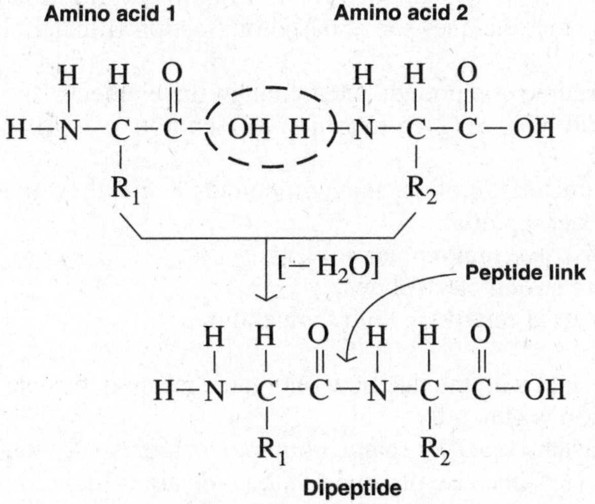

Amino acid 1   **Amino acid 2**

$$
\underset{R_1}{H-N-C-C}-(OH \quad H)-\underset{R_2}{N-C-C}-OH
$$

$$[-H_2O]$$

— Peptide link

$$
\underset{R_1}{H-N-C-C}-\underset{R_2}{N-C-C}-OH
$$

Dipeptide

Since each amino acid has *both* amino and hydroxyl groups, a *polyamide* chain containing many amino acid units can be formed. This *natural polymer* is known more commonly as a *polypeptide*. A *protein* is a molecule that contains one or more polypeptide chains. Because of its large size, the synthesis of a protein (and other biological polymers) would be impossibly slow if the process were not catalyzed by other proteins known as *enzymes*.

Addition Polymerization

Monomers that are unsaturated may undergo polymerization by undergoing addition reactions with each other. In this process, called **addition polymerization**, the double or triple bonds are reduced to single or double bonds, as illustrated below, where the "~" symbols indicate the positions where the repeating ethene units are added.

Polyethylene is an example of an addition polymer.

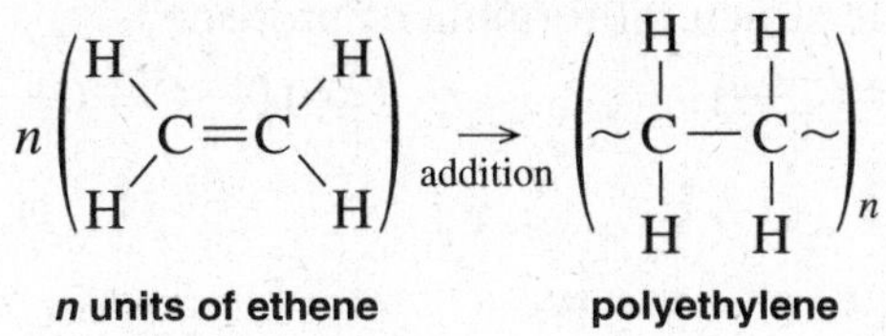

$$
n \left(\begin{array}{c} H \\ H \end{array} C=C \begin{array}{c} H \\ H \end{array} \right) \xrightarrow[\text{addition}]{} \left(\sim \underset{H}{\overset{H}{C}} - \underset{H}{\overset{H}{C}} \sim \right)_n
$$

***n* units of ethene**   **polyethylene**

END-OF-CHAPTER QUESTIONS

Some questions have the symbol "§2" in front of the question number. This symbol means that the question is based on Section II material.

1. All organic compounds must contain the element
   (1) sulfur   (2) nitrogen   (3) carbon   (4) oxygen

2. A characteristic of organic compounds is that they generally
   (1) react rapidly
   (2) dissolve in nonpolar solvents
   (3) are strong electrolytes
   (4) melt at relatively high temperatures

3. Compared with the rate of an inorganic reaction, the rate of an organic reaction is generally
   (1) faster, because organic compounds are usually ionic
   (2) faster, because organic compounds are usually molecular
   (3) slower, because organic compounds are usually ionic
   (4) slower, because organic compounds are usually molecular

4. Organic compounds that are essentially nonpolar and exhibit weak intermolecular forces have
   (1) low melting points
   (2) low vapor pressures
   (3) high conductivities in solution
   (4) high boiling points

5. The members of the alkane series of hydrocarbons are similar in that each member has the same
   (1) empirical formula   (2) general formula
   (3) structural formula   (4) molecular formula

6. Which hydrocarbon has more than one possible structural formula?
   (1) CH_4 (2) C_2H_6 (3) C_3H_8 (4) C_4H_{10}

7. Which is the structural formula of propene?
   (1) $H-C=C-H$ (2) $H-C\equiv C-H$
   (3) $H-\overset{\overset{H}{|}}{C}\equiv\overset{\overset{H}{|}}{C}-\underset{\underset{H}{|}}{\overset{\overset{H}{|}}{C}}-H$ (4) $H-\overset{\overset{H}{|}}{C}=\overset{\overset{H}{|}}{C}-\underset{\underset{H}{|}}{\overset{\overset{H}{|}}{C}}-H$

§2 **8.** Which represents benzene?

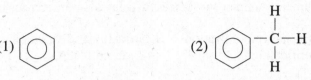

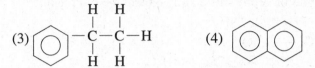

9. A compound with the molecular formula C_7H_{16} could be a member of the hydrocarbon series that has the general formula
(1) C_nH_{2n} (2) C_nH_{2n-2} (3) C_nH_{2n-6} (4) C_nH_{2n+2}

10. Which compound has the greatest possible number of isomers?
(1) butane   (2) ethane   (3) pentane   (4) propane

11. Which compound contains a triple bond?
(1) CH_4 (2) C_2H_2 (3) C_3H_6 (4) C_4H_{10}

12. Each member of the alkane series differs from the preceding member by having 1 additional carbon atom and
(1) 1 hydrogen atom   (2) 2 hydrogen atoms
(3) 3 hydrogen atoms   (4) 4 hydrogen atoms

13. Which hydrocarbon ought to have the highest normal boiling point?
(1) butene   (2) ethene   (3) pentene   (4) propene

14. Alkanes differ from alkenes in that alkanes
(1) are hydrocarbons
(2) are saturated compounds
(3) have the general formula C_nH_{2n}
(4) undergo addition reactions

15. Natural gas is composed mostly of
(1) butane   (2) gasoline
(3) methane   (4) propane

§2 **16.** Which method is commonly used to separate the components of petroleum into simpler substances by using their differences in boiling points?
(1) fractional crystallization   (2) fractional distillation
(3) esterification   (4) saponification

§2 **17.** Which equation represents a simple example of cracking?
(1) $N_2 + 3H_2 \xrightarrow{600°C} 2NH_3$
(2) $S + O_2 \longrightarrow SO_2$
(3) $C_3H_8 + 5O_2 \longrightarrow 3CO_2 + 4H_2O$
(4) $C_{14}H_{30} \xrightarrow{600°C} C_7H_{16} + C_7H_{14}$

18. How many hydrogen atoms are there in an alkene containing 20 carbon atoms?
(1) 34   (2) 20   (3) 42   (4) 40

19. Which compound is an isomer of the substance shown below?

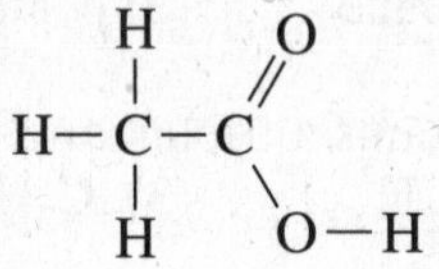

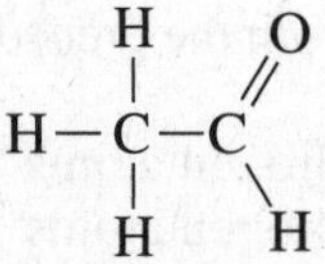

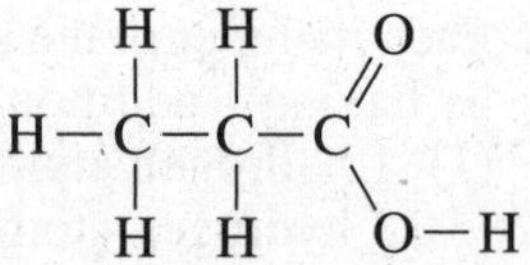

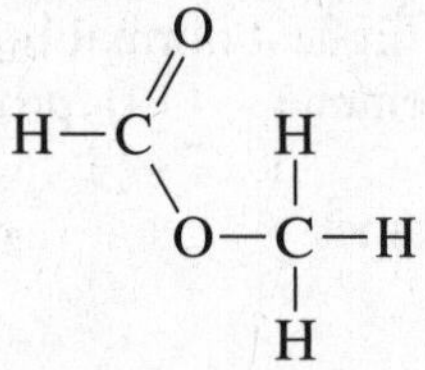

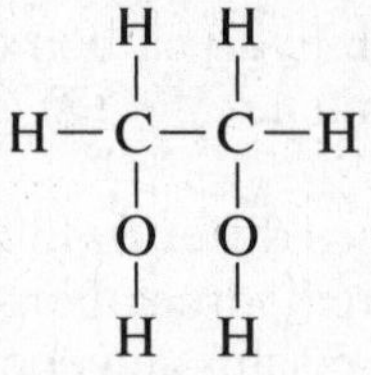

20. The functional group

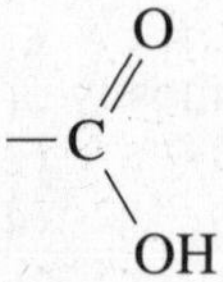

is always found in an organic
(1) acid   (2) ester   (3) ether   (4) aldehyde

21. Molecules of 1-propanol and 2-propanol have different
(1) percentage compositions   (2) molecular masses
(3) molecular formulas   (4) structural formulas

22. Which compound is an organic acid?
(1) CH_3CH_2OH (2) CH_3OCH_3
(3) CH_3COOH (4) CH_3COOCH_3

§2 **23.** Which compound is a trihydroxy alcohol?
(1) glycerol   (2) butanol   (3) ethanol   (4) methanol

24. What is the IUPAC name for a compound with the structural formula shown below?

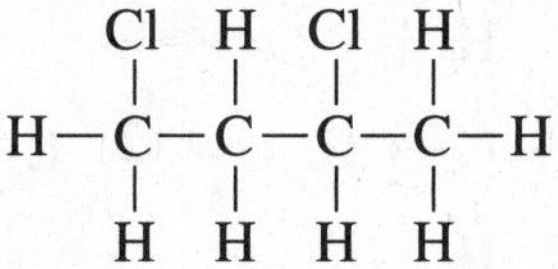

$$\begin{array}{ccccccc} & Cl & & H & & Cl & & H \\ & | & & | & & | & & | \\ H- & C & - & C & - & C & - & C & -H \\ & | & & | & & | & & | \\ & H & & H & & H & & H \end{array}$$

(1) 1,3-dichloropentane   (2) 2,4-dichloropentane
(3) 1,3-dichlorobutane   (4) 2,4-dichlorobutane

25. Which structural formula represents an aldehyde?

(1) 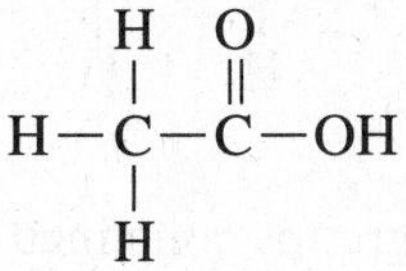
$$\begin{array}{ccc} H & O \\ | & || \\ H-C-C-OH \\ | \\ H \end{array}$$

(2) 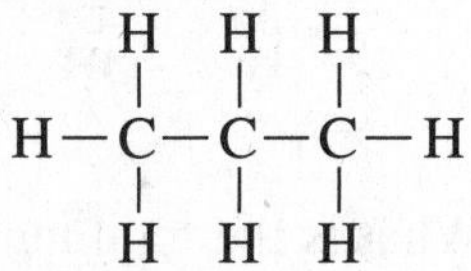
$$\begin{array}{ccc} H & H & H \\ | & | & | \\ H-C-C-C-H \\ | & | & | \\ H & H & H \end{array}$$

(3) 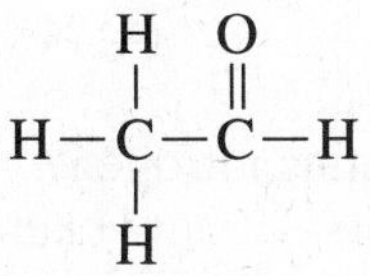
$$\begin{array}{ccc} H & O \\ | & || \\ H-C-C-H \\ | \\ H \end{array}$$

(4) 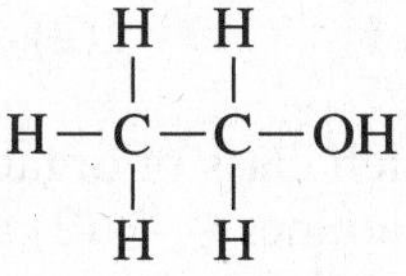
$$\begin{array}{cc} H & H \\ | & | \\ H-C-C-OH \\ | & | \\ H & H \end{array}$$

26. Which is the general formula for ketones?

(1) 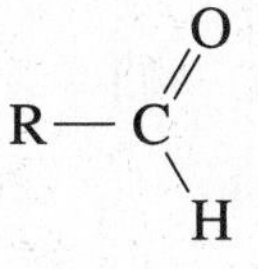
$$R-C\overset{\displaystyle O}{\underset{\displaystyle H}{\diagup\!\!\!\!\diagdown}}$$

(2) 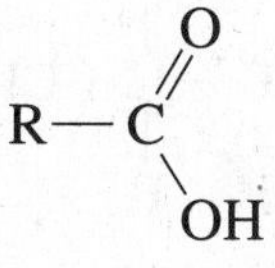
$$R-C\overset{\displaystyle O}{\underset{\displaystyle OH}{\diagup\!\!\!\!\diagdown}}$$

(3) 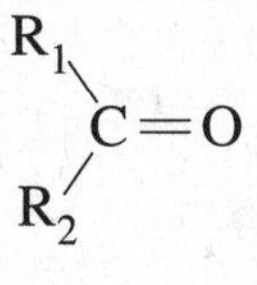
$$\begin{array}{c} R_1 \\ \diagdown \\ C=O \\ \diagup \\ R_2 \end{array}$$

(4) R_1-O-R_2

27. Which is the structural formula for diethyl ether?

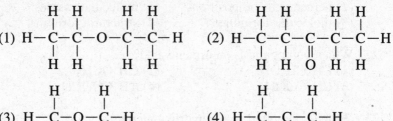

(1)

```
    H  H     H  H
    |  |     |  |
H — C — C — O — C — C — H
    |  |     |  |
    H  H     H  H
```

(2)

```
    H  H     H  H
    |  |     |  |
H — C — C — C — C — C — H
    |  |  ‖  |  |
    H  H  O  H  H
```

(3)

```
    H     H
    |     |
H — C — O — C — H
    |     |
    H     H
```

(4)

```
    H  H  H
    |  |  |
H — C — C — C — H
    |  ‖  |
    H  O  H
```

§2 **28.** Which structural formula represents a tertiary alcohol?

(1)

```
    H  H  H  H
    |  |  |  |
H — C — C — C — C — OH
    |  |  |  |
    H  H  H  H
```

(2)

```
    H  H  H  H
    |  |  |  |
H — C — C — C — C — H
    |  |  |  |
    OH H  H  H
```

(3)

```
          H
          |
    H  H—C—H  H
    |    |    |
H — C —— C —— C — H
    |    |    |
    H   OH    H
```

(4)

```
          H
          |
    H  H—C—H  H
    |    |    |
H — C —— C —— C — OH
    |    |    |
    H    H    H
```

§2 **29.** What is the total number of hydroxyl groups contained in one molecule of 1,2-ethanediol?

(1) 1   (2) 2   (3) 3   (4) 4

30. Which class of organic compounds contains nitrogen?

(1) amines   (2) aldehydes   (3) ethers   (4) ketones

31. Which is the product of the reaction between ethene and chlorine?

(1)

```
    H  H
    |  |
H — C — C — Cl
    |  |
    H  H
```

(2)

```
    H
    |
H — C — Cl
    |
    H
```

(3)

```
     H  H
     |  |
Cl — C — C — Cl
     |  |
     H  H
```

(4)

```
     H
     |
Cl — C — Cl
     |
     H
```

32. Which alcohol reacts with C_2H_5COOH to produce the ester $C_2H_5COOC_2H_5$?

(1) CH_3OH (2) C_2H_5OH (3) C_3H_7OH (4) C_4H_9OH

33. Given this equation:

$$C_6H_{12}O_6 \xrightarrow{\text{yeast enzymes}} 2C_2H_5OH + 2CO_2$$

The reaction represented by this equation is called
(1) esterification   (2) saponification
(3) fermentation   (4) polymerization

34. The reaction represented by the equation

$$nC_2H_4 \rightarrow (-C_2H_4-)_n$$

is called
(1) saponification   (2) fermentation
(3) esterification   (4) polymerization

35. The reaction

$$C_4H_{10} + Br_2 \rightarrow C_4H_9Br + HBr$$

is an example of
(1) substitution   (2) addition
(3) fermentation   (4) polymerization

36. Which physical property makes it possible to separate the components of crude oil (petroleum) by means of distillation?
(1) melting point   (2) conductivity
(3) solubility   (4) boiling point

37. In saturated hydrocarbons, carbon atoms are bonded to each other by
(1) single covalent bonds, only
(2) double covalent bonds, only
(3) alternating single and double bonds
(4) alternating double and triple bonds

38. Which formula correctly represents the product of an addition reaction between ethane and chlorine?
(1) CH_2Cl_2 (2) CH_3Cl
(3) $C_2H_4Cl_2$ (4) C_2H_3Cl

39. Which hydrocarbon is saturated?
(1) propene   (2) ethyne
(3) butene   (4) heptane

40. Which compound is an isomer of pentane?
(1) butane   (2) propane
(3) methylbutane   (4) methylpropane

41. Given the formula representing a compound:

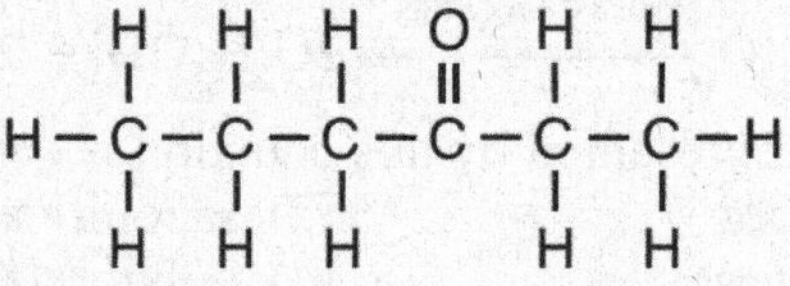

$$
\begin{array}{ccccccccccc}
 & H & & H & & H & & O & & H & & H \\
 & | & & | & & | & & \| & & | & & | \\
H- & C & - & C & - & C & - & C & - & C & - & C & -H \\
 & | & & | & & | & & & & | & & | \\
 & H & & H & & H & & & & H & & H \\
\end{array}
$$

What is an IUPAC name for this compound?
(1) ethyl propanoate   (3) 3-hexanone
(2) propyl ethanoate   (4) 4-hexanone

42. All isomers of octane have the same
(1) molecular formula   (3) physical properties
(2) structural formula   (4) IUPAC name

Constructed-Response Questions

Base your answers to questions 1 through 3 on the information below.

Many artificial flavorings are prepared using the type of organic reaction shown below.

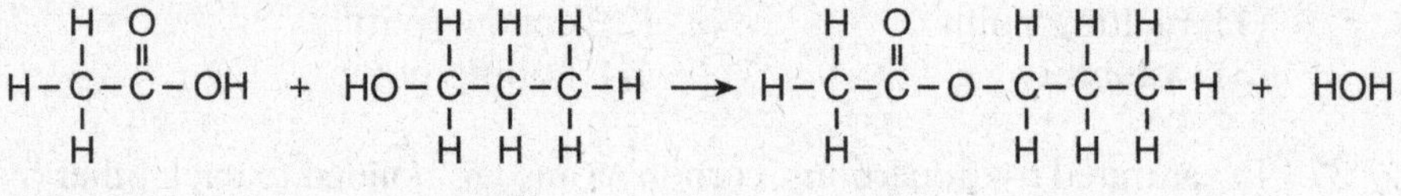

Reactant 1   Reactant 2

1. What type of organic reaction is this?

2. To what class of organic compounds does reactant 2 belong?

3. Draw the structural formula of an isomer of reactant 2.

Base your answers to questions 4–6 on the information below and on your knowledge of chemistry.

The equation below represents a reaction between propene and hydrogen bromide.

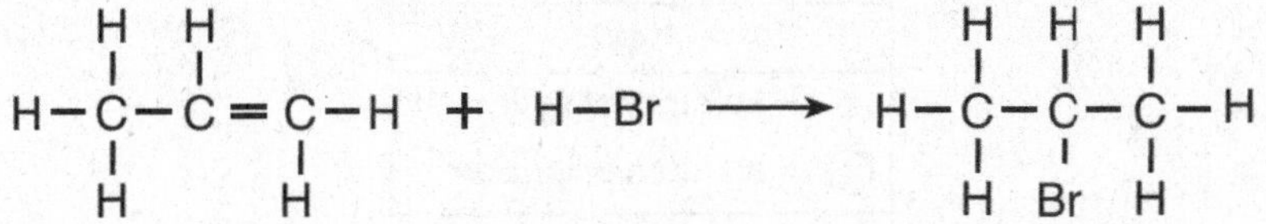

Cyclopropane, an isomer of propene, has a boiling point of –33°C at standard pressure and is represented by the formula below.

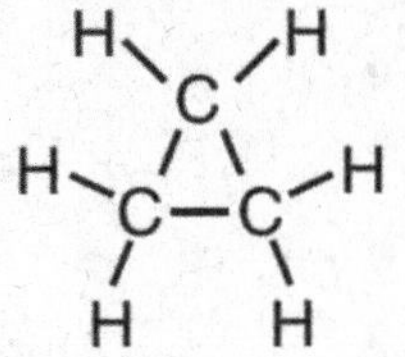

4. Explain why the reaction shown above can be classified as a synthesis reaction.

5. Identify the class of organic compounds to which the product of the reaction belongs.

6. Explain in terms of molecular formulas and structural formulas, why cyclopropane is an isomer of propene.

Base your answers to questions 7–9 on the information below and on your knowledge of chemistry.

The diagrams below represent ball-and-stick models of two molecules. (Each ball represents an atom, and each stick represents a chemical bond.)

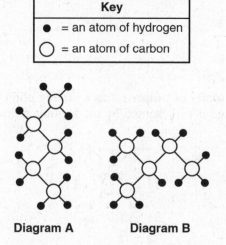

Diagram A   Diagram B

7. Draw a Lewis electron-dot diagram for an atom of the element present in all organic compounds.

8. Explain in terms of carbon-to-carbon bonds why the hydrocarbon represented in Diagram B is saturated.

9. Explain why the molecules in Diagrams A and B are isomers of each other.

The answers to these questions are found in Appendix 3.

340

Chapter Twelve

SOLUTIONS AND THEIR PROPERTIES

KEY IDEAS

This chapter focuses on the properties of the homogeneous mixtures we call solutions. The concepts of solubility, concentration, and colligative properties are explored. The chapter concludes with a discus sion of suspensions and colloidal dispersions.

KEY OBJECTIVES

At the conclusion of this chapter you will be able to:
- Define the terms *solution*, *solute*, and *solvent*.
- Provide examples of various types of solutions.
- Define the terms *miscible*, *saturated*, *unsaturated*, *solubility*, and *supersaturation*.
- Describe the factors that affect the solubility of a substance.
- Use the "solubility rules" to predict the solubility of an ionic compound in water.
- Interpret a solubility curve, and solve problems involving solubility curves.
- Describe how the concentration of a solution can be expressed with respect to the following terms: *percent*, *parts per million*, *mole fraction*, *molarity*, and *molality*, and solve problems involving these measurements of concentration.
- Describe how the solute affects the boiling point and the freezing point of a solution.
- Solve problems involving freezing point depression and boiling point elevation (colligative properties).
- Solve problems involving solutions and chemical equations.
- Define the term *electrolyte*, and indicate why solutions of electrolytes exhibit abnormal behavior.
- Distinguish among suspensions and types of colloidal dispersions.

SECTION I—BASIC
(REGENTS-LEVEL) MATERIAL

NYS REGENTS CONCEPTS AND SKILLS

Note: By the time you have finished Section I, you should have mastered the concepts and skills listed below. The Regents chemistry examination will test your knowledge of these items and your ability to apply them.

Concepts are the *basic ideas* that form the body of the Regents chemistry course (what you need to know!).

Skills are the *activities* that demonstrate your mastery of these concepts (how you show that you know them!).

Following each concept or skill is a page reference (given in parentheses) to this chapter.

12.1 Concept:
   • A solution is a homogeneous mixture of a solute dissolved in a solvent. (Page 343)
   • The solubility of a solute in a given amount of solvent is dependent on the temperature, the pressure, and the chemical natures of the solute and solvent. (Pages 344–345)

   Skills:
   • Interpret and construct solubility curves. (Pages 345–347)
   • Use solubility curves to distinguish among saturated, supersaturated, and unsaturated solutions. (Pages 345–347)
   • Apply the "like dissolves like" rule to real-world situations. (Pages 344–345)

12.2 Concept:
   The concentration of a solution may be expressed as:
   (a) molarity (M) (Pages 349–352);
   (b) percent by mass (Pages 347–348);
   (c) parts per million (ppm) (Pages 348–349).

   Skills:
   • Describe the preparation of a solution, given the molarity. (Pages 349–350)
   • Calculate solution concentrations in molarity, percent by mass, and parts per million. (Pages 348–349)

12.3 Concepts:
- The addition of a nonvolatile solute to a solvent causes the boiling point of the solvent to increase and the freezing point of the solvent to decrease. (Page 352)
- The greater the concentration of particles within the solution, the greater the effect. (Page 352)

12.4 Concepts:
- An electrolyte is a substance that releases mobile ions into an aqueous solution. As a result, the solution is capable of conducting an electric current. (Pages 352–353)
- The degree of conductivity of a solution depends on the concentration of ions within the solution. (Pages 352–353)

12.1   A SOLUTION IS . . . ?

A solution is one example of a homogeneous mixture of substances. When we think of the term *solution*, we usually picture a substance, such as table salt, well mixed or *dispersed* throughout another substance, such as water. The result is a water-clear mixture that we call a **solution**. The substance that is dissolved (the salt) is termed the **solute**, while the substance that does the dissolving (the water) is the **solvent**.

Although the solute and the solvent may be in different phases (in our example salt is solid and water is liquid), the solution assumes the phase of the *solvent*. If the solute and the solvent are in the same phase (as is the case with alcohol and water), the solvent is the substance that is present in greater quantity. Solutions in which water is the solvent are termed *aqueous*. There are times when we may be more interested in knowing the *components* of the solution than in identifying which component is the solute and which is the solvent.

Why do substances dissolve in one another? During the dissolving process, the particles of each component must be separated by—and mix with—the particles of every other component in the solution. The accompanying diagram illustrates how this may happen in a two-component solution.

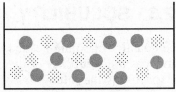

Component A   Component B   Solution of A and B

343

In the formation of a solution, (1) some of the attractive forces between the particles of each component must be overcome, and (2) the particles of all of the components must be able to attract each other with forces of similar strength.

In gases, there are essentially no attractive forces between the particles and therefore all mixtures of gases form solutions. A substance such as sugar forms a solution with water because the dipole attractions among sugar and water molecules in the solution are similar to the intermolecular attractions present in pure sugar and pure water. In contrast, oil and water do *not* form a solution because the weak London dispersion forces among the nonpolar oil molecules are not strong enough to separate the polar water molecules, which are attracted to each other by much stronger dipole forces.

Solutions exist among all three phases of matter, as the accompanying table illustrates.

Phase of Solute	Phase of Solvent	Example
Solid	Solid	Metallic alloy
Solid	Liquid	Table salt–water
Liquid	Liquid	Ethyl alcohol–water
Gas	Liquid	CO_2–water (soda)
Gas	Gas	Air

12.2 SATURATED AND UNSATURATED SOLUTIONS

When a solution of ethyl alcohol and water is prepared, the alcohol and water may be mixed in all proportions. A liquid–liquid solution of this type is said to be **miscible**. Mixtures of *gases* also form infinitely soluble solutions.

Most solutions, however, are not infinitely soluble. An example is the dissolving of table salt in water. After a certain amount of salt has been dissolved, the water will accept no more solute and any remaining salt will drop to the bottom of the vessel. This solution is said to be **saturated**. A solution that has not reached its saturation point is **unsaturated**.

12.3 SOLUBILITY

The quantity of solute needed to *just saturate* a given amount of solvent is known as the **solubility** of the solute-solvent pair. A convenient set of units for specifying the solubilities of most solids is *grams of solute per 100 grams of solvent*. For example, at 50°C, 85 grams of potassium nitrate, KNO_3, will just saturate 100 grams of H_2O. Therefore, the solubility of KNO_3 at 50°C is reported as *85 grams KNO_3 per 100 grams H_2O.*

Solubility depends on factors such as the nature of the solute and of the solvent and the temperature. If the solute is a gas, such as SO_2, its solubility is also affected by the atmospheric pressure.

Chemists frequently use the expression "like dissolves like" to indicate that the solute and the solvent need to have similar bond types in order for a solution to be formed. Substances that are ionic or consist of polar molecules will dissolve in a polar solvent such as water, and nonpolar substances (e.g., oils) will dissolve in nonpolar solvents such as benzene and toluene.

A substance such as potassium nitrate, KNO_3, that can form fairly concentrated solutions (in water) is said to be *soluble*. On the other hand, a substance such as silver iodide, AgI, that can form only very dilute solutions (in water) is termed *nearly insoluble*.

The solubilities of ionic compounds in water follow a number of general rules that are outlined in Reference Table F in Appendix 1.

TRY IT YOURSELF
Use the solubility rules in Reference Table F to classify each of the five compounds given below as *soluble*, *slightly soluble*, or *insoluble*.
(a) $NaNO_3$ (b) $BaSO_4$ (c) $(NH_4)_2S$ (d) AgI (e) $Sr(OH)_2$

ANSWERS
(a) soluble (b) insoluble (c) soluble (d) insoluble (e) slightly soluble

Solubility Curves

A *solubility curve* traces the solubility of a substance with increasing temperature. The accompanying graph displays the solubilities of $CuSO_4$ (a solid) and SO_2 (a gas) in water, as a function of temperature.

Each point on the curves represents the composition of a saturated solution at a different temperature. For example, the solubility of $CuSO_4$ is approximately 57 grams per 100 grams H_2O at 80°C, and the solubility of SO_2 is approximately 9 grams per 100 grams H_2O at 24°C.

As the temperature *increases* between 0°C and 100°C, the solubility of the solid increases, meaning that more solid is needed for a saturated solution. Most (but not all) solids increase their solubilities with increasing temperature. If the temperature of the solution is *decreased*, the solubility of the solid will decrease and any *excess* solute will drop out of the solution. Reference Table G in Appendix 1 contains a number of solubility curves for ionic solids and for gases dissolved in water.

Temperature has the opposite effect on a dissolved gas: when the temperature is *raised*, the solubility of the gas *decreases* and the some of the gas will escape from the solution. If the temperature is lowered, the solvent will be able to dissolve more gas.

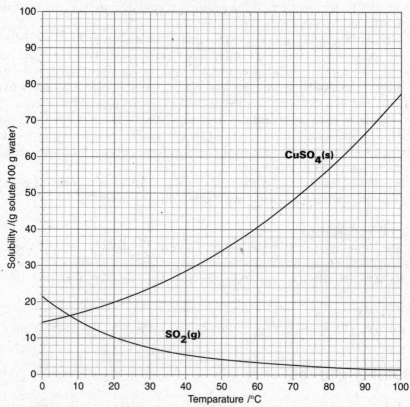

Solubility versus Temperature

Now let us solve some problems based on the solubility curves in Reference Table G.

PROBLEM
(a) At what temperature will the solubility of KNO_3 be 60 grams per 100 grams H_2O?
(b) At what temperature will the solubility be 120 grams per 100 grams H_2O?

SOLUTIONS
Refer to the solubility curve for KNO_3. (a) At 38°C the solubility of this compound is $60 g/100 g\ H_2O$. (b) At 65°C the solubility of this compound is $120 g/100 g\ H_2O$.

PROBLEM
How much $NaNO_3$ is needed to saturate 50 grams of water at 40°C?

SOLUTION
Refer to the solubility curve for $NaNO_3$. At 40°C, the solubility of this compound is 105 g/100 g H_2O. To saturate *half* the amount of water (50 g H_2O), half the amount of solute, or 53 g, will be needed.

TRY IT YOURSELF
(a) Which gas is most soluble at 0°C? At 60°C?
(b) In an experiment, 100 grams of water is saturated with NH_3 at 30°C. If the temperature is then lowered to 0°C, how much *additional* NH_3 will be needed to resaturate the solution?
(c) At what temperature do NH_4Cl and HCl have the same solubility?
(d) Which solid shows the smallest *increase* in solubility over the temperature range 0°C–100°C?

ANSWERS
(a) NH_3 (0°C); HCl (60°C)   (b) 45 g (90 g − 45 g)   (c) 58°C   (d) NaCl

Supersaturation

Occasionally, when a solution prepared at a higher temperature is cooled, the solute fails to drop out of the solution. This unstable condition is known as **supersaturation**. Eventually, the excess solute will leave the solution as it returns to its saturation point. Honey is an example of a supersaturated solution. After a period of time, the excess sugar crystallizes at the bottom of the honey jar. [The honey can be restored to its original (supersaturated) state by heating it and then allowing it to cool slowly.]

12.4   CONCENTRATIONS OF SOLUTIONS

When we prepare a solution, we are usually interested in its strength or concentration. **Concentration** is a measure of the quantity of solute dissolved in a given amount of solvent or solution. (This distinction will be made clear below.) For example, we may pick up a bottle that is labeled as *concentrated* (con) or *dilute* (dil) HCl. These terms tell us that the HCl solution is, respectively, relatively strong or relatively weak. However, the terms *concentrated* and *dilute* are not precise enough for most laboratory work.

Percent Concentration

A more precise way of expressing concentration is to specify the quantity of solute that is dissolved in one hundred parts of the solution. This type of expression is known as **percent concentration**. The percent concentration of a solution may be expressed in two ways:

• The **mass percent concentration** is defined as:

$$\text{Mass Percent Concentration} = \frac{\text{Mass of Solute}}{\text{Mass of Solution}} \times 100$$

It is essential that the quantities of the solute and solution be expressed in the same mass units.

PROBLEM
A student dissolves 25.0 grams of glucose in 475 grams of water. What is the mass percent concentration of the glucose solution?

SOLUTIONS

$$\text{Mass percent concentration} = \frac{\text{mass of solute}}{\text{mass of solution}} \times 100$$

$$= \frac{25.0\,\text{g}}{500.\,\text{g}} \times 100 = 5.00\%$$

Parts per Million (PPM)

The concentrations of very dilute solutions are often expressed in **parts per million** (ppm). In fact, scientists usually express the maximum allowable concentrations of toxic or cancer-causing substances in parts per million. For example, the maximum allowable concentration of arsenic in drinking water is 0.05 part per million, or 5 grams of arsenic per 100,000,000 grams of drinking water. We can calculate the concentration of a solution in parts per million by using the following equation:

$$\text{Parts per Million of Solute} = \frac{\text{Mass of Solute}}{\text{Mass of Solution}} \times 10^6$$

It is important to note that the numerator and denominator of this fraction must be expressed in the *same units of mass*.

PROBLEM
A certain gas has a concentration in water of 2 milligrams per 100 grams of water. What is the concentration of the gas in parts per million?

SOLUTION
There are two considerations in solving this problem. First, we must make the mass units of the solute and the solution the same. Second, we need to recognize that this solution is so dilute that the mass of the solution is essentially the mass of the water. With these considerations in mind, let us apply the equation given above and solve the problem:

$$\text{ppm of solute} = \frac{\text{mass of solute}}{\text{mass of solution}} \times 10^6$$

$$= \frac{2 \text{ mg} \cdot \dfrac{1 \text{ g}}{1000 \text{ mg}}}{100 \text{ g}} \times 10^6$$

$$= 0.00002 \times 10^6 \text{ ppm}$$

$$= 20 \text{ ppm}$$

TRY IT YOURSELF
A 2.5-gram sample of groundwater is found to contain 5.4×10^{-6} gram of Cu^{2+} ion. What is the ion concentration in parts per million?

ANSWER
2.2 ppm

In chemistry, where we are interested in the *number of particles* reacting, we measure the concentration of a solution by stating the number of *moles of solute* contained in the solution. There are three common ways to do this: molarity, molality, and mole fraction. (Mole fractions and molarity will be covered in Section 11 — Additional Material.)

Molarity

Molarity (*M*) is defined as the number of moles of solute dissolved in 1 liter of solution. In other words, molarity is a ratio between the number of moles of solute and the number of liters of solution.

$$M = \frac{n_{\text{solute}}}{V_{\text{solution (L)}}}$$

For example, if we prepare an aqueous solution by dissolving 4 moles of solute in 8 liters of solution, the concentration of this solution is $4/8 = 0.5$ mole per liter. We write this as 0.5 M (read as 0.5 *molar*).

We never specify how much *solvent* we actually use. In the example given above, 4 moles of solute is dissolved in *enough water* to make 8 liters of solution. Usually, we do this by preparing the solution in a *volumetric flask*, that is, a flask in which only the *final volume* (8 liters, in this case) is marked. The solute is added to the flask, and the solvent is poured in gradually (with constant stirring) until the volume of the solution reaches the mark. The accompanying diagram illustrates a volumetric flask.

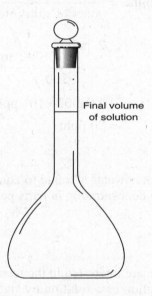

Final volume of solution

We use molarity to prepare solutions when we are more concerned about the concentrations of the solutes than about the relationship between solute and solvent.

We can solve problems involving molarity in two ways. The first method involves using the preceding molarity equation.

PROBLEM

How many moles of a solute are needed to prepare 2.5 liters of a 0.60 M solution?

SOLUTION

We rearrange the molarity equation and use the units of molarity (mol/L) to solve for the moles of solute needed:

$$n_{solute} = M \cdot V_{solution\ (L)}$$

$$= 0.60 \frac{mol}{\cancel{L}} \cdot 2.5 \cancel{L} = 1.5\ mol$$

We can also use the factor-label method (FLM) to solve molarity problems. The units of molarity (mol/L) can be used as a conversion factor. For example, if our problem involves a 0.60 molar solution, we can use either $\dfrac{0.60 \text{ mol}}{1 \text{ L}}$ or $\dfrac{1 \text{ L}}{0.60 \text{ mol}}$ to solve the problem. Let us solve the problem given above by FLM:

$$\text{mol solute} = 2.5 \; \cancel{L}_{\text{solution}} \cdot \frac{0.60 \text{ mol}_{\text{solute}}}{1 \; \cancel{L}_{\text{solution}}} = 1.5 \text{ mol}_{\text{solute}}$$

It is no accident that the two methods look essentially the same: they are entirely equivalent! Which one should you use? In solving simple problems such as the one given above, use the method that is easier for you. For more complex problems, however, FLM may prove to be your better bet.

TRY IT YOURSELF
A person wishes to prepare a 2.00 molar solution using 2.50 moles of solute. Calculate the final volume of the solution.

ANSWER
1.25 L

The next problem deals with a more complex situation involving molarity, molar mass, and volume. Problems such as this one can be solved by using the factor-label method.

PROBLEM
How many grams of urea (CH_4N_2O; molar mass = 60.1 grams per mole) are needed to make 200. milliliters of a 1.50 M solution?

SOLUTION
Let us write a solution map for this problem:

$$V_{\text{solution (mL)}} \xrightarrow{\overset{\text{unit}}{\underset{\text{conversion}}{}}} V_{\text{solution (L)}} \xrightarrow{\text{molarity}} \text{mol}_{\text{solute}} \xrightarrow{\text{molar mass}} g_{\text{solute}}$$

We can solve the problem using FLM:

$$g \text{ urea} = 200. \; \cancel{mL}_{\text{solution}} \cdot \frac{1 \; \cancel{L}_{\text{solution}}}{1000 \; \cancel{mL}_{\text{solution}}} \cdot \frac{1.50 \; \cancel{\text{mol urea}}}{1 \; \cancel{L}_{\text{solution}}} \cdot \frac{60.1 \text{ g urea}}{1 \; \cancel{\text{mol urea}}}$$

$$= 18 \text{ g urea}$$

TRY IT YOURSELF

A 3.00-molar solution of glucose ($C_6H_{12}O_6$; molar mass = 180. grams per mole) is prepared using 60.0 grams of solute. What is the final volume, in liters, of the solution?

ANSWER

0.111 L

12.5 EFFECT OF THE SOLUTE ON THE SOLVENT

When a solute is dissolved in a solvent, the particles of the solute bond with the particles of the solvent, causing the vapor pressure of the solution to be lower than the vapor pressure of the pure solvent at the same temperature. The effect of this vapor pressure lowering is to *raise* or *elevate* the boiling point of the solution and to *lower* or *depress* the freezing point of the solution in comparison to the boiling and freezing points of the pure solvent. For example, when 1 mole of glucose is dissolved in 1 kilogram of water, the boiling point of the solution is 100.52°C and the freezing point is −1.86°C. In contrast, the boiling and freezing points of pure water are 100°C and 0°C, respectively.

Boiling point elevation and freezing point depression are *colligative properties*; that is, they depend on the *number* of solute particles present and the *nature* of the solvent. Colligative properties do *not* depend on the actual *identity* of the solute. For example, if 1 mole of urea were substituted for the 1 mole of glucose in the preceding paragraph, the results would be exactly the same!

12.6 BEHAVIOR OF ELECTROLYTES IN SOLUTION

Certain substances called **electrolytes** produce ions when they dissolve in solution. Because these ions are free to move in solution, the solution conducts electricity: the greater the concentration of ions, the greater the conductivity of the solution.

Ions can be produced in solution in either of two ways.

1. When an ionic substance such as NaCl dissolves in water, the water *separates* the ions present in the NaCl crystal lattice. This process, known as *dissociation*, is represented below.

$$Na^+Cl^-(s) \xrightarrow{\text{H}_2\text{O}} Na^+(aq) + Cl^-(aq)$$

2. When a polar covalent substance such as HCl dissolves in water, ions are *created* by the interaction between HCl and H_2O molecules. This process, known as *ionization,* is represented below.

$$HCl(g) + H_2O(\ell) \rightarrow H_3O^+(aq) + Cl^-(aq)$$

When the boiling and freezing points of solutions of electrolytes are examined, it is found that they do *not* obey the simple relationship $\Delta t = k \cdot m$. The boiling points are *higher*, and the freezing points *lower*, than expected.

The reason for this abnormal behavior lies in the fact that each mole of dissolved electrolyte produces *more than 1 mole* of particles. For example, in *very dilute* aqueous solutions, 1 mole of $NaNO_3$ produces 2 moles of particles (1 mole of Na^+ and 1 mole of NO_3^-). Therefore, the boiling point elevation and freezing point depression are *twice* those for a molecular substance. Similarly, the effects of dissolving $Ca(NO_3)_2$ (1 mole produces 3 moles of particles) and $Al(NO_3)_3$ (1 mole produces 4 moles of particles) are even greater.

We need to specify that a solution of an electrolyte must be dilute because a phenomenon known as *ion-pairing* lowers the *number* of dissolved particles present in more concentrated solutions. As a result of ion-pairing, the effect that an electrolyte has on raising the boiling point and lowering the freezing point of a more concentrated solution is *less* than we would predict for a dilute solution.

SECTION II—ADDITIONAL MATERIAL

12.1A   MOLE FRACTION

Suppose a solution contains two components, *A* and *B*. The *mole fraction* (*X*) of component *A* is denoted as X_A and is defined as follows:

$$X_A = \frac{\text{moles of } A}{\text{total number of moles}} = \frac{n_A}{n_A + n_B}$$

where *n* represents the number of moles of each component present. We do not need to specify which component is the solute or the solvent. Note that the mole fraction, *X*, has no units. Why?

PROBLEM
A solution is prepared by mixing 3.0 moles of ethanol and 9.0 moles of water. Calculate the mole fraction of ethanol in this solution.

SOLUTION

$$X_{ethanol} = \frac{n_{ethanol}}{n_{ethanol} + n_{water}} = \frac{3.0 \text{ mol}}{3.0 \text{ mol} + 9.0 \text{ mol}} = \frac{3.0 \text{ mol}}{12 \text{ mol}} = 0.25$$

This result means that 25 of every 100 particles in the solution are ethanol molecules.

TRY IT YOURSELF
Calculate the mole fraction of water in this solution.

ANSWER
$X_{water} = 0.75$

12.2A MOLALITY

There are certain situations in which we must know how much solvent is present in a solution. These situations arise when we examine how a solute affects the properties of the solvent. For these instances, we use a method of measuring concentration known as *molality*.

Molality (m) is defined as the number of moles of solute dissolved in 1 kilogram of solvent. In other words, molality is a ratio between the number of moles of solute and the mass of the solvent expressed in kilograms:

$$m = \frac{n_{solute}}{kg_{solvent}}$$

For example, if we prepare an aqueous solution by dissolving 0.4 mole of solute in 0.5 kilogram of solvent, the concentration of this solution is 0.4/0.5 = 0.8 mole per kilogram. We write this as 0.8 m (read as 0.8 *molal*).

We never specify the volume of the solution we actually prepare because the volume of the solvent is not directly known to us. (We could obtain this information by measuring the *density of the solution*, but the required technique is beyond the scope of this book.)

PROBLEM
Sucrose (table sugar) has a molar mass of 342 grams per mole. What is the molality of a solution prepared by dissolving 34.2 grams of sucrose in 200. grams of water?

SOLUTION
We solve the problem by using the definition of molality and FLM:

$$m = \frac{n_{solute}}{kg_{solvent}} = \frac{34.2 \; \cancel{g \; sucrose} \cdot \dfrac{1 \; mol \; sucrose}{342 \; \cancel{g \; sucrose}}}{200. \; \cancel{g \; H_2O} \cdot \dfrac{1 \; kg \; H_2O}{1000 \; \cancel{g \; H_2O}}}$$

$$= \frac{0.100 \; mol \; sucrose}{0.200 \; kg \; H_2O} = 0.500 \; m$$

TRY IT YOURSELF
What is the molality of a solution prepared by dissolving 51 grams of NH_3 in 2.0 kilograms of H_2O?

ANSWER
1.5 m

12.3A DILUTION OF STOCK SOLUTIONS

In many laboratories solutions are frequently prepared as *concentrated stock solutions*. In order to use a *working* solution in an experiment, it is necessary to know how to dilute the stock correctly. The following problem illustrates this point.

PROBLEM
How should a 12 M stock solution be diluted in order to prepare 0.80 liter of a 3.0 M working solution?

SOLUTION
When we dilute a solution, we do *not* change the number of moles of solute present: we simply add more solvent. Therefore, since $n_{solute} = M \cdot V_{solution \; (L)}$, the following relationship must hold true:

$$M_{\substack{stock \\ solution}} \cdot V_{\substack{stock \\ solution}} = M_{\substack{working \\ solution}} \cdot V_{\substack{working \\ solution}}$$

We can now solve the problem:

$$12M \cdot V = 3.0 \; M \cdot 0.80 \; L$$
$$V = 0.20 \; L$$

This result means that we must dilute 0.20 L of the 12 M stock solution to 0.80 L with solvent. The resulting solution will have a 3.0 M concentration. Since the volume term appears on *both* sides of the equation, units *other than liters* can be used for volume provided that the *same* units appear on each side.

TRY IT YOURSELF
If 300. milliliters of a 6.00 M stock solution is diluted to produce a 2.00 M working solution, what is the volume of the working solution?

ANSWER
900. mL

12.4A SOLUTIONS AND CHEMICAL EQUATIONS

Since the molarity of a solution depends on the number of moles of solute present, we can solve chemical equation problems that include molarity.
The following example illustrates how we solve such problems.

PROBLEM
Consider this equation:

$$Cu(s) + 4HNO_3(aq) \rightarrow Cu(NO_3)_2(aq) + 2NO_2(g) + 2H_2O(\ell)$$

How many grams of $NO_2(g)$ will be produced by the complete reaction of 400. milliliters of 2.50 M HNO_3?

SOLUTION
Our solution map for this problem is as follows:

$$\text{mL HNO}_3 \xrightarrow[\text{conversion}]{\text{unit}} \text{L HNO}_3 \xrightarrow{\text{molarity}} \text{mol HNO}_3 \xrightarrow[\text{coefficients}]{\text{equation}} \text{mol NO}_2 \xrightarrow[\text{mass}]{\text{molar}} \text{g NO}_2$$

We now use our trusty friend, FLM, to solve the problem.

$$\text{g NO}_2 = 400. \text{ mL HNO}_3 \cdot \frac{1 \text{ L HNO}_3}{1000 \text{ mL HNO}_3} \cdot \frac{2.50 \text{ mol HNO}_3}{1 \text{ L HNO}_3}$$

$$\cdot \frac{2 \text{ mol NO}_2}{4 \text{ mol HNO}_3} \cdot \frac{46.0 \text{ g NO}_2}{1 \text{ mol NO}_2}$$

$$= 23.0 \text{ g NO}_2$$

TRY IT YOURSELF
In the equation given above, how many grams of Cu(s) will react completely with the quantity of HNO_3 given?

ANSWER
15.9 g

12.5A CALCULATING THE FREEZING AND BOILING POINTS OF SOLUTIONS

If a solute is *molecular* (i.e., not ionic) and is *nonvolatile* (i.e., does not evaporate with the solvent), then the changes in the boiling and freezing points of a solution bear a simple relationship to the *molality* of the solution:

$$\Delta T_b = +k_b \cdot m$$
$$\Delta T_f = -k_f \cdot m$$

where ΔT_b and ΔT_f refer to the *changes* in the boiling and freezing points, respectively. The "+" sign indicates that the boiling point of a solution is elevated; the "−" sign, that the freezing point of a solution is lowered.

In these relationships, k_b and k_f are known as the *boiling point* and *freezing point constants* and they depend on the solvent that is used. (The units of these constants are C°/m.) *The constants k_b and k_f represent the number of degrees that the boiling point of a solvent is raised (or its freezing point is lowered) when 1 mole of a molecular, nonvolatile solute is dissolved in 1 kilogram of the solvent.*

Reference Table W-3 in Appendix 2 lists freezing and boiling point data for selected solvents.

If we inspect this table, we can see that the freezing and boiling points of some solvents (e.g., cyclohexane) are very sensitive to the addition of a solute.

PROBLEM
Calculate the boiling and freezing points of a 1.50 m aqueous solution of a molecular, nonvolatile solute.

SOLUTION

$$\Delta T_b = +k_b \cdot m = +0.513 \text{ C°/m} \cdot 1.50 \text{ m} = +0.77 \text{ C°}$$
$$\Delta T_f = -k_f \cdot m = -1.86 \text{ C°/m} \cdot 1.50 \text{ m} = -2.79 \text{ C°}$$

Since the boiling point of the solution is elevated, we *add* 0.77 C° to the boiling point of pure H_2O (100°C); the solution's boiling point is 100.77°C.

Since the freezing point of the solution is lowered, we *subtract* 2.79 C° from the freezing point of pure H_2O (0°C); the solution's freezing point is −2.79°C.

TRY IT YOURSELF

A 0.80 m solution of a molecular, nonvolatile solute is prepared using benzene as a solvent.
(a) By how many degrees is the freezing point of the solvent lowered and the boiling point raised?
(b) What are the actual boiling and freezing points of the solution?

ANSWERS
(a) $\Delta T_f = -4.1$ C°; $\Delta T_b = +2.1$ C°
(b) Freezing point = 1.4°C; boiling point = 82.2°C

PROBLEM

A solution of a molecular, nonvolatile solute is prepared as follows: 90.0 grams of solute is dissolved in 250. grams of water. The freezing point of the solution is −3.72°C.
(a) How many grams of solute would be dissolved in a solution having the same concentration but containing 1 kilogram of water?
(b) What is the freezing point lowering of the solution?
(c) Based on the freezing point lowering, what is the molality of the solution?
(d) What is the molar mass of the solute in grams per mole?

SOLUTIONS

The methods employed in this problem have actually been used to calculate the molar masses of unknown compounds.
(a) Since 250. g of water represents 0.250 kg, a solution containing 1.00 kg of water—four times the amount—would need *four times* the mass of solute in order to have the same concentration. Therefore, the amount of solute needed would be 4 · 90.0 g = 360. g.
(b) The freezing point of pure water is 0°C. Since the freezing point of the solution is −3.72°C, the freezing point has been lowered by the same amount: −3.72 C°.
(c) Since $k_f = -1.86$ C°/m for H_2O, the molality of the solution is 2.00.
(d) From part (a) we know that the solution contains 360. g of solute/1.00 kg of water. From part (c) we know that the solution contains 2.00 mol of solute/1.00 kg of water. Therefore, the mass of 2.00 mol of solute is 360. g. Since the molar mass is the number of grams in 1 mol, we conclude that the molar mass of the solute is 180. g/mol.

12.6A SUSPENSIONS AND COLLOIDAL DISPERSIONS

Other examples of homogeneous mixtures include *suspensions* and *colloidal dispersions*. Whether a homogeneous mixture is a solution, a suspension, or a colloidal dispersion depends on the *size* of the particles being dispersed.

The largest particles form **suspensions**. A suspension, such as sand in water, will settle in time.

The smallest particles form *solutions*. In a solution, the solvent particles do not settle; and if the solvent is clear, a light beam passing through the solution will not be visible.

A **colloidal dispersion** is formed when particles that are larger than those in solutions but smaller than those in suspensions are dispersed in a medium. Usually, the dispersed particles range in size from 1 to 1000 nanometers (nm) and consist of very large molecules or clusters of smaller atoms or ions. In reference to colloids, the terms *dispersed substance* and *dispersing medium* are analagous to the terms *solute* and *solvent*. A colloidal dispersion will not settle, and a light beam passing through it will be clearly visible, a phenomenon known as the **Tyndall effect**. An example of the Tyndall effect is the visibility of an automobile headlight beam as it passes through fog.

The accompanying diagram shows how a solution and a colloidal dispersion affect a beam of light.

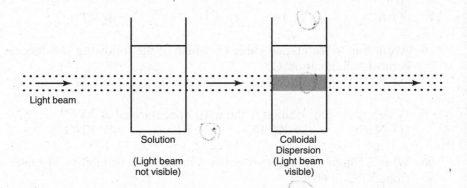

Light beam

Solution
(Light beam
not visible)

Colloidal
Dispersion
(Light beam
visible)

Many familiar mixtures of substances are actually colloidal dispersions. Some of these are listed in the accompanying table.

Colloid Type	Phase of Dispersed Substance	Phase of Dispersing Medium	Examples
Aerosol	Liquid	Gas	Fog, aerosol sprays
Aerosol	Solid	Gas	Smoke
Foam	Gas	Liquid	Whipped cream
Emulsion	Liquid	Liquid	Milk, mayonnaise
Sol	Solid	Liquid	Paint, gelatin
Solid foam	Gas	Solid	Marshmallow
Solid emulsion	Liquid	Solid	Butter, cheese

END-OF-CHAPTER QUESTIONS

Some questions have the symbol "§2" in front of the question number. This symbol means that the question is based on Section II material.

1. A solution containing 95 grams of KNO_3 dissolved in 100 grams of water at 55°C is classified as being
   (1) saturated
   (2) supersaturated
   (3) unsaturated
   (4) ultrasaturated

2. What is the maximum number of grams of NH_4Cl that will dissolve in 200 grams of water at 70°C?
   (1) 62
   (2) 70
   (3) 124
   (4) 226

3. According to Reference Table G in Appendix 1, approximately how many grams of $KClO_3$ are needed to saturate 100 grams of H_2O at 40°C?
   (1) 6
   (2) 14
   (3) 37
   (4) 47

4. According to Reference Table G, which of the following substances is most soluble at 60°C?
   (1) NH_4Cl
   (2) $NaNO_3$
   (3) NaCl
   (4) NH_3

5. Which saturated solution is the most concentrated at 20°C?
   (1) NaCl
   (2) $KClO_3$
   (3) KI
   (4) KNO_3

6. Which compound shows the *least* increase in solubility in water from 50°C to 60°C?
   (1) $KClO_3$
   (2) NaCl
   (3) KNO_3
   (4) KI

7. A solution in which an equilibrium exists between dissolved and undissolved solute must be
   (1) saturated
   (2) unsaturated
   (3) dilute
   (4) concentrated

8. According to Reference Table F in Appendix 1, which compound is most soluble in water?
   (1) $BaCO_3$
   (2) $BaSO_4$
   (3) $ZnCO_3$
   (4) $ZnSO_4$

9. A solution contains 70 grams of KNO_3 in 100 grams of water at 60°C. Approximately how many additional grams of KNO_3 are required to saturate this solution?
   (1) 36
   (2) 18
   (3) 54
   (4) 72

10. At 10.°C, 80 grams of a substance saturates 100. grams of water. The substance could be
    (1) NaCl   (2) NaNO$_3$ (3) NH$_4$Cl   (4) KClO$_3$

11. A solution is prepared by adding 20 grams of an ionic solid to 100 grams of water at 40°C. If the mixture contains undissolved solute, the solid could be
    (1) KNO$_3$ (2) NH$_4$Cl   (3) KClO$_3$ (4) NaNO$_3$

12. According to Reference Table F, which saturated solution is most concentrated?
    (1) AgC$_2$H$_3$O$_2$(aq)   (2) Ag$_2$CO$_3$(aq)
    (3) Ag$_2$S(aq)   (4) AgCl(aq)

§2 13. At which pressure would carbon dioxide gas be most soluble in 100 grams of water at a temperature of 25°C?
    (1) 1 atm   (2) 2 atm   (3) 3 atm   (4) 4 atm

14. As additional KNO$_3$(s) is added to a saturated solution of KNO$_3$ at constant temperature, the concentration of the solution
    (1) decreases   (2) increases   (3) remains the same

15. The maximum quantity of KCl that can dissolve in 150 grams of water at 40°C is closest to
    (1) 17 g   (2) 39 g   (3) 59 g   (4) 210 g

16. A crystal of solute is added to a solution, and the crystal remains undissolved. The solution is best classified as
    (1) dilute   (2) unsaturated
    (3) saturated   (4) supersaturated

§2 17. If 0.50 liter of a 12-molar solution is diluted to 1.0 liter, the molarity of the new solution is
    (1) 2.4   (2) 6.0   (3) 12   (4) 24

18. How many grams of KOH are needed to prepare 250. milliliters of a 2.00 M solution of KOH (molar mass = 56.0 g/mol)?
    (1) 1.00   (2) 2.00   (3) 28.0   (4) 112

19. How many grams of ammonium chloride (molar mass = 53.5 g/mol) are contained in 0.500 liter of a 2.00 M solution?
    (1) 10.0   (2) 26.3   (3) 53.5   (4) 107

20. What is the molarity of a solution of KNO_3 (molar mass = 101 g/mol) that contains 404 grams of KNO_3 in 2.00 liters of solution?
(1) 1.00 M   (2) 2.00 M   (3) 0.500 M   (4) 4.00 M

21. What is the molarity of a solution that contains 20. grams of $CaBr_2$ in 0.50 liter of solution?
(1) 0.50 M   (2) 0.20 M   (3) 5.0 M   (4) 10. M

22. What is the total number of grams of KCl (molar mass = 74.6 g/mol) in 1.00 liter of 0.200-molar solution?
(1) 7.46   (2) 14.9   (3) 22.4   (4) 29.8

23. What is the total number of grams of solute in 500. milliliters of 1 M CH_3COOH (molar mass = 60 g/mol)?
(1) 30.   (2) 60.   (3) 90.   (4) 120

24. Which contains the greatest number of moles of solute?
(1) 0.5 L of 0.5 M solution   (2) 0.5 L of 2 M solution
(3) 2 L of 0.5 M solution   (4) 2 L of 2 M solution

25. When a solution is diluted by the addition of solvent, which quantity does *not* change?
(1) the molarity of the solution
(2) the number of moles of solvent
(3) the mass percent of the solution
(4) the number of moles of the solute

§2 26. What is the molality of a solution containing 8.0 moles of solute dissolved in 2.4 kilograms of solvent?
(1) 0.33 m   (2) 3.3 m   (3) 1.9 m   (4) 19 m

§2 27. How many moles of solute are dissolved in 5.0 kilograms of solvent if the concentration is 2.0 molal?
(1) 0.40   (2) 2.5   (3) 10.   (4) 25

§2 28. What is the mole fraction of ethanol in a solution containing 3.00 moles of ethanol and 5.00 moles of water?
(1) 0.375   (2) 0.600   (3) 0.625   (4) 1.67

29. What is the molar concentration of a 2.00-liter solution containing 200. grams of glucose ($C_6H_{12}O_6$)?
(1) 0.555 M   (2) 1.80 M   (3) 100. M   (4) 180. M

30. 1000 grams of water contains 0.1 gram of a dissolved substance. The concentration of this solution is
    (1) 1 ppm (2) 10 ppm (3) 100 ppm (4) 1000 ppm

31. When ethylene glycol (an antifreeze) is added to water, the boiling point of the water
    (1) decreases, and the freezing point decreases
    (2) decreases, and the freezing point increases
    (3) increases, and the freezing point decreases
    (4) increases, and the freezing point increases

32. Which solution will freeze at the *lowest* temperature?
    (1) 1 mol of sugar in 500 g of water
    (2) 1 mol of sugar in 1000 g of water
    (3) 2 mol of sugar in 500 g of water
    (4) 2 mol of sugar in 1000 g of water

33. A 1-kilogram sample of water will have the highest freezing point when it contains
    (1) 1×10^{17} dissolved particles
    (2) 1×10^{19} dissolved particles
    (3) 1×10^{21} dissolved particles
    (4) 1×10^{23} dissolved particles

34. At standard pressure, an aqueous solution of sugar has a boiling point
    (1) greater than 100°C and a freezing point greater than 0°C
    (2) greater than 100°C and a freezing point less than 0°C
    (3) less than 100°C and a freezing point greater than 0°C
    (4) less than 100°C and a freezing point less than 0°C

35. Which solution containing 1 mole of solute dissolved in 1000 grams of water has the *lowest* freezing point?
    (1) KOH(aq)   (2) $C_6H_{12}O_6$(aq)
    (3) C_2H_5OH(aq)   (4) $C_{12}H_{22}O_{11}$(aq)

36. At room temperature, the solubility of which solute in water would be affected *most* by a change in pressure?
    (1) methanol   (2) sugar
    (3) carbon dioxide   (4) sodium nitrate

37. A solution that is at equilibrium must be
(1) concentrated
(2) saturated
(3) dilute
(4) unsaturated

38. What is the molarity of a solution containing 20 grams of NaOH in 500 milliliters of solution?
(1) 1 M
(2) 2 M
(3) 0.04 M
(4) 0.5 M

39. Based on Reference Table F, in Appendix 1, which of these salts is the best electrolyte?
(1) sodium nitrate
(2) magnesium carbonate
(3) silver chloride
(4) barium sulfate

40. At standard pressure when NaCl is added to water, the solution will have a
(1) higher freezing point and a lower boiling point than water
(2) higher freezing point and a higher boiling point than water
(3) lower freezing point and a higher boiling point than water
(4) lower freezing point and a lower boiling point than water

41. Which expression could represent the concentration of a solution?
(1) 3.5 g   (2) 3.5 M   (3) 3.5 mL   (4) 3.5 mol

42. Which sample, when dissolved in 1.0 liter of water, produces a solution with the highest boiling point?
(1) 0.1 mol KI
(2) 0.2 mol KI
(3) 0.1 mol $MgCl_2$
(4) 0.2 mol $MgCl_2$

43. An aqueous solution has a mass of 400 grams containing 8.5×10^{-3} gram of calcium ions. The concentration of calcium ions in this solution is
(1) 4.3 ppm   (2) 8.5 ppm   (3) 21.25 ppm   (4) 34 ppm

Constructed-Response Questions

1. A solution contains 0.161 mole of solute dissolved in 310. milliliters of solution. What is the molarity of the solution?

2. How many moles of solute are present in 35.1 milliliters of a 0.879 M solution?

3. How many grams of $Ba(OH)_2$ are needed to make 300. milliliters of a 0.692 M solution?

§2 **4.** A 250.-milliliter solution of 2.11 M H_2SO_4 is diluted to 725 milliliters. Calculate the molarity of the diluted solution.

§2 **5.** How would you prepare 325 milliliters of a 0.333 M H_2SO_4 solution, using a stock solution of 6.00 M H_2SO_4?

§2 **6.** What mass of water must be added to 3.75 moles of solute to prepare a 1.66 molal solution?

§2 **7.** How many grams of water must be added to 11.5 grams of C_2H_5OH to prepare a 0.336 molal solution?

§2 **8.** A solution is prepared by mixing 115 grams of H_2O, 31.0 grams of C_2H_5OH, and 11.2 grams of $C_6H_{12}O_6$. Calculate the mole fraction (X) of C_2H_5OH in the solution.

§2 **9.** Given this reaction:

$$2NaOH\ (aq) + H_2SO_4\ (aq) \rightarrow Na_2SO_4\ (aq) + H_2O\ (\ell)$$

What volume of a 0.750 M H_2SO_4 solution is needed to react completely with 15.0 grams of NaOH?

10. The grid below is based on the molarity concept developed in this chapter and is partially filled in. Complete the grid by using the data given in the grid.

Solute	Molar Mass of Solute ($\mathcal{M}$) (g/mol)	Moles of Solute Present (n) (mol)	Mass of Solute Present (m) (g)	Volume of Solution (V)	Molarity (M) (M)
NH_3		2.00		250. mL	
HNO_3			63.0		0.500
H_2SO_4			49.0	5.00 dm^3	
$MgCl_2$		1.00			2.00
$AlPO_4$			244	0.500 L	
$C_6H_{12}O_6$			45.0		2.00
$HC_2H_3O_2$		2.00		4000. cm^3	
KI			498		1.00
$Na_2Cr_2O_7$		3.00			9.00
$B(OH)_3$		4.00		12.0 L	
***			294	2.00 dm^3	1.50

***One of the ten compounds given above.

§2 **11.** A solution is prepared by dissolving 3.86 grams of C_2H_5OH in 325 grams of nitrobenzene (freezing point = 5.76°C; k_f = 6.87 C°/m). Calculate the freezing point of this solution.

§2 **12.** A solvent has a normal boiling point of 90.00°C. When 5.00 grams of C_2H_5OH is dissolved in 455 grams of the solvent, the boiling point of the solution is 93.33°C. Calculate the boiling point elevation constant (k_b) of the solvent.

Base your answers to questions 13–16 on the information below and on your knowledge of chemistry.

In a laboratory investigation, ammonium chloride was dissolved in water. Laboratory procedures and corresponding observations made by a student during the investigation are shown in the table below.

Dissolving $NH_4Cl(s)$ in $H_2O(\ell)$

Procedure	Observation
1. Measure the temperature of 10.0 milliliters (10.0 grams) of $H_2O(\ell)$ in a tube.	1. The temperature of the $H_2O(\ell)$ was 25.8°C
2. Add 5.0 grams of the solute, $NH_4Cl(s)$, to the $H_2O(\ell)$.	2. The $NH_4Cl(s)$ settled to the bottom of the test tube.
3. Stir the contents of the test tube for 4 minutes.	3. A small amount of $NH_4Cl(s)$ remained at the bottom of the test tube.
4. Measure the temperature of the $NH_4Cl(aq)$ solution.	4. The temperature of the solution was 11.2°C.

13. Identify *two* types of bonds in the solute.

14. State evidence from the investigation that indicates the $NH_4Cl(aq)$ solution is saturated.

15. State evidence from the investigation that indicates the process of dissolving $NH_4Cl(s)$ in water is endothermic.

16. State the observation that would be made if procedure 3 is repeated with the original temperature of the $H_2O(\ell)$ at 98°C.

The answers to these questions are found in Appendix 3.

KINETICS AND EQUILIBRIUM

KEY IDEAS

This chapter focuses on the factors that affect the speed of a reaction, the reversibility of reactions, and the way in which reversibility applies to various chemical processes.

KEY OBJECTIVES
At the conclusion of this chapter you will be able to:
• Define the term *chemical kinetics*.
• List and describe the factors that affect the rate of a reaction.
• Define the term *dynamic equilibrium*.
• Provide examples of phase and solution equilibria.
• Define the term *chemical equilibrium*.
• State Le Châtelier's principle.
• Use Le Châtelier's principle to determine the effects of changes in concentration, the addition of a common ion, volume and temperature changes, the addition of an inert gas, and the presence of a catalyst on systems in equilibrium.
• Write equilibrium-constant expressions for homogeneous and heterogeneous chemical reactions.
• Describe the effect of temperature on the equilibrium constant.
• Solve simple problems involving the equilibrium constant.

SECTION I—BASIC (REGENTS-LEVEL) MATERIAL

NYS REGENTS CONCEPTS AND SKILLS

Note: By the time you have finished Section I, you should have mastered the concepts and skills listed below. The Regents chemistry examination will test your knowledge of these items and your ability to apply them.

Concepts are the *basic ideas* that form the body of the Regents chemistry course (what you need to know!).

Skills are the *activities* that demonstrate your mastery of these concepts (how you show that you know them!).

Following each concept or skill is a page reference (given in parentheses) to this chapter.

13.1 Concept:
The rate of a chemical reaction depends on several factors: temperature, concentration, nature of reactants, surface area, and the presence of a catalyst. (Pages 371–373)

Skill:
Read and interpret potential energy diagrams with regard to:
(a) potential energies of reactants and products (Pages 372–373);
(b) activation energy (without and with a catalyst) (Page 372);
(c) heat of reaction (ΔH) (Page 372).

13.2 Concept
A catalyst provides an alternative reaction pathway that has a lower activation energy than an uncatalyzed reaction. (Pages 372–373)

13.3 Concept:
Collision theory states that a reaction is most likely to occur if reactant particles collide with the proper energy and orientation. (Pages 371–372)

Skill:
Use collision theory to explain how each of the following factors affects the rate of a reaction:
(a) temperature (Page 372);
(b) surface area (Page 372);
(c) concentration (Page 373).

13.4 Concept:
Some chemical and physical changes can reach equilibrium. (Pages 373–374)

Skill:
Identify examples of the following phenomena:
(a) solution equilibrium (a saturated solution) (Page 375);
(b) phase equilibrium (Page 374).

13.5 Concepts:
• At equilibrium the rate of the forward reaction equals the rate of the reverse reaction. (Pages 373–374)
• The measurable quantities of reactants and products remain constant at equilibrium. (Pages 373–374)

Skill:
Describe the concentrations of particles and the rates of opposing reactions in an equilibrium system. (Pages 373–374)

13.6   Concept:
Le Châtelier's principle can be used to predict the effects of stress (change in pressure, volume, concentration, and temperature) on a system at equilibrium. (Pages 375–375)

Skill:
Use Le Châtelier's principle to describe the effects of stress on a system in equilibrium. (Pages 375–378)

13.1   CHEMICAL KINETICS

Chemical kinetics is the branch of chemistry that concerns itself with the speed of chemical reactions (known as *reaction rates*) and the way in which these reactions occur (known as *reaction mechanisms*).

Let us consider this reaction:

$$4NH_3(g) + 5O_2(g) \rightarrow 4NO(g) + 6H_2O(g)$$

The study of chemical kinetics has led chemists to the conclusion that a chemical reaction such as this one occurs, not in one step, but in a series of simpler steps involving the collision of one or two particles at a time. These intermediate steps (called *elementary steps*) are usually not observed; nevertheless, they contribute to the overall reaction. Suppose that the reaction $A \rightarrow B$ consists of the three elementary steps shown below.

$$(1) \ A \xrightarrow{\text{fast}} I_1$$

$$(2) \ I_1 \xrightarrow{\text{slow}} I_2$$

$$(3) \ I_2 \xrightarrow{\text{fast}} B$$

....
(rxn) $A \rightarrow B$ (rate = ???)

Here I_1 and I_2 are intermediates that are not observed when reactant A is converted into product B, and the relative rate of each step is given above the arrow. Of the three elementary steps, (2) has the *slowest* rate and it determines the rate of the *overall* reaction ($A \rightarrow B$). In technical terms, (2) is called a *rate-limiting step*; it serves as a "bottleneck" for the other steps, much as a narrow strip of road serves as a bottleneck for traffic flow on a major highway.

Factors That Affect the Rate of a Reaction

Chemical reactions occur because the reacting particles collide with each other. The speed of a reaction depends on the number of collisions and the fraction of those collisions that are *effective*. For a collision to be effective, the particles must have enough energy and they must be *oriented* properly. Any factor that increases the number of effective collisions will serve to increase the rate of the reaction.

Nature of the Reactants

Since bonds are broken and formed in chemical reactions, the bond types of the reacting substances are an important factor. In solution, ionic compounds tend to react more quickly than covalent compounds because fewer bonds need to be rearranged. For example, the reaction between $Fe(s)$ and $O_2(g)$ (the rusting of iron) is rather slow at room temperature, whereas the ionic reaction between $AgNO_3(aq)$ and $NaCl(aq)$ is extremely rapid.

Concentrations of the Reactants

As the concentration of a reactant increases, the number of collisions increases and the rate of the reaction generally increases. For example, in the laboratory production of hydrogen gas, $Zn(s)$ reacts more rapidly with 1 M $HCl(aq)$ than with 0.5 M $HCl(aq)$.

In a system in which gases react, the pressure is directly related to the concentrations of the gases. Therefore, an increase in pressure leads to an increase in reaction rate.

Temperature

An increase in temperature increases the average kinetic energy of the particles, thereby ensuring that the number of effective collisions will increase. Therefore, an increase in temperature always increases the rate of a reaction.

Surface Area

As the surface area of the reacting particles is increased, the number of particles that collide is also increased. The result is an increase in the reaction rate. For example, when the size of $Zn(s)$ particles is reduced, the particles react more rapidly with $HCl(aq)$ to produce hydrogen gas.

Catalysts

If we refer to the potential energy diagrams on pages 115 and 116, we observe that the activation energy is like a hill that must be climbed before the reaction will conclude.

Catalysts *lower the activation energy* of a reaction by providing an alternative reaction pathway and, therefore, increase the speed of the reaction. In effect, the "hill" becomes easier to climb.

The effect of a catalyst is illustrated in the accompanying potential energy diagram. Note that *only* the activation energy is changed by the catalyst. The overall reaction and the value of ΔH, the heat of reaction, remain unchanged by its presence.

Reaction Profile With and Without Catalyst

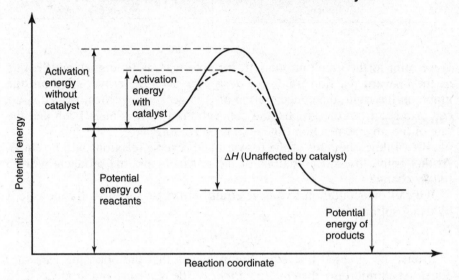

13.2 REVERSIBLE REACTIONS AND DYNAMIC EQUILIBRIUM

When we think of the term *equilibrium*, the first word that usually comes to mind is *balance*. However, balance may be achieved in a variety of ways. For example, a building remains standing, rather than falling down, because all of its stresses and strains are (hopefully) balanced. This is an example of *static equilibrium*.

In chemistry, however, *equilibrium* has a somewhat different meaning. Chemical reactions are usually *reversible*, meaning that they can proceed in *both* directions. We write a reversible reaction with a "split arrow" to indicate this reversibility:

$$A(g) \rightleftharpoons B(g)$$

The equation indicates that A can form B and that B can form A. Let us see how equilibrium can be achieved in this system by referring to the accompanying diagrams of reaction rate and concentration as functions of time.

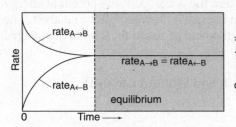 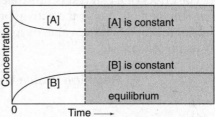

If we refer to the graph on the left, we note that, as time goes on, the rate of the forward reaction ($rate_{A \to B}$) *decreases* because, as we see from the graph on the right, the concentration of *A* is *decreasing*. Meanwhile, as we can also see from the graphs, the concentration of *B* is *increasing* and the rate of the reverse reaction ($rate_{A \leftarrow B}$) is *increasing*.

After a time, the rates of the forward and reverse reactions will be equal. At this point, the concentrations of the reactants and the products will no longer change.

We call this condition **dynamic equilibrium**, and we usually indicate it by using split arrows:

$$A(g) \rightleftharpoons B(g)$$

It must be stressed that equilibrium means that the opposing *rates* are equal. At equilibrium, the *concentrations* of the reactants and products need *not* be equal.

In Chapter 5 we studied spontaneous reactions. When a reaction is spontaneous, the change of reactants to products is observable. At equilibrium, however, we observe *neither* the forward nor the reverse reaction, and therefore neither is spontaneous.

13.3 PHASE EQUILIBRIUM

We know that phase changes are reversible, and in closed containers two or more phases may exist in dynamic equilibrium at a given temperature and pressure. For example, if water is placed in a closed container at 25°C, it will reach dynamic equilibrium with its vapor. The pressure created by the gaseous water at equilibrium is the *vapor pressure of water at* 25°C. Similarly, boiling-condensation, melting-freezing, and sublimation-deposition all represent phase equilibria.

13.4 SOLUTION EQUILIBRIUM

Solids Dissolved in Liquids

When a solid dissolves in a liquid and produces a *saturated* solution, the rate of dissolving equals the rate at which the solid leaves the solution (crystallizes). This situation represents dynamic equilibrium in the solution.

For glucose ($C_6H_{12}O_6$) and water, the equilibrium equation is

$$C_6H_{12}O_6(s) \rightleftharpoons C_6H_{12}O_6(aq).$$

For an ionic compound such as NaCl, the equilibrium equation is

$$NaCl(s) \rightleftharpoons Na^+(aq) + Cl^-(aq).$$

Solution equilibria are affected by changes in temperature.

Gases Dissolved in Liquids

In a closed system, such as a bottle of soda, equilibrium may exist between the dissolved gas (CO_2) and the gas present on top of the liquid.

For such a system the equilibrium equation is $CO_2(g) \rightleftharpoons CO_2(aq)$.

Gas-liquid equilibria are affected by changes in temperature and pressure.

13.5 CHEMICAL EQUILIBRIUM

We can now define the term *chemical equilibrium*. **Chemical equilibrium** exists when the forward and reverse rates of a chemical reaction are equal, and, as a result, the concentrations of the reactants and the products remain *constant*. For example, the equation

$$N_2(g) + 3H_2(g) \rightleftharpoons 2NH_3(g)$$

represents a chemical system in a state of equilibrium

Le Châtelier's Principle

It is possible to disturb any system at equilibrium by applying a *stress* to it. The important question is: How will the system respond to the stress?

This question was answered by the French chemist Henri Le Châtelier in 1888. **Le Châtelier's principle** states:

"When a system at equilibrium is subjected to a stress, the system shifts in order to relieve the effects of the stress and restore the equilibrium conditions as closely as possible."

This principle goes beyond chemistry. For example, our homes are usually programmed to remain at a constant temperature (the equilibrium condition). If the temperature drops suddenly because a window is opened (the stress), the heating system responds (the shift) by putting heat energy back into the system in order to try to restore the original temperature.

We note that, when a system returns to equilibrium, there is a *new equilibrium point* because the original conditions have been changed.

Let us examine how a chemical system at equilibrium reacts to a number of stresses placed upon it. The equilibrium system we will use involves the production of ammonia by the *Haber process:*

$$N_2(g) + 3H_2(g) \rightleftharpoons 2NH_3(g)$$

At equilibrium, the rate of ammonia production equals the rate at which the ammonia is decomposed into its elements, and the concentrations of nitrogen, hydrogen, and ammonia are constant.

First, however, we must define what we mean by a *chemical shift*. Any chemical reaction can proceed in only two directions: forward and reverse. When a reaction responds to a stress, either the forward or the reverse reaction will take over temporarily, and we will observe an increase in the production of products or of reactants. A forward shift is called a shift to the *right*; a reverse shift, a shift to the *left*.

Effect of Concentration

Concentrations are usually expressed in moles per liter. We express the molar concentration of a substance by placing the symbol for the substance in square brackets: for example, $[NH_3]$ means the molar concentration of NH_3.

When the concentration of a reactant or a product is changed, the system will shift in order to restore the original concentration as closely as possible.

Suppose $[N_2]$ is increased in the example of the Haber process given above. Since N_2 has been added to the system, the system will shift to the *right* in order to remove *some* of the additional N_2. As a result, $[NH_3]$ will increase, and $[H_2]$ will decrease because it is needed to react with the N_2. Eventually, a *new* equilibrium point will be established.

Suppose $[NH_3]$ is increased. The system will shift to the *left* in order to remove some of the additional NH_3. As a result, $[N_2]$ and $[H_2]$ will increase.

Suppose $[NH_3]$ is decreased. The system will shift to the *right* in order to replace some of the NH_3. As a result, both $[N_2]$ and $[H_2]$ will decrease.

A reaction in which one of the products is *continuously* removed will *never* achieve equilibrium because the reactants will be used up in the effort to replace the disappearing product. A familiar example is the laboratory preparation of hydrogen gas by reacting zinc with hydrochloric acid:

$$Zn(s) + HCl(aq) \rightarrow H_2(g) + ZnCl_2(aq)$$

Since $H_2(g)$ continuously escapes from the reaction, equilibrium is never attained. This reaction is said to go to *completion*, and we write the equation with the familiar single arrow.

Effect of a Change in Volume

The effect of a change in volume is observed in many systems containing gases. In general, *a change in volume is accomplished by changing the pressure on a gaseous system*: an increase in pressure leads to a reduction in volume, and vice versa.

Reducing the volume (by increasing the pressure) brings all of the particles into closer contact. This stress is relieved by a shift that produces a *smaller* number of particles. In contrast, an increase in volume (i.e., a decrease in pressure) leads to a shift that increases the number of particles present.

Let us again consider the Haber process:

$$N_2(g) + 3H_2(g) \rightleftharpoons 2NH_3(g)$$

The left side of the equation contains 4 moles of particles (1 mole of N_2 and 3 moles of H_2); the right side, 2 moles of particles (NH_3). A shift to the right reduces the number of particles, while a shift to the left increases the number.

Suppose the volume of this system were decreased causing the pressure to increase. The system would shift to the *right* in order to decrease the number of particles present. In this case, $[NH_3]$ would increase; $[N_2]$ and $[H_2]$ would decrease.

Let us now consider this system:

$$H_2(g) + Cl_2(g) \rightleftharpoons 2HCl(g)$$

Changing the volume would *not* cause this system to shift because each side contains the same number of particles (2 moles). A shift in either direction would not relieve the stress.

Changing the Pressure by Adding an Inert Gas

The pressure of an equilibrium system can also be changed by the addition of an inert gas. Suppose helium gas was added to the N_2-H_2-NH_3 system we have been considering. Since helium does *not* react with any of these substances, it would have *no* effect on the equilibrium of the system.

Effect of a Change in Temperature

When we raise the temperature of a system, we do so by adding heat to the system; when we lower the temperature, we remove heat from the system.

Suppose we lower the temperature of an equilibrium system. What is the effect on the system? Lowering the temperature will decrease the rates of both the forward and reverse reactions.

Le Châtelier's principle predicts, however, that the system will shift in order to relieve the stress; that is, the system will try to *replace* some of the lost heat. Since the *exothermic* reaction releases heat, it will be favored when the temperature is lowered. Why is this true? The rate of the exothermic reaction is decreased *less* than the rate of the endothermic reaction.

If, on the other hand, we raise the temperature of an equilibrium system, the rates of both reactions will increase, but the system will shift in order to *absorb* some of the added heat. Since the *endothermic* reaction absorbs heat, it is favored when the temperature is raised. In this case the rate of the endothermic reaction is increased *more* than the rate of the exothermic reaction.

Once again, we consider the Haber process (now rewritten to include the value of ΔH as part of the reaction):

$$N_2(g) + 3H_2(g) \rightleftharpoons 2NH_3(g) + 91.8 \text{ kJ}$$

We see that the forward reaction is *exothermic* and, therefore, the reverse reaction must be *endothermic*.

If the temperature of this system is raised, the reaction will shift toward the left—the endothermic reaction—and $[NH_3]$ will decrease, while $[N_2]$ and $[H_2]$ will increase.

If the temperature of this system is lowered, the reaction will shift toward the right—the exothermic reaction—and $[NH_3]$ will increase, while $[N_2]$ and $[H_2]$ will decrease.

Effect of a Catalyst

For a system in equilibrium, the presence of a catalyst increases the forward and reverse rates *equally*. Therefore, *no* shift in the system will result. If the system is *not* at equilibrium, however, a catalyst will shorten the time needed to reach equilibrium.

Summary of Effects

Direction of Stress	Result of Shift
Concentration decreases.	Some substance is replaced.
Concentration increases.	Some substance is removed.
Volume decreases (pressure increases).	Fewer particles are produced.
Volume increases (pressure decreases).	More particles are produced.
Temperature decreases.	Exothermic reaction is favored.
Temperature increases.	Endothermic reaction is favored.
An inert gas is added.	No shift occurs.
A catalyst is added.	No shift occurs.

TRY IT YOURSELF

For this equilibrium system:

$$904 \text{ kJ} + 6H_2O(g) + 4NO(g) \rightleftharpoons 4NH_3(g) + 5O_2(g)$$

what is the effect of each of the following changes?

(a) decreasing the temperature
(b) decreasing the volume
(c) increasing [NO]
(d) decreasing [H_2O]

ANSWERS

(a) a shift toward the *left*
(b) a shift toward the *right*
(c) a shift toward the *right*
(d) a shift toward the *left*

SECTION II—ADDITIONAL MATERIAL

13.1A THE COMMON-ION EFFECT

In a saturated solution of an ionic compound such as AgCl, the Ag^+(aq) and Cl^-(aq) ions are in equilibrium with the solid AgCl:

$$AgCl(s) \rightleftharpoons Ag^+(aq) + Cl^-(aq)$$

If additional Cl^-(aq) were added to this solution (perhaps by adding some NaCl), what would occur? Le Châtelier's principle predicts that the system would shift to the *left* in order to remove some of the added Cl^-(aq) ion. As a result more AgCl(s) would be produced. This shift is known as the *common-ion effect*. In this example, the addition of the common ion Cl^-(aq) *decreased* the solubility of AgCl in water because fewer Ag^+ ions were present in the saturated solution.

13.2A HETEROGENEOUS EQUILIBRIUM

An equilibrium system is *homogeneous* if all reactants and products are in a single phase (as gases or substances in aqueous solution). In a *heterogeneous* system pure solids or liquids are present; a saturated solution is an example of a heterogeneous equilibrium system. Another example of a heterogeneous equilibrium system is given by the equation

$$C(s) + O_2(g) \rightleftharpoons CO_2(g)$$

and its accompanying diagram, in which solid carbon is in equilibrium with gaseous oxygen and carbon dioxide in a sealed container.

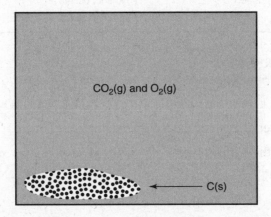

Having examined the conditions that affect systems at equilibrium, we now ask: If more *solid* carbon were added to the container shown in the diagram, how would this system be affected?

The answer is: *Not at all!* Equilibrium is affected by changes in concentration. Adding more solid does *not* change the concentration of the carbon because the solid does not occupy the entire container, as the gases do. The concentration of the solid, which is really its density, remains constant. This would be true also if a pure liquid were part of an equilibrium system.

13.3A THE EQUILIBRIUM CONSTANT (K_{eq})

When a system reaches equilibrium, it is useful to know the relative amounts of products and reactants present. Certain reactions produce hardly any products, while other reactions leave hardly any reactants remaining. The *equilibrium constant* (represented as K_{eq} or K) provides us with this information.

Defining K_{eq} for a Simple System

For a simple homogeneous system at equilibrium: $A(g) \rightleftharpoons B(g)$, the equilibrium constant expression is defined as follows:

$$K_{eq} = \frac{[B]}{[A]}$$

In defining K_{eq}, we divide the equilibrium concentration of the product, B, by the equilibrium concentration of the reactant, A. If there is more product than reactant at equilibrium, K_{eq} is greater than 1; if there is less product than reactant at equilibrium, K_{eq} is less than 1; if the amounts of product and reactant at equilibrium are approximately equal, K_{eq} is approximately 1.

PROBLEM
For the system discussed above, $[A] = 0.2$ mole per liter and $[B] = 0.02$ mole per liter at equilibrium. Calculate K_{eq} for these conditions.

SOLUTION

$$K_{eq} = \frac{[B]}{[A]} = \frac{0.02 \text{ mol/L}}{0.2 \text{ mol/L}} = 0.1$$

The small value of the equilibrium constant (0.1) confirms that, at equilibrium, there is less product than reactant.

TRY IT YOURSELF
In the system given above, if the equilibrium concentration of B is 0.15 mole per liter, what is the equilibrium concentration of A?

ANSWER
1.5 mol/L

Defining K_{eq} for a More Complicated System

For more complicated reactions, the equilibrium constant expression is itself more complicated. For the equilibrium system:

$$aM(g) + bN(g) \rightleftharpoons cP(g) + dQ(g)$$

the equilibrium constant expression is defined as

$$K_{eq} = \frac{[P]^c [Q]^d}{[M]^a [N]^b}$$

We note that, in general, we still divide the products by the reactants, but in a more complicated way: we raise each concentration to the *power of its coefficient*. A small equilibrium constant indicates more reactants at equilibrium, and a large equilibrium constant indicates more products at equilibrium.

PROBLEM
(a) Write the K_{eq} expression for this reaction: $N_2(g) + 3H_2(g) \rightleftharpoons 2NH_3(g)$.
(b) Evaluate K_{eq} if $[NH_3] = 0.2$ mole per liter, $[N_2] = 0.04$ mole per liter, $[H_2] = 0.01$ mole per liter.

SOLUTIONS

(a) $K_{eq} = \dfrac{[NH_3]^2}{[N_2][H_2]^3}$ (b) $K_{eq} = \dfrac{(0.2 \text{ mol/L})^2}{(0.04 \text{ mol/L})(0.01 \text{ mol/L})^3} = 1 \times 10^6$

The large value of the equilibrium constant confirms that, at equilibrium, there are more products than reactants. (In more complicated expressions, the units may *not* cancel. In practice, we *ignore* any unit that may be associated with an equilibrium constant.)

PROBLEM
Write the equilibrium expression for this *heterogeneous* system:

$$S(s) + O_2(g) \rightleftharpoons SO_2(g)$$

SOLUTION
The equilibrium constant expression is

$$K_{eq} = \frac{[SO_2]}{[S][O_2]}$$

Since sulfur is present as a solid, its concentration is constant and *is included* in the constant part of the equilibrium expression:

$$K_{eq} \cdot [S] = K = \frac{[SO_2]}{[O_2]}$$

When we write an equilibrium constant expression for a heterogeneous system, *we omit the concentration expressions for solids and liquids*.

TRY IT YOURSELF
Write the equilibrium expression for the system $C(s) + \dfrac{1}{2} O_2(g) \rightleftharpoons CO(g)$.

ANSWER

$$K = \frac{[CO]}{[O_2]^{\frac{1}{2}}}$$

At *constant temperature*, the equilibrium constant is truly *constant*. If we were to disturb a system at equilibrium, the point of equilibrium would shift to relieve the stress. When a new equilibrium was established, however, the value of K_{eq} would not have changed. Only a change in temperature can change K_{eq}.

The Solubility-Product Constant (K_{sp})

The *solubility-product constant* (K_{sp}) is a special type of equilibrium constant that measures the concentrations of ionic compounds in water. A compound with a small K_{sp} value is only slightly soluble in water. For example, $CaSO_4$ ($K_{sp} = 9.1 \times 10^{-6}$) is only very slightly soluble in water, and $BaSO_4$ ($K_{sp} = 9.1 \times 10^{-10}$) is even *less* soluble! In fact, the solubility of $BaSO_4$ is so low that we would classify it as *nearly insoluble* in water.

The K_{sp} expression may be developed by considering the solution equilibrium of an ionic compound AB:

$$AB(s) \rightleftharpoons A^+(aq) + B^-(aq)$$

Since AB is present as a solid, it is absorbed into the equilibrium expression:

$$K_{eq} \cdot [AB] = K_{sp} = [A^+] \cdot [B^-]$$

This expression indicates the reason for calling K_{sp} a solubility *product*; it is equal to the product of the ion concentrations in water.

PROBLEM
(a) Write the solubility equation and K_{sp} expression for $BaSO_4$.
(b) Write the solubility equation and K_{sp} expression for PbI_2.

SOLUTIONS
(a) $BaSO_4(s) \rightleftharpoons Ba^{2+}(aq) + SO_4^{2-}(aq)$; $K_{sp} = [Ba^{2+}][SO_4^{2-}]$
(b) $PbI_2(s) \rightleftharpoons Pb^{2+}(aq) + 2I^-(aq)$; $K_{sp} = [Pb^{2+}][I^-]^2$

In Appendix 2 of this book, Reference Table Y-1 lists the solubility-product constants for a number of compounds at 298.15 K.

13.4A PROBLEMS INVOLVING THE EQUILIBRIUM CONSTANT

Up to this point, the only problems we have solved in this chapter involved calculating the value of an equilibrium constant, given the equilibrium concentrations of the reactants and products. It is possible to approach problem solving from the other direction, that is, to use the equilibrium constant to calculate equilibrium concentrations. This method can become quite a complicated process that is far beyond the scope of this book. We will provide two relatively simple examples, however, to show how such problems are approached.

PROBLEM

In an experiment, hydrogen gas (H_2) and iodine gas (I_2) were placed in a container, which was then sealed and allowed to come to equilibrium with hydrogen iodide gas (HI) according to this equation:

$$H_2(g) + I_2(g) \rightleftharpoons 2HI(g)$$

At the start of the experiment, concentrations were $[H_2] = 1.00 \times 10^{-3}$ mole per liter, $[I_2] = 2.00 \times 10^{-3}$ mole per liter, $[HI] = 0.00$ mole per liter.

When equilibrium had been attained, the concentration of HI gas was 1.86×10^{-3} mole per liter.

(a) Calculate the *equilibrium concentrations* of the hydrogen and iodine gases in moles per liter.
(b) Calculate the *equilibrium constant* (K_{eq}) for this system.

SOLUTIONS

(a) We begin by constructing a table, in which all of the given information is summarized:

Reaction: $H_2(g) + I_2(g) \rightleftharpoons 2HI(g)$			
Substance	$[H_2]$ / (mol/L)	$[I_2]$ / (mol/L)	$[HI]$ / (mol/L)
Initial concentration	1.00×10^{-3}	2.00×10^{-3}	0.00
Change in concentration			
Equilibrium concentration			1.86×10^{-3}

Next, we inspect the *coefficients* of the equation. They tell us that, for every 1 mol of $H_2(g)$ that reacts, 1 mol of $I_2(g)$ will react, and 2 mol of HI(g) will be formed. *In other words, the concentrations of $H_2(g)$ and $I_2(g)$ that react must be exactly one-half of the concentration of the HI(g) that is formed.* Since *all* of the HI(g) present at equilibrium (1.86×10^{-3} mol/L) was formed in the reaction (why?), we can conclude that one-half of this concentration, 9.3×10^{-4} mol/L, of $H_2(g)$ and $I_2(g)$ reacted.

We return to our table and add this information in the appropriate spaces. We use a "$-$" sign to indicate *reacted* and a "$+$" sign to indicate *formed*.

Reaction: $H_2(g)$ + $I_2(g)$ $\rightleftharpoons$ 2HI(g)			
Substance	$[H_2]$ / (mol/L)	$[I_2]$ / (mol/L)	$[HI]$ / (mol/L)
Initial concentration	1.00×10^{-3}	2.00×10^{-3}	0.00
Change in concentration	-9.3×10^{-4}	-9.3×10^{-4}	$+1.86 \times 10^{-3}$
Equilibrium concentration			1.86×10^{-3}

Finally, we calculate each equilibrium concentration by *adding* the *change* in concentration to the *initial* concentration. Our table is now complete.

Reaction: $H_2(g)$ + $I_2(g)$ $\rightleftharpoons$ 2HI(g)			
Substance	$[H_2]$ / (mol/L)	$[I_2]$ / (mol/L)	$[HI]$ / (mol/L)
Initial concentration	1.00×10^{-3}	2.00×10^{-3}	0.00
Change in concentration	-9.3×10^{-4}	-9.3×10^{-4}	$+1.86 \times 10^{-3}$
Equilibrium concentration	7.00×10^{-5}	1.07×10^{-3}	1.86×10^{-3}

At equilibrium, the concentration of $H_2(g)$ is 7.00×10^{-5} mol/L, and the concentration of $I_2(g)$ is 1.07×10^{-3} mol/L.

(b) To calculate the equilibrium constant for this system, we write the K_{eq} expression, substitute the values that appear in the table, and perform the appropriate calculations:

$$K_{eq} = \frac{[HI]^2}{[H_2]\cdot[I_2]} = \frac{(1.86\times10^{-3})^2}{(7.00\times10^{-5})\cdot(1.07\times10^{-3})} = 46.2$$

In the next problem, we are given the equilibrium constant for the system and the initial concentrations, and are required to calculate the equilibrium concentrations of the reactant and product.

PROBLEM

In an experiment, dinitrogen tetroxide gas (N_2O_4) was placed in a 1-liter container, which was then sealed and allowed to come to equilibrium with nitrogen dioxide gas (NO_2) according to this equation:

$$N_2O_4(g) \rightleftharpoons 2NO_2(g) \qquad K_{eq} = 0.211$$

At the start of the experiment, the concentrations were $[N_2O_4] = 1.00$ mole per liter, $[NO_2] = 0.00$ mole per liter. Equilibrium is established over a period of time. What are the *equilibrium concentrations* of $NO_2(g)$ and $N_2O_4(g)$?

SOLUTION

We begin by inspecting the *coefficients* of the equation. They tell us that for every *1* mol of $N_2O_4(g)$ that react 2 mol of $NO_2(g)$ will be formed.

We know, however, that all 1.00 mol of the $N_2O_4(g)$ will *not* react because the system will reach equilibrium before that can happen. Then how many moles of $N_2O_4(g)$ *will* react? *We don't know*! Let us simply say that x mol of $N_2O_4(g)$ have reacted when equilibrium is attained. This means that $2x$ mol of $NO_2(g)$ will be formed. (Why?)

At this point we need to stress that all equilibrium calculations involve *concentrations* (in mol/L). In this example, however, the reaction container is a *1*-L vessel; therefore, the number of moles is equivalent to the number of moles per liter.

We now construct a table that summarizes our problem.

Reaction: $N_2O_4(g) \rightleftharpoons 2NO_2(g)$ $K_{eq} = 0.211$		
Substance	$[N_2O_4]/(mol/L)$	$[NO_2]/(mol/L)$
Initial concentration	1.00	0.00
Change in concentration	$-x$	$+2x$
Equilibrium concentration	$1.00 - x$	$0.00 + 2x = 2x$

We use a "−" sign to indicate *reacted* and a "+" sign to indicate *formed*.

The next step is to write the equilibrium expression for the system, substitute the equilibrium concentration values given in the table, and solve for x, $1.0 - x$, and $2x$:

$$K_{eq} = \frac{[NO_2]^2}{[N_2O_4]} = 0.211 = \frac{(2x)^2}{1.0 - x}$$

To arrive at the answer, we need to solve the quadratic equation that the equilibrium expression produces:

$$4x^2 + 0.211x - 0.211 = 0$$

When the equation is solved, we find that $x = 0.205$ mol/L.

Therefore, at equilibrium:

$$[NO_2] = 2x = 0.410 \text{ mol/L} \quad \text{and} \quad [N_2O_4] = 1 - x = 0.795 \text{ mol/L}$$

By the way, quadratic equations have *two* solutions. We *rejected* the second solution because it leads to a nonsensical answer. Can you explain why?

13.5A   APPLICATIONS OF CHEMICAL EQUILIBRIUM

The Haber Process

We have learned that ammonia is produced commercially by the reaction

$$N_2(g) + 3H_2(g) \rightleftharpoons 2NH_3(g) + heat$$

and that the formation of ammonia is favored by low temperatures and high pressures. The problem with using low temperatures is that the *rate* of ammonia formation is slow. If the temperature is raised, however, the reaction is shifted toward the *left* (the endothermic reaction), thereby decreasing the *yield* of ammonia.

In practice, an intermediate temperature (about 500°C) is used as a compromise. With a pressure of 200–1000 atmospheres and the addition of a catalyst, a reasonable rate of ammonia formation is maintained.

The Contact Process

Sulfuric acid (H_2SO_4) is one of the world's most important substances. It is prepared by a four-step method known as the *contact process*:

1. $S(s) + O_2(g) \rightleftharpoons SO_s(g) + heat energy$
Sulfur is burned to form sulfur dioxide gas.
2. $2SO_2(g) + O_2(g) \rightleftharpoons 2SO_3(g) + heat$
The sulfur dioxide is oxidized to sulfur trioxide gas.
3. $SO_3(g) + H_2SO_4(\ell) \rightarrow H_2S_2O_7(\ell)$
Sulfur trioxide dissolves readily in concentrated sufluric acid to form *fuming sulfuric acid*.
4. $H_2S_2O_7(\ell) + H_2O(\ell) \rightarrow 2H_2SO_4(\ell)$
The fuming sulfuric acid is dissolved in water to produce more sufluric acid.

The net reaction is as follows:

$$SO_3(g) + H_2O(\ell) \rightarrow H_sSO_4(\ell)$$

The sulfur trioxide must be dissolved in concentrated sulfuric acid (step 3) because it does not dissolve well in water.

The formation of SO_3 (step 2), like the formation of ammonia, is favored by high pressure and low temperature. As with the Haber process, a compromise temperature (450–575°C) is used, along with a catalyst, to maintain reasonable formation rates.

END-OF-CHAPTER QUESTIONS

Some questions have the symbol "§2" in front of the question number. This symbol means that the question is based on Section II material.

1. A 1-cubic-centimeter cube of sodium reacts more rapidly in water at 25°C than does a 1-cubic-centimeter cube of calcium at 25°C. This difference in rate of reaction is most closely associated with the different
   (1) surface areas of the metal cubes
   (2) natures of the metals
   (3) densities of the metals
   (4) concentrations of the metals

2. When a catalyst is added to a chemical reaction, there is a change in the
   (1) heat of reaction
   (2) rate of reaction
   (3) potential energy of the reactants
   (4) potential energy of the products

3. Raising the temperature speeds up the rate of a chemical reaction by increasing
   (1) the effectiveness of the collisions, only
   (2) the frequency of the collisions, only
   (3) both the effectiveness and the frequency of the collisions
   (4) neither the effectiveness nor the frequency of the collisions

4. Given the reaction

$$A + B \rightarrow C + D$$

   The reaction will most likely occur at the greatest rate if A and B represent
   (1) nonpolar molecular compounds in the solid phase
   (2) ionic compounds in the solid phase
   (3) solutions of nonpolar molecular compounds
   (4) solutions of ionic compounds

5. A student adds two 50-milligram pieces of Ca(s) to water. A reaction takes place according to the following equation:

$$Ca(s) + 2H_2O(\ell) \rightarrow Ca(OH)_2(aq) + H_2(g)$$

Which change could the student have made that would most likely have increased the rate of the reaction?
(1) used ten 10-mg pieces of Ca(s)
(2) used one 100-mg piece of Ca(s)
(3) decreased the amount of the water
(4) decreased the temperature of the water

6. In order for a chemical reaction to occur, there must always be
(1) an effective collision between reacting particles
(2) a bond that breaks in a reactant particle
(3) reacting particles with a high charge
(4) reacting particles with a high kinetic energy

7. In a gaseous system, temperature remaining constant, an increase in pressure by decreasing the volume will lead to
(1) an increase in the activation energy
(2) a decrease in the activation energy
(3) fewer particles being formed
(4) more particles being formed

8. At room temperature, which reaction would be expected to have the fastest reaction rate?
(1) $Pb^{2+}(aq) + S^{2-}(aq) \rightarrow PbS(s)$
(2) $2H_2(g) + O_2(g) \rightarrow 2H_2O(\ell)$
(3) $N_2(g) + 2O_2(g) \rightarrow 2NO_2(g)$
(4) $2KClO_3(s) \rightarrow 2KCl(s) + 3O_2(g)$

9. Given this reaction at equilibrium:

$$2CO(g) + O_2(g) \rightleftharpoons 2CO_2(g)$$

Which statement regarding this reaction is *always* true?
(1) The rates of the forward and reverse reactions are equal.
(2) The reaction occurs in an open system.
(3) The masses of the reactants and the products are equal.
(4) The concentrations of the reactants and the products are equal.

10. Which factors must be equal in a reversible chemical reaction at equilibrium?
(1) the concentrations of the reactants and products
(2) the potential energies of the reactants and products
(3) the activation energies of the forward and reverse reactions
(4) the rates of reaction of the forward and reverse reactions

11. The diagram that follows shows a bottle containing $NH_3(g)$ dissolved in water. How can the equilibrium

$$NH_3(g) \rightleftharpoons NH_3(aq)$$

be reached?

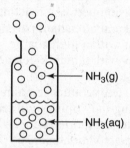

(1) Add more water.   (2) Add more $NH_3(g)$.
(3) Cool the contents.   (4) Stopper the bottle.

12. Given the reaction

$$2Na(s) + 2H_2O(\ell) \rightarrow 2Na^+(aq) + 2OH^-(aq) + H_2(g)$$

This reaction goes to completion because one of the products formed is
(1) an insoluble base   (2) a soluble base
(3) a precipitate   (4) a gas

13. In all reversible chemical reactions, equilibrium has been reached when
(1) the opposing reactions cease
(2) the molar concentrations of the reactants and products are equal
(3) any one substance leaves the field of reaction
(4) the rates of the opposing reactions are equal

14. Given the reaction

$$HC_2H_3O_2(aq) + H_2O(\ell) \rightleftharpoons H_3O^+(aq) + C_2H_3O_2^-(aq)$$

When the reaction reaches a state of equilibrium, the concentrations of the reactants
(1) are less than the concentration of the products
(2) are equal to the concentrations of the products
(3) begin decreasing
(4) become constant

15. A flask at 25°C is partially filled with water and stoppered. After a period of time the water level remains constant. Which relationship best explains this observation?
(1) The rate of condensation exceeds the rate of evaporation.
(2) The rates of condensation and evaporation are both zero.
(3) The rate of evaporation exceeds the rate of condensation.
(4) The rate of evaporation equals the rate of condensation.

16. When $AgNO_3(aq)$ is mixed with $NaCl(aq)$, a reaction occurs that tends to go to completion because
(1) a gas is formed   (2) water is formed
(3) a weak acid is formed   (4) a precipitate is formed

17. A solution in which equilibrium exists between undissolved and dissolved solute is always
(1) saturated   (2) unsaturated
(3) dilute   (4) concentrated

18. Given this system at equilibrium:

$$X_2(g) + 2Y_2(g) \rightleftharpoons 2XY_2(g) + 80 \text{ kcal}$$

The equilibrium point will shift to the right if the volume is
(1) increased and the temperature is increased
(2) increased and the temperature is decreased
(3) decreased and the temperature is increased
(4) decreased and the temperature is decreased

§2 **19.** Given this system at equilibrium:

$$PbCl_2(s) \rightleftharpoons Pb^{2+}(aq) + 2Cl^-(aq)$$

When KCl(s) is added to the system, the equilibrium shifts to the
(1) right, and the concentration of $Pb^{2+}(aq)$ ions decreases
(2) right, and the concentration of $Pb^{2+}(aq)$ ions increases
(3) left, and the concentration of $Pb^{2+}(aq)$ ions decreases
(4) left, and the concentration of $Pb^{2+}(aq)$ ions increases

20. Given this system at equilibrium:

$$H_2(g) + I_2(g) + heat \rightleftharpoons 2HI(g)$$

The equilibrium will shift to the right if there is an increase in
(1) temperature
(2) pressure
(3) concentration of $HI(g)$
(4) volume of the reaction container

21. Which system at equilibrium will shift to the right when the pressure is increased?

(1) $NaCl(s) \overset{H_2O}{\rightleftharpoons} Na^+(aq) + Cl^-(aq)$

(2) $C_2H_5OH(\ell) \overset{H_2O}{\rightleftharpoons} C_2H_5OH(aq)$

(3) $NH_3(g) \overset{H_2O}{\rightleftharpoons} NH_3(aq)$

(4) $C_6H_{12}O_6(s) \overset{H_2O}{\rightleftharpoons} C_6H_{12}O_6(aq)$

§2 **22.** Given this system at equilibrium:

$$C(s) + O_2(g) \rightleftharpoons CO_2(g)$$

As $C(s)$ is added to the system, the amount of $CO_2(g)$ will
(1) decrease   (2) increase   (3) remain the same

23. Given this system at equilibrium:

$$A(g) + B(g) \rightleftharpoons C(g)$$

If the concentration of A is increased at constant temperature, which will also increase?
(1) the rate of the forward reaction
(2) the value of the equilibrium constant
(3) the activation energy of the forward reaction
(4) the concentration of B

24. Given this system at equilibrium:

$$2SO_2(g) + O_2(g) \rightleftharpoons 2SO_3(g) + 197 \text{ kJ}$$

The amount of SO_3 will increase if there is
(1) an increase in temperature
(2) a decrease in pressure (an increase in volume)
(3) an increase in the concentration of SO_2
(4) a decrease in the concentration of O_2

25. Given this system at equilibrium:

$$2H_2(g) + O_2(g) \rightleftharpoons 2H_2O(g) + heat$$

Which concentration changes occur when the temperature of the system is increased?
(1) The $[H_2]$ decreases and the $[O_2]$ decreases.
(2) The $[H_2]$ increases and the $[O_2]$ decreases.
(3) The $[H_2]$ decreases and the $[O_2]$ increases.
(4) The $[H_2]$ increases and the $[O_2]$ increases.

26. Given this equilibrium system:

$$N_2(g) + 3H_2(g) \rightleftharpoons 2NH_3(g)$$

If the volume of the reaction vessel is decreased at constant temperature, there will be an increase in the number of moles of
(1) N_2, only   (2) H_2, only
(3) NH_3, only   (4) N_2, H_2, and NH_3

§2 **27.** For the system

$$4NH_3(g) + 5O_2(g) \rightleftharpoons 4NO(g) + 6H_2O(g)$$

the correct equilibrium expression is

(1) $K_{eq} = \dfrac{[NO]^4[H_2O]^6}{[NH_3]^4[O_2]^5}$ (2) $K_{eq} = \dfrac{[NO]^4+[H_2O]^6}{[NH_3]^4+[O_2]^5}$

(3) $K_{eq} = \dfrac{[4NO][6H_2O]}{[4NH_3][5O_2]}$ (4) $K_{eq} = \dfrac{[4NO]+[6H_2O]}{[4NH_3]+[5O_2]}$

§2 **28.** Given this solution at equilibrium:

$$CaF_2(s) \rightleftharpoons Ca^{2+}(aq) + 2F^-(aq)$$

What is the solubility product (K_{sp}) expression?
(1) $K_{sp} = [Ca^{2+}][F^-]$ (2) $K_{sp} = [Ca^{2+}][2F^-]$
(3) $K_{sp} = [Ca^{2+}][F^-]^2$ (4) $K_{sp} = [Ca^{2+}]2[F^-]$

29. Given this system at equilibrium:

$$2CO(g) + O_2(g) \rightleftharpoons 2CO_2(g)$$

The correct equilibrium expression for this system is

(1) $K_{eq} = \dfrac{[2CO][O_2]}{[CO_2]}$ (2) $K_{eq} = \dfrac{[CO]^2[O_2]}{[CO_2]^2}$

(3) $K_{eq} = \dfrac{[2CO_2]}{[2CO][O_2]}$ (4) $K_{eq} = \dfrac{[CO_2]^2}{[CO]^2[O_2]}$

§2 **30.** What is the equilibrium expression for the following reaction?

$$4Al(s) + 3O_2(g) \rightleftharpoons 2Al_2O_3(s)$$

(1) $K_{eq} = [O_2]^3$ (2) $K_{eq} = [3O_2]$

(3) $K_{eq} = \dfrac{1}{[O_2]^3}$ (4) $K_{eq} = \dfrac{1}{[3O_2]}$

§2 **31.** According to Reference Table Y-1, which compound is more soluble than $BaSO_4$?
(1) AgBr (2) CuCl (3) AgI (4) $ZnCO_3$

§2 **32.** Given this system at equilibrium:

$$H_2(g) + \frac{1}{2}O_2(g) \rightleftharpoons H_2O(g) + heat$$

The value of the equilibrium constant for this reaction can be changed by
(1) changing the pressure (2) changing the temperature
(3) adding more O_2 (4) adding a catalyst

§2 **33.** According to Reference Table Y-1, which compound is least soluble in water?
(1) AgBr (2) $BaSO_4$ (3) CuCl (4) $ZnCO_3$

§2 **34.** Which system has the following equilibrium expression?

$$K_{eq} = \frac{[A][B]^2}{[AB_2]}$$

(1) $AB_2(g) \rightleftharpoons A(g) + 2B(g)$
(2) $2AB(g) \rightleftharpoons A(g) + B_2(g)$
(3) $A(g) + 2B(g) \rightleftharpoons AB_2(g)$
(4) $A(g) + B_2(g) \rightleftharpoons 2AB(g)$

35. Given this system at equilibrium:

$$N_2(g) + 3H_2(g) \rightleftharpoons 2NH_3(g) + heat$$

Which stress on the system at equilibrium favors the production of $NH_3(g)$?
(1) decreasing the concentration of $N_2(g)$
(2) decreasing the concentration of $H_2(g)$
(3) reducing the volume of the system
(4) increasing the temperature of the system

36. Which statement best explains the role of a catalyst in a chemical reaction?
(1) A catalyst is consumed in the reaction but not regenerated.
(2) A catalyst limits the amount of reactants used.
(3) A catalyst changes the kinds of products produced.
(4) A catalyst provides an alternate reaction pathway that requires less activation energy.

37. In most aqueous reactions, as temperature increases, the effectiveness of collisions between reacting particles
(1) decreases
(2) increases
(3) remains the same
(4) reverses

38. Based on the nature of the reactants in each of the equations below, which reaction will occur at the *fastest* rate?
(1) $C(s) + O_2(g) \rightarrow CO_2(g)$
(2) $NaOH(aq) + HCl(aq) \rightarrow NaCl(aq) + H_2O(\ell)$
(3) $CH_3OH(\ell) + CH_3COOH(\ell) \rightarrow CH_3COOCH_3(aq) + H_2O(\ell)$
(4) $CaCO_3(s) \rightarrow CaO(s) + CO_2(g)$

39. Given the reaction at equilibrium:

$$A(g) + B(g) \rightleftharpoons AB(g) + \text{heat}$$

The concentration of $A(g)$ can be increased by
(1) lowering the temperature
(2) adding a catalyst
(3) increasing the concentration of $AB(g)$
(4) increasing the concentration of $B(g)$

40. Given the reaction:

$$AgCl(s) \overset{H_2O}{\rightleftharpoons} Ag^+(aq) + Cl^-(aq)$$

Once equilibrium is reached, which statement is accurate?
(1) The concentration of $Ag^+(aq)$ is greater than the concentration of $Cl^-(aq)$.
(2) The AgCl(s) will be completely consumed.
(3) The rates of the forward and reverse reactions are equal.
(4) The entropy of the forward reaction will continue to increase.

41. As the temperature of a reaction increases, it is expected that the reacting particles collide
(1) more often and with greater force
(2) more often and with less force
(3) less often and with greater force
(4) less often and with less force

42. Given the equation for a system at equilibrium:

$$N_2(g) + 3H_2(g) \rightleftharpoons 2NH_3 + \text{energy}$$

If only the concentration of $N_2(g)$ is increased, the concentration of
(1) $NH_3(g)$ increases
(2) $NH_3(g)$ remains the same
(3) $H_2(g)$ increases
(4) $H_2(g)$ remains the same

Constructed-Response Questions

1. Consider the following system at equilibrium:

$$2NO_2(g) \rightleftharpoons N_2(g) + 2O_2(g) + \text{heat}$$

A vessel contains $NO_2(g)$, $N_2(g)$, and $O_2(g)$ at equilibrium. How will each of the following stresses affect the equilibrium position?
(a) More $NO_2(g)$ is added to the vessel.
(b) Some $N_2(g)$ is removed from the vessel.
(c) The volume of the vessel is reduced to one-half its original value.
(d) Some $He(g)$ is added to the vessel.
(e) The temperature of the system is increased

§2 **2.** Write the equilibrium constant expression for each of the following systems:
(a) $4NH_3(g) + 7O_2(g) \rightleftharpoons 4NO_2(g) + 6H_2O(g)$
(b) $Fe_2O_3(s) + 3H_2(g) \rightleftharpoons 2Fe(s) + 3H_2O(g)$

§2 **3.** Consider the following system:

$$A(g) + 2B(g) \rightleftharpoons 3C(g) + 2D(g)$$

(a) Write the equilibrium constant expression for this system.

(b) The table below contains a set of equilibrium concentrations for the reactants and products when they reach equilibrium at 298 K. Use these data to calculate the value of K_{eq} for the system given above at 298 K.

Substance	Equilibrium Concentration (mol/L)
A	0.35
B	0.20
C	1.0
D	0.50

§2 **4.** In a certain experiment, 1.000 mole of $N_2(g)$ and 1.000 mole of H_2 were allowed to react at 500°C in a 1.000-liter flask. The relevant reaction was as follows:

$$N_2(g) + 3H_2(g) \rightleftharpoons 2NH_3(g)$$

After the system reached equilibrium, the flask was found to contain 0.921 mole of $N_2(g)$. Calculate the equilibrium concentrations of $H_2(g)$ and $NH_3(g)$.

For questions 5 and 6, use the given equilibrium equation at 298 K:

$$KNO_3(s) + 34.89 \text{ kJ} \overset{H_2O}{\rightleftharpoons} K^+(aq) + NO_3^-(aq)$$

5. Describe, in terms of Le Châtelier's principle, why an increase in temperature increases the solubility of KNO_3.

6. The equation indicates that KNO_3 has formed a saturated solution. Explain, in terms of equilibrium, why the solution is saturated.

Human blood contains dissolved carbonic acid, H_2CO_3, in equilibrium with carbon dioxide and water. The equilibrium system is shown below:

$$H_2CO_3(aq) \rightleftharpoons CO_2(aq) + H_2O(\ell)$$

7. Explain, using Le Châtelier's principle, why decreasing the concentration of CO_2 decreases the concentration of H_2CO_3.

The answers to these questions are found in Appendix 3.

**Chapter
Fourteen**

ACIDS AND BASES

KEY IDEAS

This chapter focuses on the ways in which acids and bases are defined and the reactions that they undergo. The strengths of acids and bases are discussed, and the important laboratory procedure of acid–base titration is introduced.

KEY OBJECTIVES
At the conclusion of this chapter you will be able to:
- Define acids and bases operationally.
- State and apply the Arrhenius definitions of acids and bases.
- Solve acid–base titration problems.
- State and apply the Brønsted–Lowry definitions of acids and bases.
- Define the term *amphiprotic* (*amphoteric*), and apply the concept.
- Define the term *conjugate acid–base pairs*, and recognize these pairs in an acid–base reaction.
- Compare the relative strengths of conjugate acid–base pairs.
- Relate ionization constants K_a and K_b to acid–base strength.
- Solve K_w problems.
- Define the terms *pH* and *pOH*, and apply these definitions.
- Describe the properties of acid–base indicators.
- Indicate which salts are likely to produce acidic, basic, or neutral solutions when dissolved in water.
- Describe the acid–base properties of metallic and nonmetallic oxides in water.
- Define the terms *Lewis acid* and *Lewis base*, and apply these concepts.

SECTION I—BASIC (REGENTS-LEVEL) MATERIAL

NYS REGENTS CONCEPTS AND SKILLS

Note: By the time you have finished Section I, you should have mastered the concepts and skills listed below. The Regents chemistry examination will test your knowledge of these items and your ability to apply them.

Concepts are the *basic ideas* that form the body of the Regents chemistry course (what you need to know!).

Skills are the *activities* that demonstrate your mastery of these concepts (how you show that you know them!).

Following each concept or skill is a page reference (given in parentheses) to this chapter.

14.1　Concept:
- An electrolyte is a substance that, when dissolved in water, forms a solution capable of conducting an electric current. (Pages 401–402)
- The ability of a solution to conduct an electric current depends on the concentration of ions. (Page 401)

14.2　Concepts
- The behaviors of many acids and bases can be explained by the Arrhenius theory. (Page 403)
- Arrhenius acids and bases are electrolytes. (Page 403)

Skill:
Given the properties of a substance, identify it as an Arrhenius acid or base. (Page 403)

14.3　Concept:
Arrhenius acids yield hydrogen ions [$H+$ (aq)] as the only positive ions in an aqueous solution. The hydrogen ion can also be written as H_3O^+ (aq) (hydronium ion). (Page 403)

14.4　Concept:
Arrhenius bases yield hydroxide ions [$OH-$ (aq)] as the only negative ions in an aqueous solution. (Page 403)

14.5　Concept:
In the process of neutralization, an Arrhenius acid and an Arrhenius base react to form a salt and water. (Page 403)

Skill:
Write simple neutralization reactions, given the reactants (acid and base). (Page 403)

14.6　Concept:
Titration is a laboratory process in which a volume of solution of known concentration is used to determine the concentration of another solution. (Pages 403–406)

Skill:
Calculate the concentration or the volume of a solution, given titration data. (Pages 403–406)

14.7 Concept:
There are alternative acid base theories. One theory states that an acid is a proton (H+) donor and a base is a proton (H+) acceptor. (Pages 406–408)

14.8 Concepts:
- The acidity or alkalinity of a solution can be measured by its pH value. (Pages 408–409)
- The relative level of acidity or alkalinity of a solution can be shown by using indicators. (Pages 409–410)

Skills:
- Identify a solution as acidic, basic, or neutral, based on the pH value of the solution. (Pages 408–409)
- Interpret changes in acid-base indicator color. (Pages 409–410)

14.9 Concept:
On the pH scale, each decrease of one unit of pH represents a tenfold increase in hydronium ion concentration. (Page 408)

14.1 OPERATIONAL DEFINITIONS OF ACIDS AND BASES

Defining something operationally means describing it by what it does. When we define acids and bases operationally, we are interested in observing their characteristic properties; we are not immediately concerned with why they exhibit these properties.

- Acids and bases are *electrolytes:* they conduct electricity when dissolved in water. (Conductivity depends on the ion concentration.)
- Dilute solutions of acids taste sour; dilute solutions of bases taste bitter.
- Aqueous solutions of bases feel slippery.
- Acids and bases cause color changes in solutions of substances known as acid–base indicators. (We will have more to say about acid–base indicators in Section 14.6.)
- Acids react with certain metals to liberate hydrogen gas:

$$Zn(s) + 2HCl(aq) \rightarrow H_2(g) + ZnCl_2(aq)$$

metal   acid   hydrogen gas

Not all metals, however, liberate hydrogen gas when treated with acid. For example, copper does not react with HCl.

- Acids and bases *neutralize* each other. If an acid reacts with a hydroxide base, a salt and water are formed:

$$HCl(aq) + NaOH(aq) \rightarrow NaCl(aq) + H_2O(\ell)$$

acid   hydroxide base   salt   water

401

The salt formed in this reaction, sodium chloride, is an ionic compound whose positive ion (Na^+) comes from the base and whose negative ion (Cl^-) comes from the acid. Common hydroxide bases include the metals of Groups 1 and 2: NaOH, KOH, LiOH, $Ca(OH)_2$, $Mg(OH)_2$.

Some common acids and the negative ions they generate in neutralization are listed in the accompanying table.

Acid	Name	Ion	Name of Salt
HCl	Hydrochloric	Cl^-	Chloride
HNO_3	Nitric	NO_3^-	Nitrate
H_2SO_4	Sulfuric	HSO_4^-	Hydrogen sulfate
		SO_4^{2-}	Sulfate
H_2CO_3	Carbonic	HCO_3^-	Hydrogen carbonate
		CO_3^{2-}	Carbonate
HBr	Hydrobromic	Br^-	Bromide
HF	Hydrofluoric	F^-	Fluoride
HNO_2	Nitrous	NO_2^-	Nitrite
H_3PO_4	Phosphoric	$H_2PO_4^-$	Dihydrogen phosphate
		HPO_4^{2-}	Hydrogen phosphate
		PO_4^{3-}	Phosphate
HOCl	Hypochlorous	ClO^-	Hypochlorite
$HClO_2$	Chlorous	ClO_2^-	Chlorite
$HClO_3$	Chloric	ClO_3^-	Chlorate
$HClO_4$	Perchloric	ClO_4^-	Perchlorate

A shorter listing of common acids and bases appears in Reference Tables K and L in Appendix 1.

PROBLEM
(a) Write the equation for the neutralization of chlorous acid by potassium hydroxide. Name the salt produced.
(b) An acid–base neutralization produces calcium nitrate as the salt. What are the names of the acid and base?

SOLUTIONS
(a) $HClO_2 + KOH \rightarrow KClO_2 + H_2O$. The salt is potassium chlorite.
(b) The acid is nitric acid; the base is calcium hydroxide.

TRY IT YOURSELF
What are the name and the formula of the salt produced by the reaction of magnesium hydroxide and chloric acid?

ANSWER
magnesium chlorate; $Mg(ClO_3)_2$

14.2 ARRHENIUS DEFINITIONS OF ACIDS AND BASES

The Swedish chemist Svante Arrhenius was the first person to propose a theory as to how solutions of electrolytes are able to conduct electric currents. He recognized that a substance such as NaCl has to dissociate in water and produce Na^+ and Cl^- ions. Arrhenius extended this idea to acids and bases, giving us the following Arrhenius definitions:

> **Arrhenius acid:** *A substance that releases H^+ ions in aqueous solution.*
> **Arrhenius base:** *A substance that releases OH^- ions in aqueous solution.*

Therefore, when HCl dissolves in water, its acidic properties are due to the H^+ ions it releases:

$$HCl(aq) \xrightarrow{H_2O} H^+(aq) + Cl^-(aq)$$

Similarly, when NaOH dissolves in water, its basic properties are due to the OH^- ions it releases:

$$NaOH(aq) \xrightarrow{H_2O} Na^+(aq) + OH^-(aq)$$

We are now able to understand what happens when an acid such as HCl neutralizes a base such as NaOH:

$$H^+(aq) + Cl^-(aq) + Na^+(aq) + OH^-(aq) \rightarrow H_2O(\ell) + Na^+(aq) + Cl^-(aq)$$
$$\text{Net reaction: } H^+(aq) + OH^-(aq) \rightarrow H_2O(\ell)$$

Neutralization is the reaction between H^+ and OH^- ions to form H_2O. The Na^+ and Cl^- ions are unchanged by the reaction and are called *spectator ions*. When the water is removed, the salt NaCl remains behind.

14.3 ACID–BASE TITRATION

Suppose we find a bottle labeled "dilute HCl" in the laboratory. A logical question would be: How dilute is the HCl? We are really asking what the molar concentration of the acid is.

The molar concentration of an unknown acid or base can readily be determined by a procedure known as *acid–base titration*. In this procedure, we perform an acid–base neutralization, proceeding as follows:

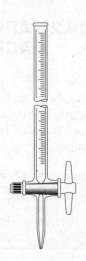

1. The volume of the unknown (acid or base) is measured carefully, and an *acid–base indicator* is added.
2. A *standard solution* (base or acid) of known concentration is slowly and accurately added to the unknown.
    (Usually, a long, narrow tube known as a *burette*, shown at the right, is used to deliver the standard solution.)
3. When the indicator changes color, this signal indicates that neutralization has occurred. This visual signal is known as the *end point* of the titration.
4. The volume of the unknown and the volume and molarity of the standard solution are recorded.
5. The molarity of the unknown is calculated.

When an acid is neutralized by a base in water, the net equation for the neutralization is

$$H^+(aq) + OH^-(aq) \rightarrow H_2O(\ell)$$

This equation indicates that *1 mole* of H^+ (acid) is needed to neutralize *1 mole* of OH^- (base).

It follows that the number of moles of H^+ and OH^- must always be equal when neutralization occurs:

$$n_{acid} = n_{base}$$

We know from Chapter 12 on solutions that, in expressions where M is the molarity, n is the number of moles, and V is the volume of the solution in liters. If we solve the equation $M = \dfrac{n}{V}$ for n, we obtain

$$n = M \cdot V.$$

Therefore, when acid–base neutralization occurs, we can write

$$\boxed{M_A \cdot V_A = M_B \cdot V_B}$$

This equation, known as the titration equation, appears also in Reference Table T in Appendix 1. The symbols M_A and M_B represent the molarities of the H^+ and OH^- ions, respectively. The volumes of the acid and base (V_A and V_B) can be measured in any units (L, mL, etc.) as long as the same units appear on both sides of the equation.

PROBLEM

In an acid–base titration, standard 0.10 M KOH is used to determine the concentration of an unknown HNO_3 solution.
(a) Write the equation for the neutralization.
(b) If 50. milliliters of KOH is needed to neutralize 200. milliliters of HNO_3, calculate the concentration of the acid.

SOLUTIONS

(a) $HNO_3(aq) + KOH(aq) \rightarrow H_2O(\ell) + KNO_3(aq)$
(b) $(M \cdot V)_{acid} = (M \cdot V)_{base}$
   $M_{acid} \cdot 200.\ mL = 0.10\ M \cdot 50.\ mL$
        $M_{acid} = 0.025\ M$

TRY IT YOURSELF

An acid–base titration reaches its end point when 100 milliliters of 0.2 M HCl is added to 50 milliliters of NaOH. What is the concentration of the NaOH?

ANSWER

0.4 M

For an acid such as H_2SO_4 or a base such as $Ba(OH)_2$, we must be careful in using the titration equation. In a 1-molar solution of H_2SO_4, for example, M_A is *2 molar* because both hydrogen atoms can ionize. Similarly, in a 0.2-molar solution of $Ba(OH)_2$, M_B is *0.4 molar* because of the presence of two OH^- ions in the formula.

PROBLEM

In an acid–base titration, standard 0.1 M H_2SO_4 is used to determine the concentration of an unknown NaOH solution.
(a) Write the equation for the neutralization.
(b) If 40 milliliters of H_2SO_4 is needed to neutralize 100 milliliters of NaOH, calculate the concentration of the base.

SOLUTIONS

(a) $H_2SO_4(aq) + 2NaOH(aq) \rightarrow 2H_2O(\ell) + Na_2SO_4(aq)$
(b) $M_A \cdot V_A = M_B \cdot V_B$
   $0.2\ M \cdot 40\ mL = M_B \cdot 100\ mL$
        $M_B = 0.08\ M$

TRY IT YOURSELF

What volume of 0.200 M H_2SO_4 is needed to neutralize 150. milliliters of 0.600 M NaOH?

ANSWER
225 mL

Lest we forget, this "Try It Yourself" problem can also be solved by the factor-label method used in Chapter 12. First we need to write the balanced equation for the neutralization reaction:

$$2NaOH(aq) + H_2SO_4(aq) \rightarrow 2H_2O(\ell) + Na_2SO_4(aq)$$

Then we construct a solution map of the problem:

$$\text{mL NaOH} \xrightarrow[\text{conversion}]{\text{volume}} \text{L NaOH} \xrightarrow{\text{molarity}} \text{mol NaOH} \xrightarrow[\text{coefficient}]{\text{equation}}$$

$$\text{mol } H_2SO_4 \xrightarrow{\text{molarity}} \text{L } H_2SO_4 \xrightarrow[\text{conversion}]{\text{volume}} \text{mL } H_2SO_4$$

The solution of this problem is given below.

$$V = 150. \text{ mL NaOH} \cdot \left(\frac{1 \text{ L NaOH}}{1000 \text{ mL NaOH}}\right) \cdot \left(\frac{0.600 \text{ mol NaOH}}{1 \text{ L NaOH}}\right)$$

$$\cdot \left(\frac{1 \text{ mol } H_2SO_4}{2 \text{ mol NaOH}}\right) \cdot \left(\frac{1 \text{ L } H_2SO_4}{0.200 \text{ mol } H_2SO_4}\right) \cdot \left(\frac{1000 \text{ mL } H_2SO_4}{1 \text{ L } H_2SO_4}\right)$$

$$= 225 \text{ mL } H_2SO_4$$

As you can see, in such cases it is simpler to rely on the general titration equation given on page 404.

14.4 BRØNSTED–LOWRY DEFINITIONS OF ACIDS AND BASES

In order to explain acid–base reactions more fully, Johannes Brønsted and T. M. Lowry extended the definitions of acids and bases as follows:

> **Brønsted–Lowry acid:**
> *Any species that can donate a proton (H+ ion) to another species, that is, a* proton donor.
> **Brønsted–Lowry base:**
> *Any species that can accept a proton (H+ ion) from another species, that is, a* proton acceptor.

(The term *species* indicates any particle: an atom, a molecule, or an ion.)

The Brønsted–Lowry definitions do not replace the Arrhenius theory; they extend it. Any substance that is an Arrhenius acid is also a Brønsted–Lowry acid; any substance that is an Arrhenius base is also a Brønsted–Lowry base.

The Brønsted–Lowry definitions explain why substances such as HCl behave as acids in water:

$$HCl(aq) + H_2O(\ell) \rightarrow H_3O^+(aq) + Cl^-(aq)$$

We see from the equation that the HCl donates a proton to the H_2O molecule, forming the H_3O^+ (hydronium) ion. (The proton forms a coordinate covalent bond with the oxygen atom.) It is the hydronium ion that is responsible for the acidic properties of acids in water solution.

If we review the equation for neutralization given in Section 14.2, we see why the OH^- ion is a Brønsted–Lowry base: it accepts a proton from the acid and forms water.

According to Arrhenius, only metallic hydroxides are bases. The Brønsted–Lowry definitions explain why a substance such as NH_3, which has no hydroxide ion, behaves as a base in water:

$$NH_3(aq) + H_2O(\ell) \rightarrow NH_4^+(aq) + OH^-(aq)$$

We see from the equation that the NH_3 molecule accepts a proton from the H_2O molecule. (The proton forms a coordinate covalent bond with the nitrogen atom.)

According to the Brønsted–Lowry definitions, every equation that has an acid must also have a base! A substance cannot donate a proton unless there is another substance that will accept it. In this sense, all such reactions are properly called *acid–base reactions*. In the reaction between HCl and H_2O, H_2O is the Brønsted–Lowry base; in the reaction between NH_3 and H_2O, H_2O is the Brønsted–Lowry acid.

PROBLEM
In the reaction $NH_3(aq) + HBr(aq) \rightarrow NH_4^+(aq) + Br^-(aq)$, identify the Brønsted–Lowry acid and the Brønsted–Lowry base.

SOLUTION
To solve this problem, we need to examine both sides of the equation. We see that NH_3 becomes NH_4^+ and HBr becomes Br^-. This could have happened only if HBr donated the proton to NH_3.

Therefore, HBr is the Brønsted–Lowry acid and NH_3 is the Brønsted–Lowry base.

TRY IT YOURSELF
In the reaction $H_2O(\ell) + NH_2^-(aq) \rightarrow NH_3(aq) + OH^-(aq)$, identify the Brønsted–Lowry acid and the Brønsted–Lowry base.

ANSWER
H_2O is the Brønsted–Lowry acid; NH_2^- is the Brønsted–Lowry base.

14.5   THE pH SCALE OF ACIDITY AND BASICITY

To measure the degree of acidity or basicity of an aqueous solution, chemists have come to rely on the **pH scale**, a scale that depends on the concentration of H^+ ions in solution. The pH scale generally runs from 0 to 14, and a change of *one* pH unit represents a *tenfold* change in the H^+ ion concentration of the solution. For example, a solution with a pH of 2 has 10 times the H^+ ion concentration of a solution with a pH of 3. Similarly, a solution with a pH of 4 has 1/100 the H^+ ion concentration of a solution with a pH of 2. The table that follows illustrates how pH is related to the degree of acidity and basicity.

pH Values and Acid-Base Character of an Aqueous Solution

pH	Solution Is:	pH	Solution Is:
0	Strongly acidic	8	Weakly basic
1	Strongly acidic	9	Moderately basic
2	Strongly acidic	10	Moderately basic
3	Moderately acidic	11	Moderately basic
4	Moderately acidic	12	Strongly basic
5	Moderately acidic	13	Strongly basic
6	Weakly acidic	14	Strongly basic
7	Neutral		

To give you an idea of how the concept of pH relates to the "real" world, examine the next table: it lists the pH values of some common fluids.

The pH Values of Some Common Fluids

Fluid	pH Value
Gastric juice (stomach)	1.0-2.0
Lemon juice	2.4
Vinegar	3.0
Grapefruit juice	3.2
Orange juice	3.5
Urine	4.8-7.5
Water (exposed to air)	5.5
Saliva	6.4-6.9
Milk	6.5
Water (pure)	7.0
Blood	7.35-7.45
Tears	7.4
Milk of magnesia	10.6
Ammonia (household)	11.5

14.6 ACID–BASE INDICATORS

Acid–base indicators are substances (usually of plant origin) that change color with changing pH. Every indicator can act either as a Brønsted–Lowry acid or base, and each form has its own specific color. The color change of an indicator takes place over a specific range of pH. Outside its pH range, an indicator cannot be used effectively since it exists only in its acidic or basic form.

Reference Table M in Appendix 1 lists some common indicators, the pH ranges over which their color changes occur, and their colors at the low and high ends of the pH range.

PROBLEM
The following three indicators appear in Reference Table M in Appendix 1: methyl orange, bromthymol blue, thymol blue. Which of these indicators can distinguish between a solution whose pH is 4.6 and a solution whose pH is 7.9?

SOLUTION
The following table lists the indicators and their colors at the two pH values:

Indicator	Color at pH 4.6	Color at pH 7.9
Methyl orange	Yellow	Yellow
Bromthymol blue	Yellow	Blue
Thymol blue	Yellow	Yellow

As we can see, only bromthymol blue will distinguish between the two solutions.

TRY IT YOURSELF
Which indicator in Reference Table M can be used to distinguish between the solutions in each of the following pairs?
(a) Solution A: pH = 2.0; solution B: pH = 5.0
(b) Solution C: pH = 5.0; solution D: pH = 9.0
(c) Solution E: pH = 7.0; solution F: pH = 10.0

ANSWERS
(a) methyl orange
(b) bromthymol blue
(c) thymol blue

SECTION II—ADDITIONAL MATERIAL

14.1A   AMPHIPROTIC (AMPHOTERIC) SUBSTANCES

We have seen that H_2O is an example of a substance that can behave as an acid in one reaction and as a base in another. Such substances are called **amphiprotic** or **amphoteric**.

14.2A   ACID–BASE EQUILIBRIA

Like other chemical reactions, acid–base reactions are equilibrium systems and therefore have both forward and reverse reactions. (Note: From now on, we will omit the phases of the reactants and products unless they have special significance.) Consider the acid HA ionizing in water:

$$HA + H_2O \rightleftharpoons H_3O^+ + A^-$$

In the forward reaction: $HA + H_2O \rightarrow H_3O^+ + A^-$, the acid, HA, donates a proton to the base, H_2O.

In the reverse reaction: $HA + H_2O \leftarrow H_3O^+ + A^-$, the H_3O^+ ion acts as an acid and donates a proton to the A^- ion, which acts as a base.

Every acid–base reaction has two acids and two bases. To summarize:

$$HA + H_2O \rightleftharpoons H_3O^+ + A^-$$
$$\text{acid} \quad\quad \text{base} \quad\quad \text{acid} \quad\quad \text{base}$$

Similarly, when a base, B, ionizes in water, it behaves as in any other acid–base reaction:

$$B + H_2O \rightleftharpoons BH^+ + OH^-$$
$$\text{base} \quad\quad \text{acid} \quad\quad \text{acid} \quad\quad \text{base}$$

PROBLEM

In the acid–base reaction $HNO_2 + H_2O \rightleftharpoons H_3O^+ + NO_2^-$, identify the acids and bases.

SOLUTION

In the forward reaction, we see that HNO_2 becomes the NO_2^- ion on the right side of the equation. Therefore, we can conclude that HNO_2 donates the proton and is the acid. It follows that the H_2O, which accepts the proton, is the base.

In the reverse reaction, we see that the H_3O^+ ion becomes H_2O on the left side of the equation. Therefore, we can conclude that the H_3O^+ ion donates the proton and is the acid. It follows that the NO_2^- ion, which accepts the proton, is the base. To summarize:

$$HNO_2 + H_2O \rightleftharpoons H_3O^+ + NO_2^-$$
$$\text{acid} \quad\quad \text{base} \quad\quad \text{acid} \quad\quad \text{base}$$

TRY IT YOURSELF

Identify the acids and bases in the following reaction:

$$NH_3 + HCO_3^- \rightleftharpoons NH_4^+ + CO_3^{2-}$$

ANSWER

$$NH_3 + HCO_3^- \rightleftharpoons NH_4^+ + CO_3^{2-}$$
$$\text{base} \quad\quad \text{acid} \quad\quad \text{acid} \quad\quad \text{base}$$

411

14.3A   CONJUGATE ACID–BASE PAIRS

If we examine all of the acid–base equilibria shown on page 411, we observe that the acid on the left side of the equation becomes the base on the right side. Similarly, the base on the left side of the equation becomes the acid on the right side. These are known as **conjugate acid–base pairs**. In the reaction

$$HNO_2 + H_2O \rightleftharpoons H_3O^+ + NO_2^-$$

$$\text{acid} \qquad \text{base} \qquad \text{acid} \qquad\qquad \text{base}$$

HNO_2 and NO_2^- are a conjugate acid–base pair, as are also H_2O and H_3O^+.

For example, we may say that HNO_2 is the conjugate acid of NO_2^-, or that NO_2^- is the conjugate base of HNO_2.

Note that a conjugate acid and its conjugate base differ by an H^+:

> **Conjugate Acid = H⁺ + Conjugate Base**

PROBLEM
(a)   What is the conjugate base of HF?
(b)   What is the conjugate base of H_2O?
(c)   What is the conjugate base of $H_2PO_4^-$?
(d)   What is the conjugate base of NO_3^-?

SOLUTIONS
A conjugate base has one less H^+ than its conjugate acid. Therefore, we will "subtract" an H^+ from each of the four compounds to determine its conjugate base:
(a) $HF - H^+ = F^-$ (b) $H_2O - H^+ = OH^-$ (c) $H_2PO_4^- - H^+ = HPO_4^{2-}$
(d)   There is no conjugate base! NO_3^- has no H^+ to donate.

TRY IT YOURSELF
(a)   What is the conjugate acid of Br^-?
(b)   What is the conjugate acid of H_2O?
(c)   What is the conjugate acid of S^{2-}?

ANSWERS
(a) $Br^- + H^+ = HBr$ (b) $H_2O + H^+ = H_3O^+$ (c) $S^{2-} + H^+ = HS^-$

The accompanying table lists a number of conjugate acid–base pairs. If we scan the list, we see that certain species (e.g., NH_3, HSO_4^-) appear as both conjugate acids and conjugate bases. These species are exhibiting amphoteric behavior. Can you find other amphoteric species in this table?

RELATIVE STRENGTHS OF CONJUGATE ACID–BASE PAIRS

Formula	Conjugate Acid Name	Formula	Conjugate Base Name
$HClO_4$	perchloric acid	ClO_4^-	perchlorate ion
HI	hydriotic acid	I^-	iodide ion
HBr	hydrobromic acid	Br^-	bromide ion
HCl	hydrochloric acid	Cl^-	chloride ion
H_2SO_4	sulfuric acid	HSO_4^-	hydrogen sulfate ion
HNO_3	nitric acid	NO_3^-	nitrate ion
H_3O^+	hydronium ion	H_2O	water
HSO_4^-	hydrogen sulfate ion	SO_4^{2-}	sulfate ion
H_3PO_4	phosphoric acid	H_2PO4^-	dihydrogen phosphate ion
HF	hydrofluoric acid	F^-	fluoride ion
HNO_2	nitrous acid	NO_2^-	nitrite ion
HCO_2H	formic (methanoic) acid	CO_2H^-	formate (methanoate) ion
$HC_2H_3O_2$	acetic (ethanoic) acid	$C_2H_3O_2^-$	acetate (ethanoate) ion
H_2CO_3	carbonic acid	HCO_3^-	hydrogen carbonate ion
$H_2PO_4^-$	dihydrogen phosphate ion	HPO_4^{2-}	hydrogen phosphate ion
NH_4^+	ammonium ion	NH_3	ammonia
HCN	hydrocyanic acid	CN^-	cyanide ion
HCO_3^-	hydrogen carbonate ion	CO_3^{2-}	carbonate ion
HPO_4^{2-}	hydrogen phosphate ion	PO_4^{3-}	phosphate ion
H_2O	water	OH^-	hydroxide ion
NH_3	ammonia	NH_2^-	amide ion

Strong acids / Weak acids — Acid strength increases ↑

Base strength increases ↓

14.4A NEUTRALIZATION (REVISITED)

At the beginning of this chapter, we said that acids and bases neutralized each other. However, we were talking about Arrhenius acids and bases, in which H^+ ions and OH^- ions combine to produce water. In the Brønsted–Lowry sense, not every acid–base reaction is a neutralization. *For acid–base neutralization to occur, the solvent of the system must be produced from its conjugate acid and base.*

We have been studying reactions in water. The conjugate acid of water is the H_3O^+ ion; the conjugate base is the OH^- ion. Therefore, neutralization involves the reaction

$$H_3O^+(aq) + OH^-(aq) \rightarrow 2H_2O(\ell)$$

Since water is a covalently bonded molecule, the reaction goes nearly to completion and removes the acidic and basic ions from the aqueous solution.

14.5A STRENGTHS OF CONJUGATE ACID–BASE PAIRS

If a conjugate acid is strong, then its conjugate base is weak; if a conjugate acid is weak, then its conjugate base is strong. This relationship follows from the Brønsted–Lowry definitions of acids and bases. If a conjugate acid is strong, it is able to donate its proton readily. Therefore, its conjugate base is not able to hold onto the proton and is weak. Similarly, a weak conjugate acid is not able to donate a proton readily because its conjugate base holds onto it tightly and is strong.

In the preceding subsection, the table of conjugate acid-base pairs lists the strengths of conjugate acids in descending order, and conjugate bases in ascending order.

PROBLEM
The following four conjugate acids are listed in descending order of strength:

$$HI > HF > H_2CO_3 > NH_3$$

List the conjugate bases of these acids, also in descending order of strength.

SOLUTION
First, we find the conjugate bases of the acids by "subtracting" H^+ from each acid on the list:

$$HI - H^+ = I^-, \qquad HF - H^+ = F^-, \qquad H_2CO_3 - H^+ = HCO_3^-,$$
$$NH_3 - H^+ = NH_2^-$$

We know that there is an inverse relationship between the strength of a conjugate acid and the strength of its conjugate base. Since HI is the strongest acid in the list, its conjugate base, I^-, must be the weakest; since NH_3 is the weakest conjugate acid, its conjugate base, NH_2^-, must be the strongest.

With this information, we can arrange the conjugate bases in descending order:

$$NH_2^- > HCO_3^- > F^- > I^-$$

TRY IT YOURSELF
If OH^- is a stronger conjugate base than HSO_4^-, what can we conclude about their conjugate acids?

ANSWER
H_2SO_4 is a stronger acid than H_2O.

14.6A   IONIZATION CONSTANTS OF ACIDS AND BASES (K_a and K_b)

Since acid–base reactions are equilibrium systems, we can write equilibrium constants for them. These constants give us information about the relative strengths of the acid and base.

In water solution, an acid ionizes according to the equation

$$HA(aq) + H_2O(\ell) \rightleftharpoons H_3O^+(aq) + A^-(aq)$$

The stronger an acid is in water, the more H_3O^+ will be produced. Therefore, the concentration of hydronium ion is a measure of the strength of the acid in water solution.

The equilibrium constant for this reaction is as follows:

$$K_{eq} = \frac{[H_3O^+][A^-]}{[HA][H_2O]}$$

In dilute solutions (which is all we concern ourselves with) the concentration of water is constant. Therefore it can be absorbed into K_{eq}, yielding the constant K_a:

$$\boxed{K_a = \frac{[H_3O^+][A^-]}{[HA]}}$$

The constant K_a is known as the ionization constant for the acid HA. If we examine the K_a expression given above, we see that the hydronium-ion concentration $[H_3O^+]$ appears in the numerator of the fraction. It follows that the stronger the acid is in water, the larger is its K_a value.

Reference Table Y-2 in Appendix 2 lists the K_a values of a number of acids in descending order. For example, nitrous acid (HNO_2), whose K_a is 7.2×10^{-4}, is a stronger acid than hypochlorous acid (HOCl), whose K_a is 2.9×10^{-8}. A **monoprotic acid** is a molecule or ion that is capable of donating *only one proton* to a Brønsted base.

If we examine Reference Table Y-2, we note that almost all of the K_a values are relatively small numbers. These low K_a values mean that these acids are relatively weak in water and the hydronium-ion concentration at equilibrium is small.

Acids that are capable of donating two or three protons are called, respectively, **diprotic** and **triprotic** acids. Collectively, they are known as **polyprotic** acids. Such acids undergo a process known as *stepwise ionization*, in which one proton is released in each step. For example, the diprotic acid carbonic acid (H_2CO_3) ionizes in *two* steps:

$$(1) \ H_2CO_3 + H_2O \rightleftharpoons H_3O^+ + HCO_3^-$$

$$(2) \ HCO_3^- + H_2O \rightleftharpoons H_3O^+ + CO_3^{2-}$$

Each step has its own distinct K_a value, written as $K_{a1}, K_{a2}, \ldots$

Reference Table Y-2 in Appendix 2 *also* lists the ionization constants for a number of diprotic and triprotic acids. Note that the K_a value for each step is *much larger* than the value for the *succeeding* step (that is, $K_{a1} \gg K_{a2} \gg \ldots$). This means that, as stepwise ionization occurs, it becomes increasingly difficult to remove a proton.

A strong acid ionizes almost completely in water:

$$HA(aq) + H_2O(\ell) \xrightarrow{\sim 100\%} H_3O^+(aq) + A^-(aq)$$

In this case, the concentration of un-ionized HA(aq) at equilibrium is nearly 0, and the value of K_a is so large that it cannot be evaluated accurately. This is the case for the first six acids ($HClO_4$, HI, HBr, HCl, H_2SO_4, and HNO_3) in the table on page 399. This also means that H_3O^+(aq) is the strongest acid that can be present in water solution. As a result, it is impossible to compare the relative strengths of the first five acids in water solution. To make such a comparison, other solvents must be used.

In water solution, a base ionizes according to the equation

$$B(aq) + H_2O(\ell) \rightleftharpoons BH^+(aq) + OH^-(aq)$$

The stronger a base is in water, the more OH^- will be produced. Therefore, the concentration of hydroxide ion is a measure of the strength of the base in water solution.

The ionization constant for a base, K_b, measures the relative strength of the base in water solution. The larger the value of K_b, the stronger the base. The expression for this constant is as follows:

$$K_b = \frac{[BH^+][OH^-]}{[B]}$$

Reference Table Y-2 in Appendix 2 lists the ionization constants for a number of weak bases. Note that all of these substances contain nitrogen, which can accept a proton from a Brønsted acid. In general, aqueous solutions of nitrogen-containing compounds are weakly basic in water. The characteristically bitter taste of coffee is due to the presence of the weak nitrogen base, caffeine ($C_8H_{10}N_4O_2$).

A strong base such as NaOH or KOH dissociates almost completely in aqueous solution:

$$NaOH \xrightarrow[H_2O]{\sim 100\%} Na^+(aq) + OH^-(aq)$$

Therefore, K_b values for these bases cannot be calculated accurately. This also means that $OH^-(aq)$ is the strongest base that can be present in water solution. As a result, it is impossible to compare the relative strengths of bases that are stronger than OH^- (such as NH_2^-) in water solution. To make such a comparison, other solvents must be used.

14.7A IONIZATION CONSTANT OF WATER (K_w)

If we examine the electrical conductivity of water, we find that water conducts poorly. This poor conductivity indicates the presence of a small amount of ions, which arise from the self-ionization of the water:

$$H_2O(\ell) + H_2O(\ell) \rightleftharpoons H_3O^+(aq) + OH^-(aq)$$

In writing an equilibrium constant for this reaction, we recall that the concentration of water is a constant and can be absorbed into K_{eq}. The ionization constant for water, K_w, can be expressed as follows:

$$K_w = [H_3O^+] \cdot [OH^-]$$

At 298 K, the value of K_w is 1.0×10^{-14}. In pure water only H_3O^+ and OH^- ions are produced; therefore, they are produced in equal amounts. We can use this fact to calculate $[H_3O^+]$ and $[OH^-]$ in pure water. If x represents $[H_3O^+]$, it must also represent $[OH^-]$:

$$1 \times 10^{-14} = K_w = [H_3O^+] \cdot [OH^-] = x \cdot x = x^2$$
$$1 \times 10^{-14} = x^2$$
$$x = 1 \times 10^{-7} \text{ M} = [H_3O^+] = [OH^-]$$

In pure water at 298 K, therefore, the concentrations of hydronium and hydroxide ions are each 1×10^{-7} M.

Suppose another substance is dissolved in the water. The fact that K_w is a constant tells us that the product of the hydronium and hydroxide ion concentrations in the water will *always* be 1.0×10^{-14} at 298 K.

If the substance is an acid, it will increase the hydronium ion concentration in the water. Therefore, the hydroxide ion concentration must *decrease* if the product is to remain at 1.0×10^{-14}. Similarly, if the substance is a base, it will increase the hydroxide ion concentration in the water. It follows that the hydronium ion concentration will *decrease*.

PROBLEM
What are $[H_3O^+]$ and $[OH^-]$ in 0.10 M HCl, a strong acid?

SOLUTION

A strong acid such as HCl ionizes completely in water:

$$HCl(aq) + H_2O(\ell) \xrightarrow{\sim 100\%} H_3O^+(aq) + Cl^-(aq)$$

At equilibrium all of the HCl is converted to $H_3O^+(aq)$. It follows that

$$[H_3O^+] = 0.10 \text{ M} = 1.0 \times 10^{-1} \text{ M}$$

Since

$$K_w = [H_3O^+] \cdot [OH^-] = 1.0 \times 10^{-14}$$

we have

$$(1.0 \times 10^{-1} \text{ M}) \cdot [OH^-] = 1.0 \times 10^{-14}$$
$$[OH^-] = 1.0 \times 10^{-13} \text{ M}$$

TRY IT YOURSELF

What are $[H_3O^+]$ and $[OH^-]$ in 0.010 M KOH, a strong base?

ANSWER

$$[OH^-] = 0.010 \text{ M} = 1.0 \times 10^{-2} \text{ M}$$
$$[H_3O^+] = 1.0 \times 10^{-12} \text{ M}$$

Let us summarize what we mean when we say that an aqueous solution is acidic, basic, or neutral (at 298 K):

Acidic: $[H_3O^+] > [OH^-]$; $[H_3O^+] > 1 \times 10^{-7}$ M and $[OH^-] < 1 \times 10^{-7}$ M

Basic: $[H_3O^+] < [OH^-]$; $[H_3O^+] < 1 \times 10^{-7}$ M and $[OH^-] > 1 \times 10^{-7}$ M

Neutral: $[H_3O^+] = [OH^-]$; $[H_3O^+] = 1 \times 10^{-7}$ M and $[OH^-] = 1 \times 10^{-7}$ M

In *every* case: $[H_3O^+] \cdot [OH^-] = K_w = 1 \times 10^{-14}$.

14.8A A MORE DETAILED LOOK AT pH AND pOH

Now let us construct a table that actually relates the pH of an aqueous solution to $[H_3O^+]$ and $[OH^-]$ at 298 K.

pH (pOH) and [H₃O⁺] ([OH⁻]) in an Aqueous Solution

pH	$[H_3O^+]$	$[OH^-]$	pOH
0	1×10^0 M	1×10^{-14} M	14
1	1×10^{-1} M	1×10^{-13} M	13
2	1×10^{-2} M	1×10^{-12} M	12
3	1×10^{-3} M	1×10^{-11} M	11
4	1×10^{-4} M	1×10^{-10} M	10
5	1×10^{-5} M	1×10^{-9} M	9
6	1×10^{-6} M	1×10^{-8} M	8
7	1×10^{-7} M	1×10^{-7} M	7
8	1×10^{-8} M	1×10^{-6} M	6
9	1×10^{-9} M	1×10^{-5} M	5
10	1×10^{-10} M	1×10^{-4} M	4
11	1×10^{-11} M	1×10^{-3} M	3
12	1×10^{-12} M	1×10^{-2} M	2
13	1×10^{-13} M	1×10^{-1} M	1
14	1×10^{-14} M	1×10^0 M	0

We note that in every case

$$[H_3O^+] \cdot [OH^-] = 1 \times 10^{-14} = K_w$$

We also note that we really don't need to know both $[H_3O^+]$ and $[OH^-]$ in order to determine how acidic or basic a solution is: either one will do. In practice, we use $[H_3O^+]$. To simplify matters even further, we can use the exponent of the hydronium ion concentration, which we call **pH** (the term stands for "power of the hydronium ion"). In our table pH values are given in the left column.

We can also use the exponent of $[OH^-]$ to develop a **pOH** scale that runs from 14 to 0, as shown in the table. A pOH less than 7 indicates a basic solution; the lower the pOH below 7, the more basic the solution. A pOH greater than 7 indicates an acidic solution; the higher the pOH above 7, the more acidic the solution. A pOH of 7 indicates a neutral solution. Since the pOH scale is based on the exponents of $[OH^-]$, a change of one pOH unit equals a tenfold change in $[OH^-]$. In our table pOH values are given in the fourth column.

Just as there is a relationship between $[H_3O^+]$ and $[OH^-]$, there is also a relationship between pH and pOH. If we inspect our table, we see that

$$\boxed{\text{pH} + \text{pOH} = 14}$$

PROBLEM

What is the pH of a solution whose $[OH^-]$ is 1×10^{-5} M? (Don't use the table!)

SOLUTION
We must find $[H_3O^+]$ in order to solve the problem.

$$[H_3O^+] \cdot (1.0 \times 10^{-5}\ M) = 1.0 \times 10^{-14}$$
$$[H_3O^+] = 1.0 \times 10^{-9}\ M$$
$$pH = 9$$

TRY IT YOURSELF
What is the pOH of the solution described above?

ANSWER
5

General Definitions of pH and pOH

Since the table on page 405 lists only concentrations of the form $1.0 \times 10^{-x}\ M$, we cannot use it to calculate the pH or pOH of *every* aqueous solution. For example, suppose $[H_3O^+]$ in an aqueous solution is 2.0×10^{-3} M. How can we calculate the pH of this solution?

The general definitions of pH and pOH, which can be used for *any* aqueous solution, are as follows:

$$pH = -\log_{10} [H_3O^+]$$
$$pOH = -\log_{10} [OH^-]$$

In our example, if $[H_3O^+]$ in an aqueous solution is 2.0×10^{-3} M, then

$$pH = -\log_{10} (2.0 \times 10^{-3}) = 2.7$$

PROBLEM
Calculate the pOH and pH of a 0.0003 M aqueous solution of KOH, a strong base.

SOLUTION
Since KOH is a strong base, it dissociates nearly 100% in aqueous solution and $[OH^-] = 0.0003$ M.

$$pOH = -\log_{10} [OH^-] = -\log_{10} (0.0003\ M) = 3.5$$
$$pH = 14 - pOH = 10.5$$

TRY IT YOURSELF
Calculate the pH and pOH of 0.04 M HCl, a strong acid.

ANSWER
pH = 1.4; pOH = 12.6

14.9A HYDROLYSIS OF SALTS IN AQUEOUS SOLUTIONS

If we test an aqueous NaCl solution, we find that its pH is 7. Since NaCl is the product of the neutralization of HCl by NaOH, we are not surprised that the solution is neutral.

However, not all salts produce neutral solutions when dissolved in water. For example, the salt NH_4Cl produces an acidic solution; that is, its pH is less than 7. The salt K_2CO_3 produces a basic solution; its pH is greater than 7. To learn why this occurs, we must look at the acids and bases that produce the salt.

If the acid and base are both strong, the neutralization is complete and the salt will produce a neutral solution. We recall that HI, HBr, HCl, HNO_3, and H_2SO_4 are considered strong acids. The Group 1 hydroxides (LiOH, NaOH, KOH, RbOH, and CsOH) are considered strong bases. Any of these combinations (strong acid and strong base) will produce a salt whose aqueous solution is neutral.

Let us analyze why the salt produced by the interaction of a *strong acid and a weak base* produces an *acidic* solution when dissolved in water. The salt NH_4Cl is produced by the reaction of HCl (a strong acid) and NH_3 (a weak base). When this salt is dissolved in water, NH_4^+ and Cl^- ions are present. If we examine the conjugate acid-base pairs table on page 413, we find that:

- The Cl^- ion is a *weaker base* than a molecule of H_2O. Therefore, Cl^- ions have no tendency to accept protons from water molecules, and no additional OH^- ions will be produced in the solution.
- The NH_4^+ ion is a *stronger acid* than a molecule of H_2O. Therefore, NH_4^+ ions can donate protons to water molecules, and additional H_3O^+ ions will be produced in the solution.

As a result, the solution contains a slight excess of H_3O^+ ions and is acidic. We call this process, in which NH_4^+ ions interact with H_2O molecules, **hydrolysis**, and we say that NH_4^+ ions *hydrolyze* in water, but Cl^- ions do not.

Using this analysis, we can conclude that *any combination of a strong acid and a weak base will produce a salt whose aqueous solution is acidic.*

We can perform a similar analysis for salts that form from the interaction of a *weak acid and a strong base*. The salt $NaC_2H_3O_2$ (sodium acetate) is formed from the interaction of $HC_2H_3O_2$ (acetic acid, a weak acid) and

NaOH (sodium hydroxide, a strong base). In aqueous solution, Na^+ and $C_2H_3O_2^-$ ions are present. A Na^+ ion will not hydrolyze because it is neither a conjugate acid nor a conjugate base. However, if we examine the table on p. 399, we find that the $C_2H_3O_2^-$ ion is a stronger base than a molecule of H_2O. Therefore, $C_2H_3O_2^-$ ions will hydrolyze in water by accepting protons from water molecules, and additional OH^- ions will be produced in the solution.

As a result, the solution contains an excess of OH^- ions and is basic. Using this analysis, we can conclude that *any combination of a weak acid and a strong base will produce a salt whose aqueous solution is basic*.

PROBLEM

Using the analyses described above, explain why an aqueous solution of KCl (derived from the strong acid HCl and the strong base KOH) will be *neutral*.

SOLUTION

The aqueous solution contains K^+ and Cl^- ions. A K^+ ion does not hydrolyze in water because it is neither a conjugate acid nor a conjugate base. A Cl^- ion does not hydrolyze in water because it is a weaker base than a molecule of H_2O. As a result, no additional H_3O^+ or OH^- ions will be produced and the solution will be neutral.

What will happen if a weak acid reacts with a weak base? We cannot easily predict the outcome; the salt may produce an acidic, basic, or neutral solution in water. To answer the question, we would need to know more about the relative strengths of the acid and base that react. This is the stuff of more advanced chemistry courses!

PROBLEM

For each of the following salts, identify its parent acid and base. Then indicate the kind of aqueous solution—acidic, basic, or neutral—that the salt ought to produce.
(a) LiF   (b) RbI   (c) $Al(NO_3)_3$ (d) $MgCl_2$

SOLUTIONS

	Parent Acid	Parent Base	Result
(a)	HF (weak)	LiOH (strong)	Basic solution
(b)	HI (strong)	RbOH (strong)	Neutral solution
(c)	HNO_3 (strong)	$Al(OH)_3$ (weak)	Acidic solution
(d)	HCl (strong)	$Mg(OH)_2$ (weak)	Acidic solution

TRY IT YOURSELF

For each salt, tell whether it will dissolve in water to produce an acidic, basic, or neutral solution.
(a) NaBr   (b) $CuSO_4$ (c) K_3PO_4

ANSWERS
(a) NaBr(aq) is neutral. (b) $CuSO_4$(aq) is acidic. (c) K_3PO_4(aq) is basic.

14.10A ACID–BASE PROPERTIES OF OXIDES

Certain oxides can combine *directly* with water to produce an acid or a base. The metallic oxides of Groups 1 and 2 form bases:

$$K_2O(s) + H_2O\ (\ell) \rightarrow 2KOH(aq)$$
$$CaO(s) + H_2O\ (\ell) \rightarrow Ca(OH)_2(aq)$$

Such substances are called *basic oxides*.
The nonmetallic oxides of Groups 14, 15, 16, and 17 form acids:

$$CO_2(g) + H_2O\ (\ell) \rightarrow H_2CO_3(aq)$$
$$P_2O_5(g) + 3H_2O\ (\ell) \rightarrow 2H_3PO_4(aq)$$
$$SO_3(g) + H_2O\ (\ell) \rightarrow H_2SO_4(aq)$$
$$Cl_2O_7(g) + H_2O\ (\ell) \rightarrow 2HClO_4(aq)$$

Such substances are called *acidic oxides*. Other oxides may be acidic, basic, or even amphoteric.

14.11A LEWIS DEFINITIONS OF ACIDS AND BASES

The Brønsted–Lowry definition of acid–base behavior, which extended the Arrhenius definition, was in turn extended by the Lewis definition of acid–base behavior (formulated by G.N. Lewis in the early 1920s). The Brønsted–Lowry theory classifies acids and bases in terms of how they interact with protons; the Lewis theory, in terms of how they interact with electron pairs. The Lewis definitions are as follows:

> **Lewis acid:** *Any species that can accept a pair of electrons from another species, that is, an* electron-pair acceptor.
>
> **Lewis base:** *Any species that can donate a pair of electrons to another species, that is, an* electron-pair donor.

In the Lewis system, H^+ is an acid because it can accept a pair of electrons from a Lewis base such as OH^-, which can donate the pair of electrons, as shown on the next page.

$$H^+ \quad + \quad :\overset{..}{\underset{..}{O}}-H^- \quad \longrightarrow \quad \overset{\textstyle H}{\underset{}{:\overset{..}{O}-H}}$$

Lewis Acid	Lewis Base	Lewis Acid–Base
(electron-pair	(electron-pair	Product
acceptor)	donor)	

We might be tempted to ask what all the shouting is about since H^+ and OH^- are very nicely defined in terms of the Brønsted–Lowry model. A second example, however, will make the distinction more apparent.

In Chapter 9 we discovered that the compound BF_3 was stable even though boron contains only six electrons in its outer principal energy level. When BF_3 combines with NH_3, boron forms a coordinate covalent bond with nitrogen. In terms of the Lewis definitions, BF_3 is a Lewis acid and NH_3 is a Lewis base, as shown below.

$$\overset{\textstyle F}{\underset{\textstyle F}{F-B}} \quad + \quad \overset{\textstyle H}{\underset{\textstyle H}{:N-H}} \quad \longrightarrow \quad \overset{\textstyle F\ H}{\underset{\textstyle F\ H}{F-B:N-H}}$$

Lewis Acid	Lewis Base	Lewis Acid–Base
(electron-pair	(electron-pair	Product
acceptor)	donor)	

What is remarkable is that BF_3 cannot be considered a Brønsted–Lowry acid because it has no proton to donate! The Lewis definitions of acids and bases play important roles in understanding reactions that occur by means of shifting electron-pairs. This statement is particularly true in organic chemistry.

END-OF-CHAPTER QUESTIONS

Some questions have the symbol "§2" in front of the question number. This symbol means that the question is based on Section II material.

1. Which substance is classified as a salt?
   (1) $Ca(OH)_2$ (2) $C_2H_4(OH)_2$
   (3) CCl_4 (4) $CaCl_2$

2. Water containing dissolved electrolyte conducts electricity because the solution contains mobile
   (1) electrons   (2) molecules   (3) atoms   (4) ions

3. What is the name for the sodium salt of the acid $HClO_2$? (*Hint:* See the table on page 402.)
   (1) sodium chlorite   (2) sodium chloride
   (3) sodium chlorate   (4) sodium perchlorate

4. According to the Arrhenius theory, when an acid substance is dissolved in water it will produce a solution containing only one kind of positive ion. To which ion does the theory refer?
   (1) acetate   (2) hydrogen   (3) chloride   (4) sodium

5. Which equation represents a neutralization reaction?
   (1) $H^+(aq) + OH^-(aq) \rightarrow H_2O\ (\ell)$
   (2) $Ag^+(aq) + I^-(aq) \rightarrow AgI(s)$
   (3) $Zn(s) + 2HCl(aq) \rightarrow ZnCl_2(aq) + H_2(g)$
   (4) $NaCl(aq) + AgNO_3(aq) \rightarrow NaNO_3(aq) + AgCl(s)$

6. When substance X is dissolved in water, the only positive ions in the solution are hydronium ions. Substance X could be
   (1) $NaOH$ (2) NaH (3) HNO_3 (4) NH_3

7. Which substance can act as an Arrhenius acid in aqueous solution?
   (1) NaI (2) HI (3) LiH (4) NH_3

8. Which species is classified as an Arrhenius base?
   (1) CH_3OH (2) $LiOH$ (3) NH_3 (4) BF_3

9. A water solution contains 0.50 mole of HCl. How much KOH should be added to the HCl solution to exactly neutralize it?
   (1) 1.0 mole   (2) 2.0 moles   (3) 0.25 mole   (4) 0.50 mole

10. A 30-milliliter sample of HCl is completely neutralized by 10. milliliters of a 1.5 M NaOH solution. What is the molarity of the HCl solution?
    (1) 0.25   (2) 0.50   (3) 1.5   (4) 4.5

11. How much water is formed when 1.0 mole of HCl reacts completely with 1.0 mole of NaOH?
    (1) 1.0 mole   (2) 2.0 moles   (3) 0.50 mole   (4) 0.25 mole

12. What volume of a 0.200 M NaOH solution is needed to exactly neutralize 40.0 milliliters of a 0.100 M H_2SO_4 solution?
    (1) 10.0 mL   (2) 20.0 mL   (3) 40.0 mL   (4) 80.0 mL

§2 **13.** Which ion is the conjugate base of H_2SO_4?
    (1) SO_3^{2-} (2) S^{2-} (3) HSO_3^- (4) HSO_4^-

§2 **14.** According to the table on page 413, ammonia can act as
    (1) a Brønsted acid, only
    (2) a Brønsted base, only
    (3) either a Brønsted acid or a Brønsted base
    (4) neither a Brønsted acid nor a Brønsted base

§2 **15.** Which of the following Brønsted acids has the strongest conjugate base?
    (1) HF   (2) HCl   (3) HBr   (4) HI

§2 **16.** Which of the following is the *weakest* Brønsted acid?
    (1) HBr   (2) H_2S (3) H_2O (4) NH_3

17. According to the Brønsted–Lowry theory, an acid is any species that can
    (1) donate a proton   (2) accept a proton
    (3) donate an electron   (4) accept an electron

18. In the reaction

$$HSO_4^- + H_2O \rightarrow H_3O^+ + SO_4^{2-}$$

the HSO_4^- ion is
    (1) a proton donor   (2) an amphiprotic substance
    (3) a proton acceptor   (4) a base

§2 **19.** According to the Brønsted–Lowry theory, neutralization occurs in aqueous solution when there is a reaction between
    (1) H_3O^+ and H_2O (2) OH^- and H_2O
    (3) H_3O^+ and OH^- (4) H_2O and H_2O

20. In the reaction

$$NH_2^- + HOH \rightleftharpoons NH_3 + OH^-$$

the two Brønsted–Lowry bases are
    (1) OH^- and NH_2^- (2) HOH and NH_3
    (3) OH^- and HOH   (4) OH^- and NH_3

§2 **21.** The ionization constants (K_a values) of four acids are shown below. Which K_a represents the *weakest* of these acids?
    (1) $K_a = 1.0 \times 10^{-5}$ (2) $K_a = 1.0 \times 10^{-4}$
    (3) $K_a = 7.1 \times 10^{-3}$ (4) $K_a = 1.7 \times 10^{-2}$

22. According to the table on page 413, which 0.1 M acid solution contains the highest concentration of H_3O^+ ions?
(1) H_2S (2) H_3PO_4 (3) HNO_2 (4) HF

23. The diagram below illustrates an apparatus used to test the conductivities of various solutions.

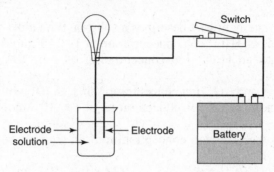

When the switch is closed, which of the following 1–molar solutions will cause the bulb to glow most brightly?
(1) ammonia   (2) acetic acid
(3) carbonic acid   (4) sulfuric acid

§2 **24.** A solution of hydrochloric acid contains
(1) fewer hydronium ions than chloride ions
(2) fewer hydroxide ions than hydronium ions
(3) more hydronium ions than chloride ions
(4) more hydroxide ions than hydronium ions

§2 **25.** As the H_3O^+ ion concentration of a solution increases at 298 K, K_w
(1) decreases   (2) increases   (3) remains the same

§2 **26.** The OH^- ion concentration is greater than the H_3O^+ ion concentration in a water solution of
(1) CH_3CH_2OH (2) $Ba(OH)_2$
(3) HCl   (4) H_2SO_4

§2 **27.** The K_w value for a sample of water at 1 atmosphere and 298 K will be most likely to change when there is an increase in the
(1) concentration of H^+ ions
(2) concentration of OH^- ions
(3) pressure
(4) temperature

§2 **28.** Given:

$$K_w = [H^+][OH^-] = 1 \times 10^{-14} \text{ at 298 K.}$$

What is the concentration of H^+ in pure water at 298 K?
(1) 1×10^{-7} mol/L   (2) 1×10^7 mol/L
(3) 1×10^{-14} mol/L   (4) 1×10^{14} mol/L

§2 **29.** Which statement best describes a solution with a pH of 3?
(1) It has an H_3O^+ ion concentration of 1×10^3 mol/L and is acidic.
(2) It has an H_3O^+ ion concentration of 1×10^{-3} mol/L and is acidic.
(3) It has an H_3O^+ ion concentration of 1×10^3 mol/L and is basic.
(4) It has an H_3O^+ ion concentration of 1×10^{-3} mol/L and is basic.

§2 **30.** What is the OH^- ion concentration of an aqueous solution that has a pH of 11?
(1) 1.0×10^{-11} mol/L   (2) 1.0×10^{-3} mol/L
(3) 3.0×10^{-1} mol/L   (4) 11×10^{-1} mol/L

§2 **31.** What is the pOH of a solution whose OH^- ion concentration is 1×10^{-3} M?
(1) 14   (2) 11   (3) 3   (4) 7

§2 **32.** The H_3O^+ ion concentration of an aqueous solution is 1×10^{-4} mole per liter. This solution has a pH of
(1) 4 and a pOH of 4   (2) 4 and a pOH of 10
(3) 10 and a pOH of 4   (4) 10 and a pOH of 10

§2 **33.** An aqueous solution has a hydronium ion concentration of 6.0×10^{-9} mole per liter at 25°C. The pH of this solution is closest to
(1) 5.8   (2) 6.0   (3) 8.2   (4) 9.0

34. If an aqueous solution turns phenolphthalein pink, the solution will turn litmus
(1) red   (2) purple   (3) blue   (4) violet

35. An indicator was used to test a water solution with a pH of 12. Which indicator color would be observed?
(1) colorless with litmus
(2) red with litmus
(3) colorless with phenolphthalein
(4) pink with phenolphthalein

36. According to Reference Table M, at what pH would the indicators bromthymol blue, thymol blue, and methyl orange all appear as yellow?
(1) 1.9   (2) 2.9   (3) 4.7   (4) 9.8

§2 **37.** When K_2CO_3 is dissolved in water, the resulting solution turns litmus paper
(1) red, and is acidic   (2) blue, and is acidic
(3) red, and is basic   (4) blue, and is basic

§2 **38.** Which salt will hydrolyze in water to produce a basic solution?
(1) BaI_2 (2) $NaNO_2$ (3) $CaCl_2$ (4) K_2SO_4

§2 **39.** Oxides of nonmetals react with water to form
(1) acids   (2) bases   (3) salts   (4) anhydrides

§2 **40.** Which of the following is the best classification for the compound BF_3 in this reaction: $BF_3 + NH_3 \rightarrow BF_3NH_3$?
(1) Lewis acid   (2) Lewis base
(3) Arrhenius acid   (4) Arrhenius base

41. Which chemical equation represents the reaction of an Arrhenius acid with an Arrhenius base?
(1) $HC_2H_3O_2 + NaOH(aq) \rightarrow NaC_2H_3O_2(aq) + H_2O(\ell)$
(2) $C_3H_8(g) + 5\ O_2(g) \rightarrow 3\ CO_2(g) + 4\ H_2O(\ell)$
(3) $Zn(s) + 2\ HCl(aq) \rightarrow ZnCl_2(aq) + H_2(g)$
(4) $BaCl_2(aq) + Na_2SO_4(aq) \rightarrow BaSO_4(s) + 2\ NaCl(aq)$

42. When the pH of a solution changes from a pH of 5 to a pH of 3, the hydronium ion concentration is
(1) 0.01 of the original content
(2) 0.1 of the original content
(3) 10 times the original content
(4) 100 times the original content

43. A sample of $Ca(OH)_2$ is considered to be an Arrhenius base because it dissolves in water to produce
(1) Ca^{2+} ions as the only positive ion in solution
(2) H_3O^+ ions as the only positive ions in solution
(3) OH^- ions as the only negative ions in solution
(4) H^- ions as the only negative ions in solution

44. Which reaction occurs when hydrogen ions react with hydroxide ions to form water?
(1) substitution
(2) saponification
(3) ionixation
(4) neutralization

45. Which of these 1 M solutions has the highest pH?
(1) CH_3OH
(2) NaOH
(3) HCl
(4) NaCl

46. Phenolphthalein is pink in an aqueous solution having a pH of
(1) 5
(2) 2
(3) 7
(4) 12

47. The pH of a solution is 7. When acid is added to the solution, the hydronium ion concentration becomes 100 times greater. What is the pH of the new solution?
(1) 1
(2) 5
(3) 9
(4) 14

Constructed-Response Questions

§2 **1.** The substance methylamine (CH_3NH_2) can act as a Brønsted–Lowry base in water.
(a) Write the reaction between methylamine and water. [*Hint*: Note the structural similarity of methylamine to ammonia (NH_3).]
(b) Write the K_b expression for methylamine.

§2 **2.** (a) Calculate the pH of a solution in which $[OH^-] = 2.00 \times 10^{-4}$ M.
(b) Calculate the pOH of a solution in which $[H_3O^+] = 5.00 \times 10^{-3}$ M.

Base your answers to questions 3 through 6 on the information and data table below.

A titration setup was used to determine the unknown molar concentration of a solution of NaOH. A 1.2 M HCl solution was used as the titration standard. The following data were collected.

	Trial 1	Trial 2	Trial 3	Trial 4
Amount of HCl Standard Used	10.0 mL	10.0 mL	10.0 mL	10.0 mL
Initial NaOH Buret Reading	0.0 mL	12.2 mL	23.2 mL	35.2 mL
Final NaOH Buret Reading	12.2 mL	23.2 mL	35.2 mL	47.7 mL

3. Calculate the volume of NaOH solution used to neutralize 10.0 mL of the HCl standard solution in trial 3.

4. According to Reference Table M in Appendix 1, which indicator would be most appropriate in determining the end point of this titration? Give one reason for choosing this indicator.

5. Calculate the average molarity of the unknown NaOH solution for all four trials.

6. Explain why it is better to use the average data from multiple trials rather than the data from a single trial to calculate the results of the titration.

Base your answers to questions 7–10 on the information below and on your knowledge of chemistry.

A NaOH(aq) solution and an acid–base indicator are used to determine the molarity of an HCl(aq) solution. A 25.0-milliliter sample of HCl(aq) is exactly neutralized by 15.0 milliliters of 0.20 M NaOH(aq).

7. Identify the laboratory process mentioned in this passage.

8. Write the complete equation for the neutralization reaction that occurs.

9. Based on the data given, the calculated molarity of the HCl(aq) solution should be expressed to what number of significant figures?

10. Using the data, determine the concentration of the HCl(aq).

Base your answers to questions 11–13 on the information below and on your knowledge of chemistry.

A company produces a colorless vinegar that is 5.0% $HC_2H_3O_2$ in water. Using thymol blue as an indicator, a student titrates a 15.0-milliliter sample of the vinegar with 43.1 milliliters of 0.30 M NaOH(aq) solution until the acid is neutralized.

11. Based on Reference Table M in Appendix 1, what is the color of the indicator in the vinegar solution before any base is added?

12. Identify the negative ion in the NaOH(aq) used in this titration.

13. Using the titration data, determine the molarity of the $HC_2H_3O_2$ in the vinegar sample.

The answers to these questions are found in Appendix 3.

REDUCTION-OXIDATION (REDOX) AND ELECTROCHEMISTRY

KEY IDEAS

This chapter focuses on the roles of oxidation and reduction in chemistry. In particular, the production of electrical energy by electrochemical cells is examined, and a number of important commercial applications are presented.

KEY OBJECTIVES

At the conclusion of this chapter you will be able to:

• Formally define the terms *oxidation* and *reduction*.
• Define the term *redox reaction*, and identify these reactions.
• Write oxidation and reduction half-reactions for redox equations.
• Balance redox equations by the half-reaction and ion-electron methods.
• Predict when a redox reaction will occur spontaneously.
• Draw and label a simple voltaic cell with a porous barrier and with a salt bridge.
• Describe electron and ion movements in a simple voltaic cell.
• Define the terms *standard electrode potential, standard reduction potential,* and *standard oxidation potential*.
• Use standard potentials to calculate the potential difference of a voltaic cell operating under standard conditions.
• Define *electrolysis*, and draw and label a simple electrolytic cell.
• Describe the electrolysis of fused salts, water, and brine.
• Describe the principles of electroplating.
• Apply redox and electrochemistry to real-world applications.

SECTION I—BASIC (REGENTS-LEVEL) MATERIAL

NYS REGENTS CONCEPTS AND SKILLS

Note: By the time you have finished Section I, you should have mastered the concepts and skills listed below. The Regents chemistry examination will test your knowledge of these items and your ability to apply them.

Concepts are the *basic ideas* that form the body of the Regents chemistry course (what you need to know!).

Skills are the *activities* that demonstrate your mastery of these concepts (how you show that you know them!).

Following each concept or skill is a page reference (given in parentheses) to this chapter.

15.1 Concept:
An oxidation-reduction (redox) reaction involves the transfer of electrons (e^-). (Page 437)

Skill:
Determine a missing product in a balanced redox equation. (Pages 437–438)

15.2 Concept:
Reduction is the gain of electrons. (Page 436)

15.3 Concept:
A half-reaction can be written to represent reduction. (Page 436)

Skill:
Write and balance reduction half-reactions for free elements and their monatomic ions. (Page 436)

15.4 Concept:
Oxidation is the loss of electrons. (Page 436)

15.5 Concept:
A half reaction can be written to represent oxidation. (Page 436)

Skill:
Write and balance oxidation half-reactions for free elements and their monatomic ions. (Page 436)

15.6 Concept:
In a redox reaction the number of electrons lost is equal to the number of electrons gained. (Pages 437–438)

15.7 Concepts:
• Oxidation numbers (states) can be assigned to atoms and ions. (Page 436)
• Changes in oxidation numbers indicate that oxidation or reduction has occurred. (Page 437)

15.8 Concepts:
• An electrochemical cell can be either voltaic or electrolytic. (Page 441)
• In an electrochemical cell, oxidation occurs at the anode and reduction at the cathode. (Page 442)

Skill:
Compare and contrast voltaic and electrolytic cells.
(Pages 442, 443–444)

15.9 Concept:
A voltaic cell spontaneously converts chemical energy to electrical energy. (Pages 441–442)

Skills:
• Given a redox equation, identify the parts of a voltaic cell:
    (a) anode (Page 441);
    (b) cathode (Page 441);
    (c) salt bridge (Page 441).
• Identify the direction of electron flow in a voltaic cell.
    (Page 441)
• Use an activity series to determine whether a redox reaction is spontaneous. (Pages 438–441)

15.10 Concept:
An electrolytic cell requires electrical energy to produce chemical change. This process is known as electrolysis. (Pages 438–441)

Skills:
• Given a redox equation, identify the parts of an electrolytic cell:
    (a) anode (Page 443);
    (b) cathode (Page 443).
• Identify the direction of electron flow in an electrolytic cell.
    (Page 443)

15.1 WHAT ARE OXIDATION AND REDUCTION?

Most of us are already acquainted with the term *oxidation*. Historically, *oxidation* means the combination of a substance with oxygen, as in the following example:

$$2Mg + O_2 \rightarrow 2MgO$$

Oxygen is called the *oxidizing agent*, and magnesium is the substance that is oxidized.

Historically, *reduction* is associated with the removal of oxygen from a substance, as in the following example:

$$CuO + H_2 \rightarrow Cu + H_2O$$

Hydrogen is called the *reducing agent*, and copper(II) oxide is the substance that is reduced.

15.2 FORMAL DEFINITIONS OF OXIDATION AND REDUCTION

At this point, you should review the section on assigning oxidation numbers (Chapter 3, pages 62–63). Remember that charges are written so that the sign is placed *after* the number, as in 2+, and oxidation numbers are written so that the sign of the number is placed *before* the number itself, as in +2. Oxidation and reduction arise out of the competition for electrons in a chemical reaction.

Let us examine the oxidation of magnesium by oxygen, this time including the oxidation number for each atom in the equation:

$$2Mg^0 + O^0_2 \rightarrow 2Mg^{2+}O^{2-}$$

We see that magnesium has increased its oxidation number from 0 to 2+. We know from Chapter 9 on chemical bonding that charges arise from the loss or gain of electrons. We can conclude that each magnesium atom lost two electrons when it became oxidized, and we can write this oxidation as a half-reaction:

$$Mg^0 \rightarrow Mg^{2+} + 2e^- \quad \text{(oxidation half-reaction)}$$

The formal definition of **oxidation** is as follows:

> ### Oxidation: *a loss of electrons*

Now let us examine the reduction of copper(II) oxide by hydrogen and also include the oxidation numbers for each element in the equation:

$$Cu^{2+}O^{2-} + H^0_2 \rightarrow Cu^0 + H^{+1}_2O^{-2}$$

We note that the reduction of copper(II) oxide does not involve the oxygen because the oxidation number of oxygen remains unchanged. The oxidation number of copper, however, has changed: it has been decreased from +2 to 0. We conclude that the copper gained two electrons when it was reduced, and we can also write this reduction as a half-reaction:

$$Cu^{2+} + 2e^- \rightarrow Cu^0 \quad \text{(reduction half-reaction)}$$

The formal definition of **reduction** is as follows:

> ### Reduction: *a gain of electrons*

As a result of these formal definitions, we have expanded what we mean by oxidation and reduction. For example, the presence of oxygen is not even required. Any atom that loses one or more electrons is oxidized, and any atom that gains electrons is reduced.

Not only are oxidation and reduction opposite processes, but also one cannot occur without the other! If an atom loses electrons, there must be another atom that will gain them. We call all reactions involving oxidation and reduction **redox reactions**.

In Section 15.1, in our example involving the oxidation of magnesium, we called oxygen the *oxidizing agent*. We see that its oxidation number decreases; therefore, it has been reduced. This is a general rule: *In a redox reaction, the oxidizing agent is reduced.*

In our example involving copper(II) oxide and hydrogen, we called hydrogen the *reducing agent*. We see that its oxidation number has been increased; therefore, it has been oxidized. This is also a general rule: *In a redox reaction, the reducing agent is oxidized.*

15.3   REDOX EQUATIONS

Is this (unbalanced) equation:

$$NH_3 + O_2 \rightarrow NO + H_2O$$

a redox equation? To answer this question, we rewrite the equation and include the oxidation number for each element. Then, we look for changes that occurred in the oxidation numbers as a result of the reaction: an increase means oxidation, and a decrease means reduction.

$$N^{-3}H^{+1}_3 + O^0_2 \rightarrow N^{+2}O^{2-} + H^+_2O^{-2}$$

In this reaction, nitrogen changes its oxidation number from -3 to $+2$; its oxidation number increases, and therefore nitrogen has been oxidized. Oxygen changes its oxidation number from 0 to -2; its oxidation number decreases, and therefore oxygen has been reduced. Hydrogen, whose oxidation number is not changed, is neither oxidized nor reduced. This is indeed a redox equation. As with all other chemical reactions, mass, energy, and electric charge are *conserved* within redox reactions.

PROBLEM
Indicate whether each of the following equations is a redox equation. If it is, write the half-reactions representing oxidation and reduction.
(a) $HNO_3 + I_2 \rightarrow HIO_3 + NO_2 + H_2O$ (unbalanced)
(b) $HCl + NaOH \rightarrow H_2O + NaCl$

SOLUTIONS
(a) $H^{+1}N^{+5}O_3^{-2} + I^0_2 \rightarrow H^{+1}I^{+5}O_3^{-2} + N^{+4}O_2^{-2} + H^{+1}_2O^{-2}$
   N changes its oxidation number from $+5$ to $+4$; it is reduced.
   $N^{+5} + e^- \rightarrow N^{+4}$ (reduction half-reaction)
   I changes its oxidation number from 0 to $+5$; it is oxidized.

437

$I^0 \rightarrow I^{+5} + 5e^-$ (oxidation half-reaction)

This equation represents a redox equation.

(b) $H^{+1}Cl^{-1} + Na^+O^{2-}H^{+1} \rightarrow H^{+1}_2O^{-2} + Na^+Cl^-$

No atom changes its oxidation number.

This equation does *not* represent a redox equation.

TRY IT YOURSELF

Indicate whether each of the following unbalanced equations is a redox equation.

(a) $Cu + HNO_3 \rightarrow Cu(NO_3)_2 + NO_2 + H_2O$

(b) $CH_4 + O_2 \rightarrow CO_2 + H_2O$

ANSWERS

Both (a) and (b) are redox equations.

15.4   SPONTANEOUS REDOX REACTIONS

The accompanying diagram represents a simple laboratory experiment. A strip of silver-gray zinc metal is placed in a beaker containing a (blue) solution of aqueous Cu^{2+} ions, and a strip of red copper metal is inserted into a second beaker containing a (colorless) solution of aqueous Zn^{2+} ions. In the beaker at the left, we will observe that the solution loses its blue color and the zinc strip becomes coated with copper metal. We observe no reaction in the beaker at the right.

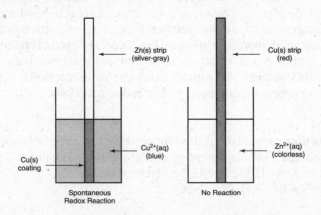

The loss of the blue color and the appearance of the metallic deposit indicate that Cu^{2+} ion is being converted to $Cu(s)$ by reduction. The half-reaction that occurs is

$$Cu^{2+}(aq) + 2e^- \rightarrow Cu(s)$$

The two electrons are supplied by the Zn(s), which is oxidized to Zn^{2+}. The half-reaction that occurs is

$$Zn(s) \rightarrow Zn^{2+}(aq) + 2e^-$$

The net equation for this reaction is

$$Zn(s) + Cu^{2+}(aq) \rightarrow Zn^{2+}(aq) + Cu(s)$$

This redox reaction is clearly spontaneous under laboratory conditions; therefore, the reverse reaction—between Cu(s) and Zn^{2+}(aq)—cannot be spontaneous. This is why we see no reaction in the beaker on the right side of the diagram.

Like all redox reactions, this reaction involves a competition for electrons. Here, Zn(s) loses electrons more easily than Cu(s), and Cu^{2+}(aq) gains electrons more easily than Zn^{2+}(aq), as we observe.

How can we predict whether another, similar experiment will yield a spontaneous redox reaction? The *Activity Series* in Reference Table J in Appendix 1 lists a number of *metals* in (descending) order of their abilities to be *oxidized*. The table also lists a number of *nonmetals* in (descending) order of their abilities to be *reduced*. Both series are based on H_2, known as the *hydrogen standard*.

For example, in Reference Table J, manganese (Mn) is listed higher than nickel (Ni). This means that the oxidation half-reaction

$$Mn(s) \rightarrow Mn^{2+}(aq) + 2e^-$$

is more likely to occur than the oxidation half-reaction

$$Ni(s) \rightarrow Ni^{2+}(aq) + 2e^-$$

The spontaneous reaction that will occur between Mn/Mn^{2+} and Ni/Ni^{2+} is

$$Mn(s) + Ni^{2+}(aq) \rightarrow Mn^{2+}(aq) + Ni(s)$$

Notice that the oxidation of Mn to Mn^{2+} has *forced the reduction* of Ni^{2+} to Ni. We can always use Reference Table J and a simple rule to determine how a spontaneous redox reaction between two metals (including H_2) will occur: *The metal that is listed higher in the table will be oxidized, and the metal that is listed lower in the table will be reduced.*

PROBLEM

Use Reference Table J to write the balanced spontaneous redox reaction between the following pairs of metals and their ions:

$$Au(s)/Au^{3+}(aq)$$

$$Sn(s)/Sn^{2+}(aq)$$

SOLUTION

We see that Sn appears higher on Reference Table J than Au. Therefore, Sn(s) will be oxidized and Au^{3+}(aq) will be reduced. The half-reactions are

$$Sn(s) \rightarrow Sn^{2+}(aq) + 2e^-$$

$$Au^{3+}(aq) + 3e^- \rightarrow Au(s)$$

In order to combine these two half-reactions, we must be certain that the number of electrons lost by Sn(s) is equal to the number of electrons gained by Au^{3+}(aq). This is accomplished by multiplying the first half-reaction by 3 and the second half-reaction by 2. As a result, 6 electrons will be transferred during the reaction. The final *balanced* redox equation is

$$3Sn(s) + 2Au^{3+}(aq) \rightarrow 3Sn^{2+}(aq) + 2Au(s)$$

TRY IT YOURSELF

Use Reference Table J to write the balanced spontaneous redox reaction between the following pairs of metals and their ions:

$$Fe(s)/Fe^{2+}(aq)$$

$$Pb(s)/Pb^{2+}(aq)$$

ANSWER

$$Fe(s) + Pb^{2+}(aq) \rightarrow Fe^{2+}(aq) + Pb(s)$$

Pairs of nonmetals listed in Reference Table J (F_2, Cl_2, Br_2, and I_2) are handled similarly, *except* that the higher nonmetal is *reduced* and lower one *oxidized*.

PROBLEM

Use Reference Table J to write the balanced spontaneous redox reaction between the following pairs of metals and their ions:

$$Br_2(\ell)/Br^-(aq)$$

$$Cl_2(g)/Cl^-(aq)$$

SOLUTION

Since Cl_2 is higher than Br_2, Cl_2(g) will be reduced to Cl^-(aq), and Br^-(aq) will be oxidized to $Br_2(\ell)$ The relevant half reactions are

$$Cl_2(g) + 2e^- \rightarrow 2Cl^-(aq)$$

$$2Br^-(aq) \rightarrow Br_2(\ell) + 2e^-$$

The balanced redox equation is

$$Cl_2(g) + 2Br^-(aq) \rightarrow 2(Cl^-(aq) + Br_2(\ell)$$

15.5 ELECTROCHEMICAL CELLS

An **electrochemical cell** is a device that relates electricity to oxidation-reduction reactions. There are two types of electrochemical cells: *voltaic cells* and *electrolytic cells*.

Voltaic Cells

A **voltaic cell** (also known as a *Galvanic cell*) uses a spontaneous redox reaction to provide a source of electrical energy. It is designed so that the oxidation and reduction half-reactions occur in separate half-cells that are connected to one another.

Under standard laboratory conditions, electrochemical cells are constructed to operate at 298 K and 100 kilopascals (nearly 1 atmosphere). All solutions have a 1 M concentration, and any gases present have a partial pressure of 100 kilopascals.

A simple design for a voltaic cell involving the zinc-copper reaction described at the beginning of Section 13.6 is shown in the accompanying diagram.

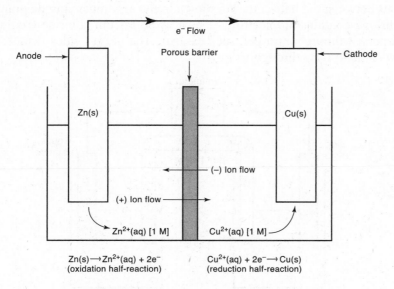

$Zn(s) \rightarrow Zn^{2+}(aq) + 2e^-$
(oxidation half-reaction)

$Cu^{2+}(aq) + 2e^- \rightarrow Cu(s)$
(reduction half-reaction)

Net Reaction: $Zn(s) + Cu^{2+}(aq) \rightarrow Cu(s) + Zn^{2+}(aq)$

Oxidation occurs in the left half-cell. The electrons that are released travel through the external wire and enter the Cu(s). Reduction occurs in the right half-cell. The metal strips are called *electrodes*. The electrode at which oxidation occurs is the *anode*; the electrode at which reduction occurs, the *cathode*. *The direction of the electron flow in a voltaic cell is always from the anode (where oxidation occurs) to the cathode (where reduction occurs).*

As the reaction progresses, an electrical imbalance in the half-cells will stop the voltaic cell from functioning. However, the *porous barrier* allows the migration of positive and negative ions between the half-cells, keeping them electrically neutral. In the diagram, the directions of the positive and negative ion flows are shown.

If we connect the external wire to an external circuit, we will have a usable source of electrical energy; if it were feasible to use this voltaic cell commercially, we would call it a *battery*! *The anode of a voltaic cell is the negative terminal* because electrons are flowing out of it; *the cathode is the positive terminal* because electrons are flowing into it.

After a period of time, the redox reaction of the cell reaches equilibrium and the cell no longer operates. When this occurs in a battery, we say that the battery is dead!

Voltaic Cells with Salt Bridges

We can also build a voltaic cell whose half-cells are completely separated, but we must provide a device known as a *salt bridge* to allow for the flow of ions between the half-cells. As shown in the accompanying diagram, the salt bridge contains an electrolyte, such as potassium chloride (KCl), that is dispersed throughout a gel, such as agar. The gel provides firmness but allows ions to flow through it.

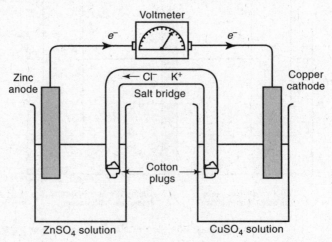

Since Zn^{2+}(aq) is produced in the left half-cell, negative ions are required to keep the solution electrically neutral. Therefore, Cl^- ions flow into this half-cell. At the same time, Cu^{2+}(aq) is being used up in the right half-cell, which requires positive ions to keep the solution electrically neutral. Therefore, K^+ ions flow into this half-cell.

Electrolytic Cells and Electrolysis

Electrolytic cells are electrochemical cells that are used to force nonspontaneous redox reactions to occur. Unlike voltaic cells, they do not generate electrical energy—they use it. In practice, an electric current is passed through the substance inside the cell. The process is called *electrolysis* (meaning "to break apart using electricity").

Operation of an Electrolytic Cell

The redox reaction

$$2NaCl \rightarrow 2Na + Cl_2$$

is *not spontaneous*.

To make this reaction occur, we must perform the electrolysis on the fused or molten salt, $NaCl(\ell)$. In the liquid phase, NaCl is a "soup" of freely moving Na^+ and Cl^- ions. The accompanying diagram illustrates how an electrolytic cell operates.

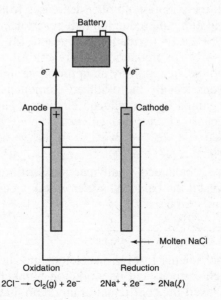

443

The source of electricity establishes positive and negative electrodes. Ions of opposite charge migrate to these electrodes, and the half-reaction occurs. Because oxidation occurs at the positive electrode, this is the anode of the electrolytic cell. Similarly, the negative electrode, at which reduction occurs, is the cathode.

SECTION II—ADDITIONAL MATERIAL

15.1A   BALANCING REDOX EQUATIONS BY THE HALF-REACTION METHOD

Redox equations may be very difficult to balance simply by inspection. A good case in point is this redox equation:

$$Cu + HNO_3 \rightarrow Cu(NO_3)_2 + NO + H_2O$$

Redox equations can be balanced, however, by identifying the elements that are oxidized and reduced and then writing their half-reactions. The seven steps in this **half-reaction method**, with the equation above used as an example, are as follows:

1. Rewrite the equation with its oxidation numbers.
   $$Cu^0 + H^{1+}N^{5+}O_3{}^{2-} \rightarrow Cu^{2+}(N^{5+}O_3{}^{2-})_2 + N^{2+}O^{2-} + H_2^{+1}O^{2-}$$
2. Identify the elements that are oxidized and reduced.
   Cu is oxidized; its oxidation number increases from 0 to +2.
   N is reduced; its oxidation number decreases from +5 to +2.
3. Write the oxidation and reduction half-reactions.
   $Cu^0 \rightarrow Cu^{2+} + 2e^-$ (oxidation half-reaction)
   $N^{5+} + 3e^- \rightarrow N^{2+}$ (reduction half-reaction)
4. Multiply each half-reaction by an appropriate number so that the number of electrons lost by the oxidized element is equal to the number of electrons gained by the reduced element. (This "balancing" step is necessary because *electric charge must be conserved*.)
   $3 \cdot (Cu^0 \rightarrow Cu^{2+} + 2e^-) = 3Cu^0 \rightarrow 3Cu^{2+} + 6e^-$
   $2 \cdot (N^{5+} + 3e^- \rightarrow N^{2+}) = 2N^{5+} + 6e^- \rightarrow 2N^{2+}$
5. Add the two "balanced" half-reactions, eliminating the electrons. We call the result the balanced *skeleton* redox equation.
   $3Cu^0 \rightarrow 3Cu^{2+} + 6e^-$
   $2N^{5+} + 6e^- \rightarrow 2N^{2+}$
   $\overline{\phantom{3Cu^0 + 2N^{5+} \rightarrow 3Cu^{2+} + 2N^{2+}}}$
   $3Cu^0 + 2N^{5+} \rightarrow 3Cu^{2+} + 2N^{2+}$
6. Insert the coefficients from the skeleton into the original equation by matching each element and its oxidation number. There is one exception: Do *not* insert the coefficient of any item that appears in more than *one* place in the equation.

$$3Cu^0 + \_H^{1+}N^{+5}O_3^{2-} \rightarrow 3Cu^{2+}(H^{+1}N^{+5}O_3^{-2})_2 + 2N^{2+}O^{2-} + \_H_2^{1+}O^{2-}$$

Since N^{+5} appears twice, we do *not* insert the coefficient for it.

$$3Cu + \_HNO_3 \rightarrow 3Cu(NO_3)_2 + 2NO + \_H_2O$$

7. Balance the rest of the equation (the nonredox part) by *inspection*.

$$3Cu + 8HNO_3 \rightarrow 3Cu(NO_3)_2 + 2NO + 4H_2O$$

TRY IT YOURSELF

Balance each of the following redox equations by the half-reaction method:
(a) $Cu + HNO_3 \rightarrow Cu(NO_3)_2 + NO_2 + H_2O$
(b) $HNO_3 + I_2 \rightarrow HIO_3 + NO_2 + H_2O$

ANSWERS

(a) $Cu + 4HNO_3 \rightarrow Cu(NO_3)_2 + 2NO_2 + 2H_2O$
(b) $10HNO_3 + I_2 \rightarrow 2HIO_3 + 10NO_2 + 4H_2O$

15.2A   BALANCING REDOX EQUATIONS BY THE ION–ELECTRON METHOD

Another technique, known as the **ion–electron method**, is useful for balancing *ionic* redox equations. The redox equation between copper and nitric acid:

$$Cu + HNO_3 \rightarrow Cu(NO_3)_2 + NO + H_2O$$

can be written as an ionic equation, stripped of all the items that do not actually participate in oxidation-reduction:

$$Cu + NO_3^- \rightarrow Cu^{2+} + NO \quad \text{(acidic solution)}$$

We need to indicate that the solution is *acidic*, because HNO_3 is not written out.

The ion–electron method used to balance this ionic redox equation is simpler than balancing by the half-reaction method because we do not have to assign oxidation numbers for individual elements. We balance for both mass and charge in a step-by-step fashion:

1. Separate the equation into half-reactions by selecting similar items on both sides of the arrow.

$$Cu \rightarrow Cu^{2+}$$
$$NO_3^- \rightarrow NO$$

2. Balance the half-reactions for all atoms that are neither hydrogen nor oxygen. The Cu and N atoms are already balanced.

3. Balance the half-reactions for oxygen by adding one H_2O molecule for each O atom needed. Add the H_2O molecules to the appropriate side of the half-reaction.

$$Cu \rightarrow Cu^{2+}$$
$$NO_3^- \rightarrow NO + 2H_2O$$

4. Balance the half-reactions for hydrogen by adding one H^+ ion for each H atom needed. Add the H^+ ions to the appropriate side of the half-reaction. (Remember: Acid is present.)

$$Cu \rightarrow Cu^{2+}$$
$$4H^+ + NO_3^- \rightarrow NO + 2H_2O$$

5. Balance the half-reactions for electric charge by adding electrons (e^-) to the appropriate side of the half-reaction.

$$Cu \rightarrow Cu^{2+} + 2e^- \quad \text{(oxidation half-reaction)}$$
$$3e^- + 4H^+ + NO_3^- \rightarrow NO + 2H_2O \quad \text{(reduction half-reaction)}$$

6. Multiply the half-reactions in order to conserve electric charge. (The number of electrons lost must equal the number of electrons gained.)

$$3 \cdot (Cu \rightarrow Cu^{2+} + 2e^-) = 3Cu \rightarrow 3Cu^{2+} + 6e^-$$
$$2 \cdot (3e^- + 4H^+ + NO_3^- \rightarrow NO + 2H_2O) =$$
$$6e^- + 8H^+ + 2NO_3^- \rightarrow 2NO + 4H_2O$$

7. Add the half-reactions, but eliminate the electrons. If H_2O and/or H^+ appears on *both* sides of the equation, subtract the smaller amount from each side of the equation.
The resulting equation is now balanced.

$$3Cu + 8H^+ + 2NO_3^- \rightarrow 3Cu^{2+} + 2NO + 4H_2O$$

PROBLEM
Balance the following ionic redox equation by the ion-electron method:

$$NO_3^- + I_2 \rightarrow HIO_3 + NO_2 \quad \text{(acidic solution)}$$

SOLUTION
We will balance the equation step by step according to the method given above.

1. $NO_3^- \rightarrow NO_2$
   $I_2 \rightarrow HIO_3$

2. $NO_3^- \rightarrow NO_2$
   $I_2 \rightarrow 2HIO_3$

3. $NO_3^- \rightarrow NO_2 + H_2O$
   $6H_2O + I_2 \rightarrow 2HIO_3$

4. $2H^+ + NO_3^- \rightarrow NO_2 + H_2O$

 $6H_2O + I_2 \rightarrow 2HIO_3 + 10H^+$

5. $e^- + 2H^+ + NO_3^- \rightarrow NO_2 + H_2O$

 $6H_2O + I_2 \rightarrow 2HIO_3 + 10H^+ + 10e^-$

6. $10 \cdot (e^- + 2H^+ + NO_3^- \rightarrow NO_2 + H_2O)$

 $1 \cdot (6H_2O + I_2 \rightarrow 2HIO_3 + 10H^+ + 10e^-)$

7. $20H^+ + 10NO_3^- + 6H_2O + I_2 \rightarrow 2HIO_3 + 10H^+ + 10NO_2 + 10H_2O$

 This reduces to

 $10H^+ + 10NO_3^- + I_2 \rightarrow 2HIO_3 + 10NO_2 + 4H_2O$

If the reaction should occur in *basic* solution, we balance it as though it were in acid solution. Then, at the last step, we add enough OH^- ions (to both sides of the equation) to remove all of the H^+ present. Each $H^+ - OH^-$ combination creates one H_2O molecule.

PROBLEM

Balance the following ionic redox equation, which occurs in basic solution:

$$NO_2^- + Al \rightarrow AlO_2^- + NH_3$$

SOLUTION

If this reaction were balanced in *acid* solution, it would look like this:

$$2H_2O + NO_2^- + 2Al \rightarrow 2AlO_2^- + NH_3 + H^+$$

To remove one H^+, we add one OH^- to each side of the equation:

$$OH^- + 2H_2O + NO_2^- + 2Al \rightarrow 2AlO_2^- + NH_3 + H^+ + OH^-$$

The equation becomes

$$OH^- + 2H_2O + NO_2^- + 2Al \rightarrow 2AlO_2^- + NH_3 + H_2O$$

which reduces to

$$OH^- + H_2O + NO_2^- + 2Al \rightarrow 2AlO_2^- + NH_3$$

TRY IT YOURSELF

Balance each of the following redox equations by the ion-electron method:

(a) $Fe^{2+} + MnO_4^- \rightarrow Mn^{2+} + Fe^{3+}$ (acid solution)

(b) $IO_3^- + H_2S \rightarrow I_2 + SO_3^{2-}$ (basic solution)

ANSWERS

(a) $5Fe^{2+} + MnO_4^- + 8H^+ \rightarrow Mn^{2+} + 5Fe^{3+} + 4H_2O$

(b) $6IO_3^- + 5H_2S + 4OH^- \rightarrow 3I_2 + 5SO_3^{2-} + 7H_2O$

15.3A HALF-CELL POTENTIALS AND CELL VOLTAGE

Now that we have "constructed" a voltaic cell from zinc and copper, our next question is: How good is it? We measure the energy output of electrical devices in relation to the amount of charge they transfer. This quantity is known as *potential difference* or *voltage*, and its SI unit is the volt (V).

Under standard conditions, the zinc-copper cell would have a maximum voltage of 1.10 volts. As the cell continued to operate, however, its voltage would drop and finally become 0 when the cell reached equilibrium. Table Z, *Standard Electrode Potentials*, in Appendix 2 can be used to calculate the *maximum* standard voltage of a voltaic cell. Before we perform these calculations, however, we need to know more about the numerical values given in Table Z.

For each reduction half-reaction in the table, a **standard reduction potential** (which is one type of *standard electrode potential*) has been assigned. Its symbol is either $\mathscr{E}°$ or $E°$, and it is also measured in volts. (We will use the script form, $\mathscr{E}°$, in this book.) Note that some of the values in Table Z are positive, some are negative, and one is equal to 0. *The more positive a reduction potential is, the more easily the reduction half-reaction associated with it can occur.*

We could just as well have built a similar table based on oxidation half-reactions and assigned another type of standard electrode potential, known as a **standard oxidation potential**, to each half-reaction. How would these two tables compare? Since an oxidation half-reaction is the reverse of a reduction half-reaction, *the standard oxidation potential is simply the standard reduction potential with its sign reversed*. The following example illustrates this principle:

Half-Reaction	$\mathscr{E}°$	
$Cu^{2+} + 2e^- \rightarrow Cu(s)$	$+0.34$ V	(reduction potential)
$Cu(s) \rightarrow Cu(aq) + 2e^-$	-0.34 V	(oxidation potential)

In this book, we will use standard *reduction* potentials exclusively. The maximum standard cell voltage ($\mathscr{E}°_{cell}$) is calculated by subtracting the standard reduction potential for the half-cell that is reduced from the standard reduction potential for the half-cell that is oxidized:

$$\mathscr{E}°_{cell} = \mathscr{E}°_{reduced} - \mathscr{E}°_{oxidized}$$

$$\mathscr{E}°_{cell} = \mathscr{E}°_{cathode} - \mathscr{E}°_{anode}$$

In Table Z, we find the following listings for the copper-zinc electrochemical cell that we constructed above:

Half-Reaction	$\mathcal{E}°$
$Cu^{2+}(aq) + 2e^- \rightarrow Cu(s)$	$+0.34$ V
$Zn^{2+}(aq) + 2e^- \rightarrow Zn(s)$	-0.76 V

In this case, Cu^{2+} is reduced since its reduction potential is more positive than that of Zn^{2+}. Therefore, we need to reverse the sign of the reduction potential for Zn (in order to convert it into an oxidation potential) and to add the two values:

$$\mathcal{E}°_{cell} = \mathcal{E}°_{Cu^{2+}|Cu} - \mathcal{E}°_{Zn|Zn^{2+}} = (+0.34 \text{ V}) - (-0.76 \text{ V}) = +1.10 \text{ V}$$

Note that we indicate a reduction or oxidation potential by writing a "shorthand" version of the half-reaction as a subscript of $\mathcal{E}°$. For example, the notation "$Cu^{2+}|Cu$" is an abbreviated way of writing the reduction half-reaction $Cu^{2+}(aq) + 2e^- \rightarrow Cu(s)$, and $Zn|Zn^{2+}$ is an abbreviated way of writing the oxidation half-reaction $Zn(s) \rightarrow Zn^{2+}(aq) + 2e^-$.

The positive sign of $\mathcal{E}°_{cell}$ is not simply a result of the arithmetic: it tells us that the cell's redox reaction is spontaneous as written.

PROBLEM
Calculate $\mathcal{E}°_{cell}$ for the following:

Half-Reaction	$\mathcal{E}°$
$Sn^{2+}(aq) + 2e^- \rightarrow Sn(s)$	-0.14 V
$Ag^+(aq) + e^- \rightarrow Ag(s)$	$+0.80$ V

SOLUTION
The half-cell that will be reduced will have the more positive $\mathcal{E}°$. This is the $Ag^+|Ag$ half-cell. Therefore:

$$\mathcal{E}°_{cell} = \mathcal{E}°_{Ag^+|Ag} - \mathcal{E}°_{Sn|Sn^{2+}} = (0.80 \text{ V}) - (-0.14 \text{ V}) = 0.94 \text{ V}$$

The fact that Sn(s) loses two electrons, whereas Ag^+ gains only one electron, is ignored when $\mathcal{E}°_{cell}$ is calculated.

TRY IT YOURSELF
Using the values given in Reference Table Z, calculate $\mathcal{E}°_{cell}$ for an electrochemical cell that has this redox reaction:

$$Fe(s) + Pb^{2+}(aq) \rightarrow Fe^{2+}(aq) + Pb(s)$$

ANSWER
$\mathcal{E}°_{cell} = +0.31$ V

15.4A   THE STANDARD HYDROGEN HALF-CELL

The $\mathcal{E}°$ value assigned to each half-reaction (and its accompanying half-cell) in Reference Table Z is determined by connecting that half-cell to a standard hydrogen half-cell and measuring the voltage of this special electrochemical cell.

The *standard hydrogen half-cell* consists of hydrogen gas (at a partial pressure of 100 kPa) bubbled into a solution of a strong acid, such as HCl, that yields a $[H_3O^+]$ equal to 1 M. This half-cell operates at a temperature of 298 K and has the following reversible half-reactions:

$$2H^+ (aq, 1\ M) + 2e^- \rightleftharpoons H_2 (g, 100\ kPa)$$

By convention, $\mathcal{E}°$ for either half-reaction of the standard hydrogen half-cell is set at exactly 0.00 volt. An illustration of the special voltaic cell used to measure $\mathcal{E}°$ for a half-cell follows.

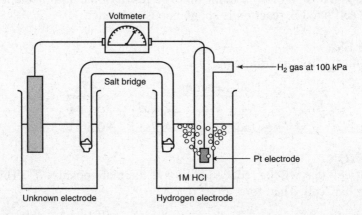

If the half-cell is *reduced* (that is, it causes the standard hydrogen half-cell to be *oxidized*), its $\mathcal{E}°$ is reported as a *positive* value.

If the half-cell is *oxidized* (that is, it causes the standard hydrogen half-cell to be *reduced*), it is reported as a *negative* value.

PROBLEM

A half-cell consisting of Zn(s) and Zn^{2+}(aq, 1 M) is connected to a standard hydrogen half-cell.
(a) Using Reference Table Z, determine each of the half-reactions that occurs in the cell.
(b) Write the overall redox reaction for this electrochemical cell.
(c) Calculate $\mathcal{E}°_{cell}$.

SOLUTION

(a) Since $\mathscr{E}°$ for Zn has a *negative* value, its half-cell will undergo *oxidation* when the cell operates. Therefore, the two half-reactions are

$$Zn(s) \rightarrow Zn^{2+}(aq) + 2e^-$$
$$2H^+(aq) + 2e^- \rightarrow H_2(g)$$

(b) Overall redox reaction: $Zn(s) + 2H^+(aq) \rightarrow Zn^{2+}(aq) + H_2(g)$

(c) $\mathscr{E}°_{cell} = \mathscr{E}°_{H^+|H_2} - \mathscr{E}°_{Zn|Zn^{2+}} = (+0.00 \text{ V}) - (-0.76 \text{ V}) = +0.76 \text{ V}$

TRY IT YOURSELF

Repeat the problem given above for a half-cell consisting of Cu(s) and $Cu^{2+}(aq, 1 \text{ M})$.

ANSWERS

(a) $H_2(g) \rightarrow 2H^+(aq) + 2e^-$
$Cu^{2+}(aq) + 2e^- \rightarrow Cu(s)$

(b) Overall redox reaction: $H_2(g) + Cu^{2+}(aq) \rightarrow 2H^+(aq) + Cu(s)$

(c) $\mathscr{E}°_{cell} = \mathscr{E}°_{Cu^{2+}|Cu} - \mathscr{E}°_{H_2|H^+} = (+0.34 \text{ V}) - (+0.00 \text{ V}) = +0.34 \text{ V}$

15.5A ELECTROLYSIS OF WATER AND AQUEOUS NaCl (BRINE)

Pure water is a very poor conductor of electricity. We can electrolyze water, however, by adding an electrolyte such as H_2SO_4. The reactions that occur at the electrodes are as follows:

Anode (oxidation): $2H_2O(\ell) \rightarrow O_2(g) + 4H^+(aq) + 4e^-$
Cathode (reduction): $4H_2O(\ell) + 4e^- \rightarrow 2H_2(g) + 4OH^-(aq)$

Net reaction: $6H_2O(\ell) \rightarrow 2H_2(g) + O_2(g) + 4H^+(aq) + 4OH^-(aq)$

When the H^+ and OH^- ions come into contact, 4 moles of H_2O are formed $[4H^+(aq) + 4OH^-(aq) \rightarrow 4H_2O(\ell)]$, and the net reaction reduces to

$$2H_2O(\ell) \rightarrow 2H_2(g) + O_2(g)$$

If an aqueous solution of NaCl (known as brine) is electrolyzed, the reactions that occur at the electrodes are as follows:

Anode (oxidation): $2Cl^-(aq) \rightarrow Cl_2(g) + 2e^-$
Cathode (reduction): $2H_2O(\ell) + 2e^- \rightarrow H_2(g) + 2OH^-(aq)$

Net reaction: $2H_2O(\ell) + 2Cl^-(aq) \rightarrow H_2(g) + Cl_2(g) + 2OH^-(aq)$

As we can see, chlorine gas is produced at the positive electrode but sodium metal is *not* produced at the negative electrode; water reacts and produces OH^- ions and H_2 gas instead. The explanation of why brine electrolyzes as it does is quite complicated and lies beyond the scope of this book.

15.6A ELECTROPLATING

Electroplating is a process in which an electric current is used to deposit a layer of metal, such as silver, on the object to be plated. An electroplating cell is a special adaptation of the electrolytic cell. The accompanying diagram illustrates how electroplating is accomplished.

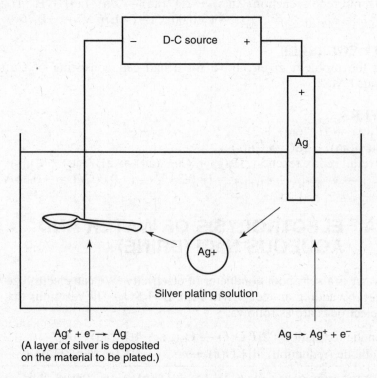

Ag$^+$ + e$^-$ → Ag
(A layer of silver is deposited on the material to be plated.)

Ag → Ag$^+$ + e$^-$

The plating solution is a salt of the metal. A bar of metal serves as the anode (positive electrode) and provides the positive ions needed for plating. The material to be plated acts as the cathode (negative electrode) and receives the positive ions that are reduced to the metallic layer of electroplate.

15.7A ADDITIONAL APPLICATIONS OF REDOX AND ELECTROCHEMISTRY

Reduction of Metals

Most metals do not occur freely in nature; they occur in the oxidized state (i.e., as positive ions). A metal that belongs to Group 1 or 2 is recovered by electrolysis of its fused salt. Other metals are recovered by reduction of their ores; the method used depends on the activity of the metal and the nature of the ore.

In the production of chromium metal, chromium(III) oxide is fairly stable and is treated with aluminum metal, which is a relatively strong reducing agent:

$$2Al + Cr_2O_3 \rightarrow Al_2O_3 + 2Cr$$

Metals such as zinc and iron are extracted by the reduction of their oxides. Carbon, in the form of coke, or carbon monoxide is used as the reducing agent:

$$ZnO + C + heat \rightarrow Zn + CO$$
$$Fe_2O_3 + 3CO + heat \rightarrow 2Fe + 3CO_2$$

Electrolytic Purification of Metals

Electrolytic methods are useful in purifying metals. For example, a cell similar to the electroplating cell shown in Section 15.6A can be used to purify copper from its ores. The anode is made from impure copper, and purified copper (more than 99.5 percent purified) is collected at the cathode. The reactions are as follows:

$$Cathode: Cu^{2+}(aq) + 2e^- \rightarrow Cu(s)$$
$$Anode: Cu(s) \rightarrow Cu^{2+}(aq) + 2e^-$$

The impurities, iron and zinc, are also oxidized at the anode and enter the solution as $Fe^{2+}(aq)$ and $Zn^{2+}(aq)$. Their electrode potentials are too negative, however, to allow their reduction at the cathode. Other impurities, such as silver and gold, remain in their metallic states and are valuable by-products of this process.

Preventing the Corrosion of Metals

Corrosion occurs when a metal is attacked slowly by elements in its environment. In many cases, the metal ceases to be useful. Corrosion is a redox reaction, and agents such as moisture may contribute to the process.

Metals may be protected from corrosion in a variety of ways. Aluminum oxidizes rapidly, but the oxide adheres tightly to the metal and forms a self-protective coating that shields the metal underneath from further oxidation. Zinc also produces a self-protective oxide coating.

When a metal such as iron oxidizes, however, the oxide (rust) flakes off and allows the oxidation of fresh metal. Iron can be protected by plating it with a corrosion-resistant metal such as chromium or nickel or by coating it with paint, oil, or porcelain.

Another protection technique is to coat the iron with zinc—a process known as *galvanizing*. If the coating is broken, exposing fresh iron and zinc, the more active zinc will be oxidized first and will produce a protective oxide coating.

END-OF-CHAPTER QUESTIONS

Some questions have the symbol "§2" in front of the question number. This symbol means that the question is based on Section II material.

1. How many moles of electrons are needed to reduce 1 mole of Cu^{2+} to $Cu(s)$?
   (1) 1   (2) 2   (3) 3   (4) 4

2. A reducing agent is a substance that
   (1) gains protons   (2) loses protons
   (3) gains electrons   (4) loses electrons

3. Metals located near the top of Reference Table J tend to have
   (1) high ionization energies and high electronegativities
   (2) high ionization energies and low electronegativities
   (3) low ionization energies and high electronegativities
   (4) low ionization energies and low electronegativities

4. Which half-reaction correctly represents oxidation?
   (1) $Mg + 2e^- \rightarrow Mg^{2+}$
   (2) $Mg^{2+} + 2e^- \rightarrow Mg$
   (3) $Mg^{2+} \rightarrow Mg + 2e^-$
   (4) $Mg \rightarrow Mg^{2+} + 2e^-$

5. In the reaction

$$Pb + 2Ag^+ \rightarrow Pb^{2+} + 2Ag$$

   the Ag^+ is
   (1) reduced, and the oxidation number changes from $+1$ to 0
   (2) reduced, and the oxidation number changes from $+2$ to 0
   (3) oxidized, and the oxidation number changes from 0 to $+1$
   (4) oxidized, and the oxidation number changes from $+1$ to 0

6. In the reaction

$$2Al(s) + 3Cu^{2+}(aq) \rightarrow 2Al^{3+}(aq) + 3Cu(s)$$

Al(s)
(1) gains protons   (2) loses protons
(3) gains electrons   (4) loses electrons

7. In the reaction

$$4Zn + 10HNO_3 \rightarrow 4Zn(NO_3)_2 + NH_4NO_3 + 3H_2O$$

the zinc is
(1) reduced, and the oxidation number changes from 0 to +2
(2) oxidized, and the oxidation number changes from 0 to +2
(3) reduced, and the oxidation number changes from +2 to 0
(4) oxidized, and the oxidation number changes from +2 to 0

8. All redox reactions involve
(1) the gain of electrons, only
(2) the loss of electrons, only
(3) both the gain and the loss of electrons
(4) neither the gain nor the loss of electrons

§2 **9.** In the reaction

$$2Fe^{2+} + Cl_2 \rightarrow 2Fe^{3+} + 2\,Cl^-$$

which is the oxidizing agent?
(1) Fe^{2+} (2) Cl_2 (3) Fe^{3+} (4) Cl^-

§2 **10.** In the reaction

$$3Cu + 8HNO_3 \rightarrow 3Cu(NO_3)_2 + 2NO + 4H_2O$$

the reducing agent is
(1) Cu   (2) N^{5+} (3) Cu^{2+} (4) N^{2+}

11. Which is a redox reaction?
(1) $CaCO_3 \rightarrow CaO + CO_2$
(2) $NaOH + HCl \rightarrow NaCl + H_2O$
(3) $2NH_4Cl + Ca(OH)_2 \rightarrow 2NH_3 + 2H_2O + CaCl_2$
(4) $2H_2O \rightarrow 2H_2 + O_2$

12. Given the reaction

$$3Ag + Au^{3+} \rightarrow 3Ag^+ + Au$$

Which equation correctly represents the oxidation half-reaction?
(1) $3Ag + 3e^- \rightarrow 3Ag^+$ (2) $3Ag \rightarrow 3Ag^+ + 3e^-$
(3) $Au^{3+} + 3e^- \rightarrow Au$ (4) $Au^{3+} \rightarrow Au + 3e^+$

13. When the redox equation

$$\_\_Cr^{3+}(aq) + \_\_Mn(s) \rightarrow \_\_Mn^{2+}(aq) + \_\_Cr(s)$$

is completely balanced, the coefficient of $Cr^{3+}(aq)$ will be
(1) 1   (2) 2   (3) 3   (4) 4

§2 **14.** When the reaction

$$\_\_Cu(s) + \_\_HNO_3(aq) \rightarrow \_\_Cu(NO_3)_2(aq) + \_\_NO_2(g) + \_\_H_2O(\ell)$$

is completely balanced using smallest whole numbers, the coefficient of $HNO_3(aq)$ will be
(1) 1   (2) 2   (3) 3   (4) 4

§2 **15.** When the reaction

$$\_\_K_2Cr_2O_7 + \_\_HCl \rightarrow \_\_KCl + \_\_CrCl_3 + \_\_Cl_2 + \_\_H_2O$$

is completely balanced using smallest whole numbers, the coefficient of Cl_2 will be
(1) 1   (2) 2   (3) 3   (4) 4

§2 **16.** The redox reaction below occurs in *acid* solution:

$$\_\_Cu(s) + \_\_NO_3^- \rightarrow \_\_Cu^{2+} + \_\_NO_2$$

When the equation in correctly balanced, the coefficient of NO_3^- is
(1) 1   (2) 2   (3) 3   (4) 4

17. According to Reference Table J, which will reduce Mg^{2+} to $Mg(s)$?
(1) Fe(s)   (2) Ba(s)   (3) Pb(s)   (4) Ag(s)

18. According to Reference Table J, which of the following nonmetals is most easily reduced?
(1) F_2 (2) Cl_2 (3) Br_2 (4) I_2

19. According to Reference Table J, which element will react spontaneously with Al^{3+}?
(1) Cu   (2) Au   (3) Li   (4) Ni

§2 **20.** According to Reference Table Z in Appendix 2, which ion can be both an oxidizing agent and a reducing agent?
(1) Fe^{2+} (2) Cu^{2+} (3) Al^{3+} (4) Cr^{3+}

21. Which metal will react spontaneously with H_2SO_4?
(1) Au   (2) Ag   (3) Cu   (4) Mg

22. Which will oxidize $Zn(s)$ to Zn^{2+}, but will *not* oxidize $Pb(s)$ to Pb^{2+}?
(1) Al^{3+} (2) Au^{3+} (3) Co^{2+} (4) Mg^{2+}

§2 **23.** Which half-reaction is the arbitrary standard used in the measurement of the standard electrode potentials in Reference Table Z in Appendix 2?
(1) $2H^+ + 2e^- \rightarrow H_2(g)$
(2) $2H_2O + 2e^- \rightarrow 2OH^- + H_2(g)$
(3) $F_2(g) + 2e^- \rightarrow 2F^-$
(4) $Li^+ + e^- \rightarrow Li(s)$

§2 **24.** Given the reaction

$$2Fe^{3+} + 2I^- \rightarrow 2Fe^{2+} + I_2$$

The net potential difference ($\mathscr{E}°$) for the overall reaction is
(1) 1.00 V (2) 1.31 V (3) 2.08 V (4) 0.23 V

25. In a voltaic cell composed of two half-cells, *ions* are allowed to flow from one half-cell to another by means of
(1) electrodes (2) an external conductor
(3) a voltmeter (4) a salt bridge

§2 **26.** What is the voltage of a voltaic cell that has reached equilibrium?
(1) 1 (2) greater than 1
(3) between 0 and 1 (4) 0

Base your answers to questions 27 and 28 on the diagram below, which represents a voltaic cell at 298 K. The equation that accompanies the diagram represents the net cell reaction.

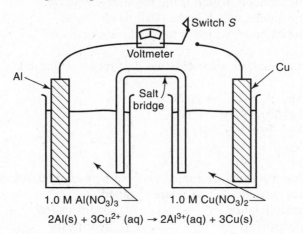

$$2Al(s) + 3Cu^{2+}\ (aq) \rightarrow 2Al^{3+}(aq) + 3Cu(s)$$

§2 **27.** When switch S is closed, the maximum potential difference ($\mathscr{E}°$) for the cell will be
(1) 1.32 V   (2) 2.00V   (3) -1.32 V   (4) -2.00 V

28. When switch S is closed, electrons in the external circuit will flow from
(1) Al to Al^{3+} (2) Al to Cu
(3) Cu to Al   (4) Cu to Cu^{2+}

§2 **29.** In order for a redox reaction to be spontaneous, the potential ($\mathscr{E}°$) for the overall reaction must be
(1) greater than 0   (2) 0
(3) between 0 and -1 (4) less than -1

Base your answers to questions 30 and 31 on the diagram below, which represents the electroplating of a metal fork with Ag(s). (An electroplating cell is a special kind of *electrolytic* cell.)

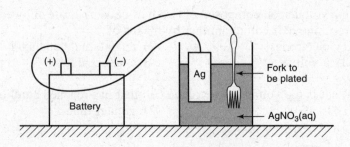

§2 **30.** Which part of the electroplating system is provided by the fork?
(1) the anode, which is the negative electrode
(2) the cathode, which is the negative electrode
(3) the anode, which is the positive electrode
(4) the cathode, which is the positive electrode

§2 **31.** Which equation represents the half-reaction that takes place at the fork?
(1) $Ag^+ + NO_3^- \rightarrow AgNO_3$
(2) $AgNO_3 \rightarrow Ag^+ + NO_3^-$
(3) $Ag^+ + e^- \rightarrow Ag(s)$
(4) $Ag(s) \rightarrow Ag^+ + e^-$

32. If redox reactions are forced to occur by use of an externally applied electric current, the procedure is called
(1) neutralization   (2) esterification
(3) electrolysis   (4) hydrolysis

33. In an electrolytic cell, a negative ion will be attracted to the
    (1) positive electrode and oxidized
    (2) positive electrode and reduced
    (3) negative electrode and oxidized
    (4) negative electrode and reduced

34. Which half-reaction occurs at the cathode in an electrolytic cell in which an object is being plated with copper?
    (1) $Cu(s) \rightarrow Cu^{2+} + 2e^-$
    (2) $Cu(s) + 2e^- \rightarrow Cu^{2+}$
    (3) $Cu^{2+} \rightarrow Cu(s) + 2e^-$
    (4) $Cu^{2+} + 2e^- \rightarrow Cu(s)$

§2 35. Which metal is *not* obtained from its ore by electrolytic reduction?
    (1) Na   (2) Li   (3) Au   (4) K

36. In which substance does chlorine have an oxidation number of +1?
    (1) Cl_2 (2) HCl   (3) HClO   (4) $HClO_2$

37. Which statement is true for any electrochemical cell?
    (1) Oxidation occurs at the anode, only.
    (2) Reduction occurs at the anode, only.
    (3) Oxidation occurs at both the anode and the cathode.
    (4) Reduction occurs at both the anode and the cathode.

38. A diagram of a chemical cell and an equation are shown below.

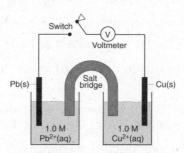

$$Pb(s) + Cu^{2+}(aq) \longrightarrow Pb^{2+}(aq) + Cu(s)$$

When the switch is closed, electrons will flow from
    (1) the Pb(s) to the Cu(s)
    (2) the Cu(s) to the Pb(s)
    (3) the $Pb^{2+}(aq)$ to the Pb(s)
    (4) the $Cu^{2+}(aq)$ to the Cu(s)

39. When a neutral atom undergoes oxidation, the atom's oxidation state
(1) decreases as it gains electrons
(2) decreases as it loses electrons
(3) increases as it gains electrons
(4) increases as it loses electrons

40. Given the equation:

$$C(s) + H_2O(g) \rightarrow CO(g) + H_2(g)$$

Which species undergoes reduction?
(1) $C^0(s)$ (2) H^{1+} (3) C^{2+} (4) $H_2^0(g)$

41. Where do reduction and oxidation occur in an electrolytic cell?
(1) Both occur at the anode.
(2) Both occur at the cathode.
(3) Reduction occurs at the anode, and oxidation occurs at the cathode.
(4) Reduction occurs at the cathode, and oxidation occurs at the anode.

Constructed-Response Questions

1. Zinc metal can be reacted with sulfuric acid to produce hydrogen gas according to the balanced equation

$$Zn(s) + H_2SO_4(aq) \rightarrow ZnSO_4(aq) + H_2(g)$$

The balanced *redox* equation for this reaction is

$$Zn(s) + 2H^+(aq) \rightarrow Zn^{2+}(aq) + H_2(g)$$

(a) Write the oxidation and reduction half-reactions for this equation.

When copper metal [Cu(s)] is mixed with sulfuric acid, *no* generation of hydrogen gas occurs. Using Reference Table J in Appendix 1:

(b) Explain why zinc reacts, but copper will not.
(c) Write the balanced redox equation that *can* occur between the pairs:

$$Cu(s)/Cu^{2+}(aq) \text{ and } H_2(g)/H^+(aq)$$

§2 **2.** Balance each of the following redox equations:
(a) $KOH + Zn + H_2O + KNO_3 \rightarrow K_2Zn(OH)_4 + NH_3$
(b) $H_2SO_4 + FeSO_4 + KClO_3 \rightarrow Fe_2(SO_4)_3 + KCl + H_2O$
(c) $H_3AsO_4 + Zn + HNO_3 \rightarrow AsH_3 + Zn(NO_3)_2 + H_2O$
(d) $H_2O + P_4 + LiOH \rightarrow PH_3 + LiH_2PO_2$
(e) $Zn(s) + NO_3^-(aq) \rightarrow Zn^{2+}(aq) + NH_4^+(aq)$ (acidic solution)
(f) $MnO_4^-(aq) + CN^-(aq) \rightarrow MnO_2(s) + CNO^-(aq)$ (basic solution)

§2 **3.** (a) Write the half-reaction for the reduction of 1.000 mole of $Cu^{2+}(aq)$ to $Cu(s)$.
(b) How many moles of electrons are needed in this half-reaction?
(c) If the electric charge on 1.000 mole of electrons is 96,470 coulombs, calculate the total charge that is needed to reduce all of the $Cu^{2+}(aq)$.
(d) One ampere of electric current is equal to an electric charge flow of one coulomb per second. If the electric charge you calculated in part was supplied over a period of 5.000 hours, how much electric current was needed for the reduction?

Base your answers to questions 4–6 on the diagram of a voltaic cell provided below and on your knowledge of chemistry.

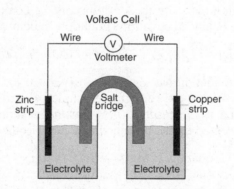

4. On the diagram, indicate with one or more arrows the direction of electron flow through the wire.

5. Write an equation for the half-reaction that occurs at the zinc electrode.

6. Explain the function of the salt bridge.

Base your answers to questions 7–9 on the information below and on your knowledge of chemistry.

One type of voltaic cell, called a mercury battery, uses zinc and mercury(II) oxide to generate an electric current. Mercury batteries were used because of their miniature size even though mercury is toxic. The overall reaction for a mercury battery is given in the equation below.

$$Zn(s) + HgO(s) \rightarrow ZnO(s) + Hg(\ell)$$

7. Determine the change in oxidation number of the zinc while the battery operates.

8. Compare the number of moles of electrons lost to the number of moles of electrons gained during the reaction.

9. Using the information in the passage, state *one* risk and *one* benefit of using a mercury battery.

Base your answers to questions 10–12 on the information below and on your knowledge of chemistry.

Early scientists defined oxidation as a chemical reaction in which oxygen is combined with another element to produce the oxide of the element. An example of oxidation based on this definition is the combustion of methane represented by the balanced equation (Equation 1) below.

$$\text{Equation 1: } CH_4(g) + 2O_2(g) \rightarrow CO_2(g) + 2H_2O(g)$$

The definition of oxidation has since been expanded to include many reactions that do not involve oxygen. One example of oxidation based on the expanded definition is the reaction between magnesium ribbon and powdered sulfur represented by the balanced equation (Equation 2) below.

$$\text{Equation 2: } Mg(s) + S(s) \rightarrow MgS(s)$$

10. Based on the information given in the passage, state why early scientists classified the reaction represented by Equation 1 as oxidation.

11. Determine the change in the oxidation number of carbon in the reaction represented by Equation 1.

12. Write a balanced half-reaction equation for the oxidation that occurs in the reaction represented by Equation 2.

The answers to these questions are found in Appendix 3.

| Chapter Sixteen | **THE CHEMISTRY LABORATORY** |

KEY IDEAS

This chapter is a review of the laboratory skills you have acquired and the laboratory activities you have performed during your chemistry course. Included in the chapter are listings of safety procedures and of basic skills and activities, illustrations of common laboratory equipment, and a guide for preparing laboratory reports.

KEY OBJECTIVES

At the conclusion of this chapter you will have reviewed the following information:

- Which safety procedures are basic in the laboratory.
- Which measuring devices are commonly used in the laboratory.
- Which basic laboratory skills you should be familiar with.
- Which equipment is common to most laboratories.
- Which laboratory activities are commonly employed in a chemistry laboratory.
- How colors are used to identify various substances.
- What general guidelines should be observed in constructing a laboratory report.

16.1   INTRODUCTION

This chapter is quite different from the 15 chapters that precede it. You cannot learn laboratory skills by reading a book any more than you can learn to drive a car by reading a book! Learning about the laboratory can take place only in the laboratory.

In this chapter, we will briefly review the activities that you should have performed and the skills that you should have acquired during the course of the year. You should supplement this chapter by referring to your laboratory manual and/or notebook.

16.2 SAFETY PROCEDURES

While each chemistry laboratory has its own procedures to ensure safety, a number of rules for students are common to every laboratory. Included are the following:

- Never be in a laboratory without a teacher.
- Do not bring any food into the laboratory.
- If your hair is long, tie it back to avoid catching fire.
- Always wear safety goggles in the laboratory.
- Do not handle equipment until you receive instructions.
- Always be alert when you handle chemicals or work with open flames.
- Learn and follow the correct procedures for diluting acids, heating substances, mixing chemicals, and working with glass.
- Never get "creative" and mix chemicals "to see what will happen."

You can undoubtedly add to this list. The point is that you must practice safety procedures at all times!

16.3 USING MEASURING DEVICES

Most of the work in a chemistry laboratory involves measurement. Devices that you may use include a balance, ruler, graduated cylinder, thermometer, burette, and eudiometer tube.

   When you measure the level of a liquid, as with a burette or graduated cylinder, your eye should be level with the *bottom* of the meniscus of the liquid, and the reading should be taken at that point, as illustrated in the accompanying diagram.

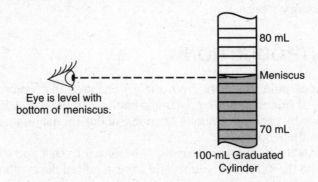

Eye is level with
bottom of meniscus.

80 mL

Meniscus

70 mL

100-mL Graduated
Cylinder

In general, when you use any measuring device that has a ruled scale, such as a ruler, balance, or graduated cylinder, you may estimate to $\frac{1}{10}$ of the smallest division present. For example, in the preceding diagram, the smallest divisions on the cylinder are 1 milliliter each. Therefore, you read this measurement as 75.7 milliliters.

16.4   BASIC LABORATORY SKILLS

During your chemistry laboratory course, you acquired a number of basic skills. You should be familiar with the following skills:

- How to adjust a (Bunsen) burner.
- How to cut, bend, and fire-polish glass tubing.
- How to separate a mixture with a funnel and filter paper.
- How to dilute acids and other liquids safely.
- How to remove solids from containers.
- How to pour liquids safely.
- How to heat materials in test tubes, beakers, and flasks safely.

16.5   IDENTIFICATION OF COMMON LABORATORY APPARATUS

The illustrations on pages 466 and 467 show the laboratory equipment that you use most often during your study of chemistry. You should review how each item is used in the laboratory.

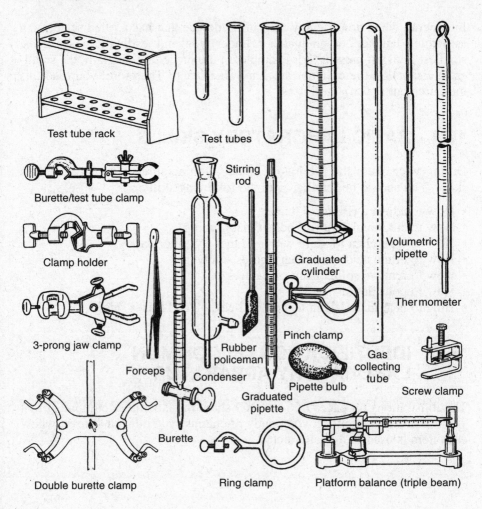

Test tube rack

Test tubes

Burette/test tube clamp

Stirring rod

Clamp holder

Graduated cylinder

Volumetric pipette

3-prong jaw clamp

Thermometer

Forceps

Condenser

Rubber policeman

Pinch clamp

Gas collecting tube

Screw clamp

Graduated pipette

Pipette bulb

Burette

Double burette clamp

Ring clamp

Platform balance (triple beam)

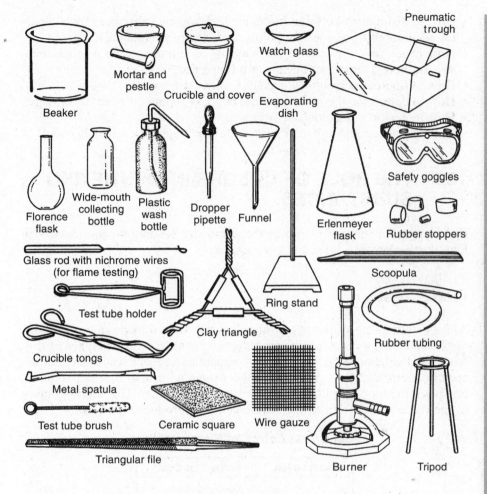

Beaker

Mortar and pestle

Crucible and cover

Watch glass

Evaporating dish

Pneumatic trough

Florence flask

Wide-mouth collecting bottle

Plastic wash bottle

Dropper pipette

Funnel

Erlenmeyer flask

Safety goggles

Rubber stoppers

Glass rod with nichrome wires (for flame testing)

Test tube holder

Clay triangle

Ring stand

Scoopula

Rubber tubing

Crucible tongs

Metal spatula

Test tube brush

Ceramic square

Wire gauze

Triangular file

Burner

Tripod

16.6   BASIC LABORATORY ACTIVITIES

The following is a list of other laboratory activities that are common to most courses in chemistry. You should use your laboratory manual and notebook to review these activities.

- How to construct a heating or cooling curve and to interpret points on the curve.
- How to measure the heat of a simple chemical reaction.
- How to identify exothermic and endothermic processes using temperature measurements.
- How to determine the solubility of a substance at various temperatures and to construct a solubility curve for that substance.

- How to distinguish between inorganic and organic substances by comparing their solubilities, melting points, and electrical conductivities.
- How to identify the ions of such metals as lithium, sodium, potassium, strontium, and copper by means of a flame test.
- How to find the percent (by mass) of water in a crystal.
- How to determine the molar volume of a gas generated in an experiment.
- How to perform an acid–base titration.
- How to prepare and collect a gas such as hydrogen or carbon dioxide.

16.7 THE ROLE OF COLORS IN IDENTIFYING SUBSTANCES

Colors play an important role in the chemistry laboratory, particularly in identifying elements, ions, and compounds.

Flame Tests

All of us are aware of the beautiful colors that fireworks produce. The ions of certain elements produce the characteristic colors of various types of fireworks. In the laboratory, we generally prepare an aqueous solution of certain specific elements, dip a platinum or Nichrome wire loop into the solution, and place the loop into the flame of a laboratory burner. The resulting color of the flame is used to identify the element that produced the color.

Flame Test Colors of Elemental Ions

Elemental Ion	Flame Test Color
Li^+	Deep red (crimson)
Na^+	Yellow
K^+	Pale violet
Ca^{2+}	Orange-red
Sr^{2+}	Red
Ba^{2+}	Yellow-green
Cu^{2+}	Blue-green

Sometimes the flame test of a mixture of Na^+ ions and K^+ ions in aqueous solution will show the yellow color of Na^+, but not the pale violet color of K^+ because of the difference in the intensities of the two colors. In this case, using a special type of glass, known as cobalt blue glass, absorbs the yellow color and allows the pale violet color to be observed in the flame.

Colors of Ions in a Solid and/or in Aqueous Solution

Many students notice that solid copper(II) sulfate is blue in color, as is its aqueous solution. In certain cases, a (positive) metallic ion or a polyatomic ion containing a transition metal will yield a solid or aqueous solution that is colored.

Colors of Solids and Aqueous Solutions of Ions

Ion	Associated Color
Cu^+	Green
Cu^{2+}	Blue
Fe^{2+}	Yellow-green (depending on the anion)
Fe^{3+}	Orange-red (depending on the anion)
Co^{2+}	Pink
Cr^{3+}	Violet ($Cr(NO_3)_3$) to Green ($CrCl_3$)
Ni^{2+}	Green
Mn^{2+}	Pink
MnO_4^-	Purple
CrO_4^{2-}	Yellow
$Cr_2O_7^{2-}$	Orange
$FeSCN^{2+}$	Deep red
$CoCl_4^{2-}$	Blue

Note that various ions such as Al^{3+}, Li^+, Na^+, K^+, Mg^{2+}, Ca^{2+}, Ba^{2+}, Pb^{2+}, Pb^{4+}, Sr^{2+}, and Zn^{2+} are colorless in aqueous solution but most (though not all) of their solid compounds are white. A notable exception is PbI_2, which is an intensely yellow solid.

Colors of Elements and Certain Compounds

Several elements and compounds can be readily identified by their characteristic colors.

Element or Compound and Associated Color

Elemental or Compound	Associated Color
F_2	Pale yellow gas
Cl_2	Green-yellow gas
Br_2	Deep red liquid
I_2	Metallic gray solid; violet vapor
S_8	Yellow solid
Cu	Red metallic solid
Au	Yellow metallic solid
NO_2	Brown gas

In addition, the sulfides of transition metals tend to be black solids. Metals that consist of very small particles appear black because they reflect light poorly.

16.8 GUIDELINES FOR LABORATORY REPORTS

Although each teacher has his or her own rules for keeping a laboratory notebook, there are a number of general guidelines for laboratory reports that you should bear in mind.

• Make your reports concise. A wordy report tends to obscure important details.
• Arrange the information in your reports logically. There should be an introduction, a materials and methods section, a section for observations and results, and a section for discussion and conclusions.
• Never erase data that you consider unacceptable. Draw a single line through questionable material. (It is entirely possible that "bad" data may turn out to be correct!)
• Whenever possible, present your data and results in the form of tables and graphs.
• List your observations clearly.
• Be sure that your conclusions follow directly and logically from your observations.

END-OF-CHAPTER QUESTIONS

1. Which procedure represents the safest technique to use for diluting a concentrated acid?
   (1) Add the acid to the water quickly.
   (2) Add the water to the acid quickly.
   (3) Add the acid slowly to the water with steady stirring.
   (4) Add the water slowly to the acid with steady stirring.

2. Which procedure is the safest to follow when using an open flame to heat the contents of a test tube that contains a flammable mixture?
   (1) Cork the test tube and then heat it gently near the bottom only.
   (2) Heat the open test tube gently near the bottom only.
   (3) Cork the test tube and place it in a beaker of water, then heat the water in the beaker.
   (4) Place the open test tube in a beaker of water, then heat the water in the beaker.

3. The diagram below represents a portion of a triple beam balance. If the beams are in balance, with the riders in the position shown, what is the total mass, in grams, of the object being massed?

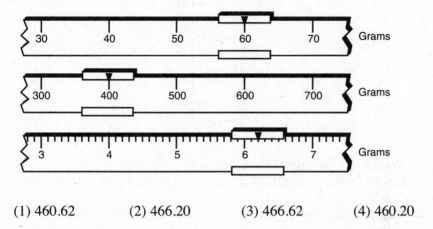

(1) 460.62   (2) 466.20   (3) 466.62   (4) 460.20

4. A student has to measure the diameter of a test tube (as shown in the diagram below) in order to calculate the tube's volume.

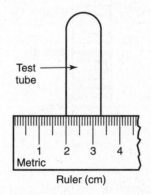

Based on the diagram, the tube's diameter is closest to
(1) 1.25 cm   (2) 2.32 cm   (3) 3.25 cm   (4) 12.5 cm

5. The diagram below represents a portion of a burette.

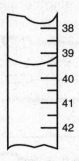

   What is the reading of the meniscus?
   (1) 39.2 mL   (2) 39.5 mL   (3) 40.7 mL   (4) 40.9 mL

6. The diagram below shows a section of a 100-milliliter graduated cylinder:

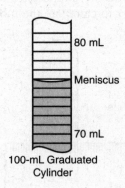

100-mL Graduated
Cylinder

   In order to read the volume of the liquid most accurately, the eye of the student should be
   (1) level with the meniscus   (2) above the meniscus
   (3) at the bottom of the cylinder   (4) at the top of the cylinder

7. As a result of dissolving a salt in water, a student found that the temperature of the water increased. From this observation alone, the student should conclude that the dissolving of the salt
   (1) produced an acid solution   (2) produced a basic solution
   (3) was endothermic   (4) was exothermic

8. A Bunsen burner flame is sooty black and mixed with an orange-yellow color. Which is the probable reason for this condition?
(1) No oxygen is mixing with the gas.
(2) No gas is mixing with the oxygen.
(3) Insufficient oxygen is mixing with the gas.
(4) Insufficient gas is mixing with the oxygen.

9. A student obtained the following data in determining the solubility of $NaNO_3$:

Temperature (°C)	Solubility (g $NaNO_3$/100 g H_2O)
0	73
10	80
20	88
30	97
40	105
50	115
60	124
70	134
80	145

Which set of coordinates graphically present the data in the table most clearly?

(1)
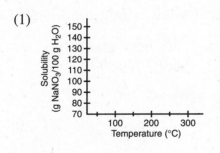

(2)
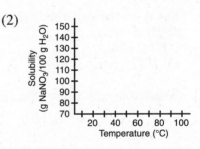

(3)
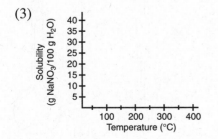

(4)
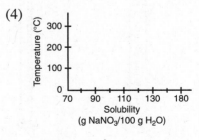

10. The volume of an acid required to neutralize exactly 15.00 milliliters of a base could be measured most precisely if it were added to the base solution from a

(1) 100-mL graduate   (2) 125-mL Erlenmeyer flask
(3) 50-mL burette   (4) 50-mL beaker

11. Which diagram represents a graduated cylinder?

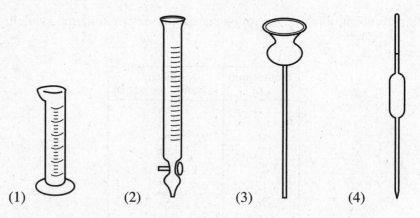

(1)   (2)   (3)   (4)

12. Refer to the experimental data below.

Mass of empty graduated cylinder   =   141 g
Mass of graduated cylinder and distilled water   =   163 g
Volume of distilled water   =   25.3 mL

Based on the experimental data collected, what is the density of the distilled water?

(1) 1.0 g/mL   (2) 0.253 g/mL
(3) 0.87 g/mL   (4) 1.15 g/mL

13. Which diagram below represents a pipette?

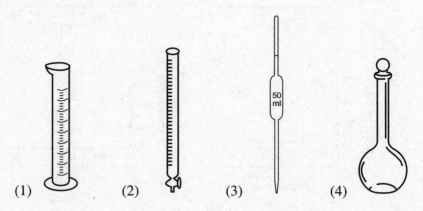

50 ml

(1)   (2)   (3)   (4)

14. Which piece of laboratory apparatus would most likely be used to evaporate a 1-milliliter sample of a solution to dryness?
    (1) a volumetric flask (2) a burette
    (3) a pipette (4) a watch glass

15. Which diagram below represents an Erlenmeyer flask?

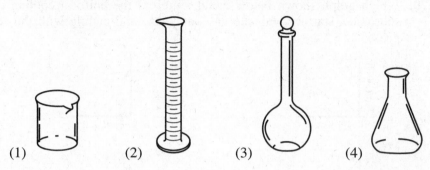

    (1)   (2)   (3)   (4)

16. In a laboratory experiment, a student collects a gas by water displacement. Which piece of apparatus is best used to measure the volume of the gas as it is generated?
    (1) a pipette (2) a volumetric flask
    (3) a eudiometer tube (4) a beaker

17. The heat energy given off by an acid solution as it is neutralized by a basic solution is best measured by using a
    (1) pipette (2) pH meter
    (3) calorimeter (4) test tube

18. In order to measure the solubility of a compound in water at 20°C, a student prepared a saturated solution of the compound at that temperature and then evaporated it to dryness. The student's experimental data are given below.

    Mass of the saturated solution before evaporation = 18.8 g
    Mass of the solid remaining after evaporation   =   8.8 g

    Based on the data and on Reference Table G in Appendix 1, the compound could have been
    (1) $NaNO_3$ (2) $KClO_3$ (3) $NaCl$ (4) NH_4Cl

19. The process of filtration is performed in the laboratory to
    (1) form precipitates
    (2) remove water from solutions
    (3) separate dissolved particles from the solvent
    (4) separate insoluble substances in an aqueous mixture

20. Which graph shown below could represent the uniform cooling of a substance, starting with the gaseous phase and ending with the solid phase?

    (1)

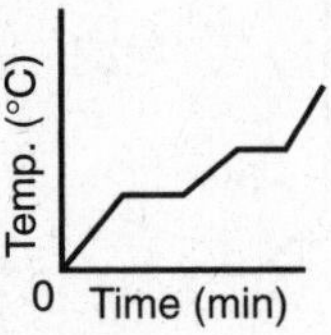

    (2)

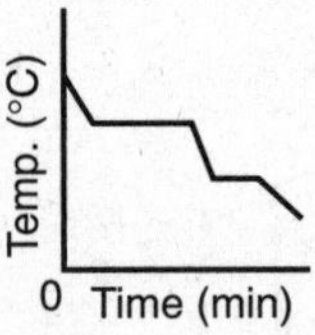

    (3)

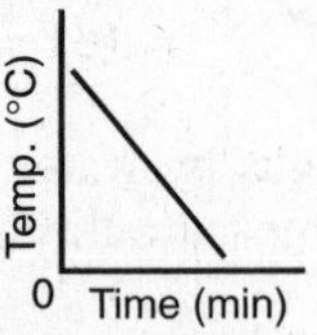

    (4)

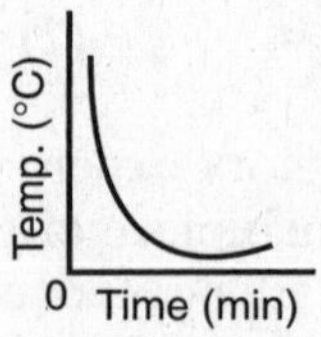

21. A 1.20-gram sample of a hydrated salt is heated to a constant mass of 0.80 gram. What was the percent by mass of water contained in the original sample?
    (1) 20   (2) 33   (3) 50   (4) 67

22. A student obtained the following data while cooling a substance. The substance was originally in the liquid phase at a temperature below its boiling point.

Time (min)	0.5	1.0	1.5	2.0	2.5	3.0	3.5	4.0	4.5	5.0	5.5	6.0
Temperature (°C)	70.	63	57	54	53	53	53	53	53	52	51	48

What is the freezing point of the substance?
    (1) 70.°C   (2) 59°C   (3) 53°C   (4) 48°C

Base your answers to questions 23 and 24 on the table below, which shows the data collected during the heating of a 5.0-gram sample of a hydrated salt.

Mass of Salt (g)	Heating Time (min)
5.0	0.0
4.1	5.0
3.1	10.
3.0	15.
3.0	30.
3.0	60.

23. After 60. minutes, how many grams of water appear to remain in the salt?
(1) 0.00   (2) 2.0   (3) 1.9   (4) 0.90

24. What is the percent of water in the original sample?
(1) 82.   (2) 60.   (3) 30.   (4) 40.

25. Which piece of glassware is best for transferring water to an Erlenmeyer flask?
(1) beaker
(2) test tube
(3) volumetric flask
(4) condenser

26. In an experiment, a student determined the normal boiling points of four unknown liquids. The collected data were organized into the table below.

Unknown	Normal Boiling Point (°C)
A	10
B	33
C	78
D	100

Which unknown liquid has the weakest attractive forces between its molecules?
(1) A   (2) B   (3) C   (4) D

27. A student investigated samples of four different substances in the solid state. The table below is a record of the properties observed when each solid was tested.

Property	Sample I	Sample II	Sample III	Sample IV
Melting point	High	Low	High	Low
Solubility in water	Soluble	Insoluble	Insoluble	Soluble
Stability at high temperature	Stable	Decomposes	Stable	Stable
Electrical conductivity	Good	Poor	Poor	Good

According to the tabulated results, which of the solids investigated had the characteristics most closely associated with those of an organic compound?
(1) Sample I
(2) Sample II
(3) Sample III
(4) Sample IV

28. A 9.90-gram sample of a hydrated salt is heated to a constant mass of 6.60 grams. What was the percent, by mass, of water contained in the sample?
(1) 66.7 (2) 50.0 (3) 33.3 (4) 16.5

29. The data below were obtained during the determination of the percent of water in $BaCl_2 \cdot 2H_2O$.

	Mass
Empty crucible	45.12 g
Crucible + $BaCl_2 \cdot 2H_2O$	75.82 g
Crucible + contents after first heating	72.60 g
Crucible + contents after second heating	71.34 g
Crucible + contents after third heating	71.34 g

The percent of water in $BaCl_2 \cdot 2H_2O$ may be expressed as

(1) $\dfrac{30.70}{75.82} \times 100\%$

(2) $\dfrac{30.70}{45.12} \times 100\%$

(3) $\dfrac{71.34}{75.82} \times 100\%$

(4) $\dfrac{4.48}{30.70} \times 100\%$

30. The following procedures are carried out during a laboratory activity to determine the mass, in grams, of $CuSO_4$ in a sample of $CuSO_4 \cdot 5H_2O$:

Step 1. Determine the mass, in grams, of the crucible and $CuSO_4 \cdot 5H_2O$.
Step 2. Determine the mass, in grams, of the crucible and $CuSO_4$.
Step 3. Determine the mass, in grams, of $CuSO_4 \cdot 5H_2O$.
Step 4. Determine the mass, in grams, of the empty crucible.
Step 5. Determine the mass, in grams, of $CuSO_4$.

Which sequence of steps would produce the desired result using only *three* weighings?
(1) 1, 3, 4, 2, 5   (2) 2, 4, 1, 5, 3
(3) 3, 4, 1, 2, 5   (4) 4, 1, 3, 2, 5

31. All of the following are good laboratory safety practices EXECPT
(1) wearing eye goggles
(2) placing unused chemicals back into their original containers
(3) adding acid to water when diluting an acid
(4) allowing heated glass to cool before touching it

32. A student observes that an aqueous solution of a compound is blue. Which ion could the solution contain?
(1) Zn^{2+}
(2) Na^+
(3) Al^{3+}
(4) Cu^{2+}

33. In order to identify the metallic element in an ionic compound, a student places a small amount of the compound onto a wire loop and places the loop into a flame. If the flame is pale violet, the compound most likely contains
(1) K^+
(2) Na^+
(3) Li^+
(4) Ca^{2+}

34. A student wishes to prepare one liter of a 0.2-molar aqueous solution. Which piece of laboratory apparatus will yield the most precise final volume?
(1) beaker
(2) graduated cylinder
(3) volumetric flask
(4) buretette

35. When a flame test on a solid was performed, a student observed a bright orange-yellow flame. When the student used cobalt blue glass, a pale violet color flame was observed. The solid most likely contained
(1) sodium and potassium
(2) lithium and strontium
(3) copper and barium
(4) calcium and zinc

Constructed-Response Questions

Five pieces of laboratory equipment are listed below.

A. graduated cylinder
B. triple-beam balance
C. beaker
D. filter paper
E. Nichrome loop

Match laboratory activities 1–4 with the equipment in the list above. A choice may be used once, more than once, or not at all

1. used to measure 100 mL of NaCl(aq)

2. used in flame tests to insert a substance into the flame

3. used to separate sand from water

4. used to measure the mass of a solid

The answers to these questions are found in Appendix 3.

GLOSSARY

absolute zero The lowest possible temperature, written as 0 K or $-273.15°C$.

accuracy The closeness of a measurement to an accepted value; see also **precision**.

acid See **Arrhenius acid**; **Brønsted–Lowry acid**; **Lewis acid**.

acid ionization constant (K_a) A constant whose value indicates the relative strength of an acid in aqueous solution.

activated complex The intermediate state between reactants and products in a chemical reaction; the peak of the potential energy diagram.

activation energy The minimum energy needed to initiate a reaction.

addition polymerization The joining of unsaturated monomers by a series of addition reactions.

addition reaction The process in which a substance reacts across a double or triple bond in an organic compound.

alcohol An organic compound containing one or more hydroxyl ($-OH$) groups.

aldehyde An organic compound containing a carbonyl group with at least one hydrogen atom attached to the carbonyl carbon.

alkali metal Any Group 1 element, excluding hydrogen.

alkaline earth element Any Group 2 element.

alkane A hydrocarbon containing only single bonds between adjacent carbon atoms.

alkene A hydrocarbon containing one double bond between two adjacent carbon atoms.

alkyl group An open-chained hydrocarbon less one hydrogen atom; for example, CH_3 = methyl group, C_2H_5 = ethyl group. Unspecified alkyl groups are designated by the letter R.

alkyne A hydrocarbon containing one triple bond between two adjacent carbon atoms.

allotrope A specific form of an element that can exist in more than one form; graphite and diamond are allotropes of the element carbon.

alloy A solid metallic solution.

alpha decay The radioactive process in which an alpha particle is emitted.

alpha particle (α) A helium-4 nucleus.

amide A functional group that combines an amino and a carbonyl group; $-CONH_2$.

amine A hydrocarbon derivative containing an amino group.

amino acid An organic compound containing at least one amino group and one carboxyl group.

amino group An ammonia molecule less one hydrogen atom; $-NH_2$.

ampere (A) The SI unit of electric current.

amphiprotic Pertaining to a substance that can act as a Brønsted–Lowry acid or base by donating or accepting H^+ ions.

amphoteric See **amphiprotic**.

anhydrous Pertaining to a compound from which the water of crystallization has been removed.

anode The electrode at which oxidation occurs.

aqueous Pertaining to a solution in which water is the solvent.

aromatic hydrocarbon Any ring hydrocarbon whose electronic structure is related to that of benzene.

Arrhenius acid Any substance that releases H^+ ions in water.

Arrhenius base Any substance that releases OH^- ions in water.

atmosphere (atm) A unit of pressure equal to 101.3 kilopascals.

atom The basic unit of an element.

atomic mass The weighted average of the masses of the naturally occurring isotopes of an element.

atomic mass unit (u) One-twelfth the mass of a carbon-12 atom.

atomic number The number of protons in the nucleus of an atom; the atomic number defines the element.

atomic radius An estimate of the size of an atom.

Avogadro's hypothesis Equal volumes of gases, measured at the same temperature and pressure, contain equal numbers of particles.

Avogadro's law At constant temperature and pressure, the volume of an ideal gas is directly proportional to the number of gas particles present; $\dfrac{V_1}{n_1} = \dfrac{V_2}{n_2}$.

Avogadro's number (N_A) The number of particles in 1 mole; 6.02×10^{23}.

base See **Arrhenius base**; **Brønsted–Lowry base**; **Lewis base**.

base ionization constant (K_b) A constant whose value indicates the relative strength of a base in aqueous solution.

battery A commercial voltaic cell.

benzene C_6H_6; the parent hydrocarbon of all aromatic compounds.

beta decay The radioactive process in which a beta particle is emitted.

beta (−) particle (β^-) An electron.

beta (+) particle (β^+) A positron.

binary compound A compound containing two elements.

binding energy The energy released when a nucleus is assembled from its nucleons.

boiling The transition of liquid to gas; boiling occurs when the vapor pressure of a liquid equals the atmospheric pressure above the liquid.

boiling point The temperature at which boiling occurs; the temperature at which the liquid and vapor phases of a substance are in equilibrium.

boiling point elevation The increase in the boiling point of a solvent due to the presence of solute particles.

bond energy The energy needed to break a chemical bond.

Boyle's law At constant temperature and mass, the pressure of an ideal gas is inversely proportional to its volume; $P_1V_1 = P_2V_2$.

breeder reactor A fission reactor that generates its own nuclear fuel.

bright-line spectrum The lines of visible light emitted by elements as electrons fall to lower energy levels.

Brønsted–Lowry acid A substance that can donate H^+ ions.

Brønsted–Lowry base A substance that can accept H^+ ions.

burette A long, graduated tube used to deliver a controlled volume of liquid precisely.

calorie A quantity of energy; 1 calorie is approximately equal to 4.2 joules.

carbonyl group The functional group characteristic of aldehydes and ketones; $>C=O$.

carboxyl group The functional group characteristic of organic acids; $-COOH$.

catalyst A substance that speeds a chemical reaction by lowering the activation energy of the reaction.

cathode The electrode at which reduction occurs.

cathode rays A stream of electrons emitted at the negative electrode when electricity is passed through a gas at very low pressure.

Celsius (C) scale The temperature scale on which the freezing and boiling points of water (at 1 atm) were originally set at 0 and 100, respectively. The Celsius scale is now defined in terms of the Kelvin scale: $C = K - 273.15$.

chain reaction A chemical or nuclear reaction in which one step supplies energy or reactants for the next step.

Charles's law At constant pressure and mass, the volume of an ideal gas is directly proportional to the Kelvin temperature; $\dfrac{V_1}{T_1} = \dfrac{V_2}{T_2}$.

chemical bond The stabilizing of two atoms by sharing or transferring electrons.

chemical change See **chemical reaction**.

chemical energy The part of internal energy that is associated with the bonds and intermolecular attractions of substances.

chemical equation A shorthand listing of reactants, products, and molar quantities in a chemical reaction.

chemical equilibrium The state in which the rates of the forward and reverse reactions are equal.

chemical family See **group**.

chemical property A property that describes the composition and reactivity of a substance.

chemical reaction A process in which one or more substances are converted into other substances.

chromatography The process in which the components of certain mixtures are separated because of their differences in solubility.

cis-trans isomerism A form of stereoisomerism across the double bond of an alkene. In the *cis* isomer, two identical atoms lie on the *same side* of the molecule across the double bond. In the *trans* isomer, the two atoms lie on *opposite sides* of the molecule across the double bond.

coefficient A number in a chemical equation that indicates how many particles of a reactant or product are required or formed in the reaction.

colligative property A property that depends on the number of particles present rather than the type of particle; see also **boiling point elevation**; **freezing point depression**.

combined gas law At constant mass, the product of the pressure and volume divided by the Kelvin temperature is a constant; $\dfrac{P_1 V_1}{T_1} = \dfrac{P_2 V_2}{T_2}$.

common ion effect An equilibrium shift caused by the addition of an ion present in a reaction.

compound A combination of two or more elements with a fixed composition by mass.

concentrated Pertaining to a solution that contains a relatively large quantity of solute.

concentration The "strength" of a solution; the quantity of solute relative to the quantity of solvent.

condensation The change of gas to liquid.

condensation polymerization The joining of monomers by a series of dehydration reactions.

conjugate acid-base pair Two particles that differ by a single H^+ ion; HCl and Cl^- are a conjugate acid-base pair.

constitutional isomers Isomers that differ because their atoms are connected in different orders.

control rod The part of a fission reactor that controls the rate of fission by absorbing neutrons.

coordinate covalent bond A single covalent bond in which the pair of electrons is supplied by one atom.

coulomb (C) The SI unit of electric charge.

covalent bond A chemical bond formed by the sharing of electrons.

cracking The process of breaking large hydrocarbon molecules into smaller ones in order to increase the yield of compounds such as gasoline.

critical temperature The temperature above which a gas cannot be liquefied, regardless of the applied pressure.

crystal A solid whose particles are arranged in a regularly repeating pattern.

crystallization The process in which a solute separates from its solution; also called *precipitation*.

Dalton's law of partial pressures The sum of the partial pressures of the gases in a container is equal to the total pressure of all of the gases in the container; $P_i = X_i \cdot P_t$. See also **mole fraction**.

decomposition A reaction in which a compound forms two or more simpler substances.

density Mass per unit volume; $d = \dfrac{m}{v}$.

deposition The direct transition from gas to solid.

deuterium The isotope of hydrogen with a mass number of 2.

diatomic molecule A neutral particle consisting of two atoms; Br_2 and CO are diatomic molecules.

diffusion The movement of one substance through another.

dihydroxy alcohol An organic compound with two hydroxyl groups.

dilute (adjective) Pertaining to a solution that contains a relatively small quantity of solute; (verb) to reduce the concentration of a solution by adding solvent.

dipole An unsymmetrical charge distribution in a neutral molecule.

dipole-dipole attraction The attractive force between two oppositely charged dipoles of neighboring polar molecules.

dispersion forces The attractive forces between neighboring nonpolar molecules; also called *London dispersion forces*.

dissociation The separation of an ionic compound in solution into positive and negative ions.

distillation The simultaneous boiling of a liquid and condensation of its vapor.

double bond A covalent bond in which two pairs of electrons are shared by two adjacent atoms.

ductility The property of a substance that allows it to be drawn into a wire; metallic substances possess ductility.

dynamic equilibrium The state in which the rates of opposing processes are equal; see also **chemical equilibrium**; **phase equilibrium**; **solution equilibrium**.

effusion The escape of a gas from a small porous opening; see also **Graham's law of effusion**.

Einstein's mass-energy relationship The equivalence of mass and energy; $E = mc^2$.

electrochemical cell A voltaic cell or an electrolytic cell.

electrode A conductor in an electrochemical or electrolytic cell that serves as the site of oxidation or reduction.

electrode potential A measurement, expressed in volts (V), that indicates the relative ease with which an oxidation or reduction half-reaction can occur. Electrode potentials are determined with respect to a standard hydrogen half-cell; see also **standard hydrogen half-cell**.

electrolysis A nonspontaneous redox reaction driven by an external source of electricity.

electrolyte A substance whose aqueous solution conducts electricity.

electrolytic cell A device for carrying out electrolysis.

electron The elementary unit of negative charge.

electron affinity The energy change that occurs when an atom or ion gains an electron.

electron-dot diagram See **Lewis structure**.

electronegativity The measure of an atom's attraction for a bonded pair of electrons.

electroplating The use of an electric current to deposit a layer of metal on a negatively charged object.

element A substance all of whose atoms have the same atomic number.

elementary reactions A series of simpler reactions that are the "building blocks" of a more complex reaction.

empirical formula A formula in which the elements are present in the smallest whole-number ratio; NO_2 is an empirical formula, but C_2H_4 is not.

endothermic reaction A reaction that absorbs energy; ΔH is positive for an endothermic reaction.

end point The point in a titration that signals that equivalent quantities of reactants have been added.

energy A quantity related to an object's capacity to do work.

enthalpy change (ΔH) The heat energy absorbed or released by a system at constant pressure.

entropy (S) The measure of the randomness or disorder of a system.

entropy change (ΔS) An increase or decrease in the randomness of a system.

equilibrium See **dynamic equilibrium**.

ester The organic product of esterification.

esterification The reaction of an acid with an alcohol to produce an ester and water.

ethanoic acid CH_3COOH; acetic acid.

ethanol CH_3CH_2OH; ethyl (grain) alcohol.

ethene C_2H_4; ethylene; the parent of the alkene family of hydrocarbons.

ether An organic compound containing the arrangement $R-O-R$.

ethyne C_2H_2; acetylene; the parent of the alkyne family of hydrocarbons.

evaporation The surface transition of liquid to gas.

excited state A condition in which one or more electrons in an atom are no longer in the lowest possible energy state.

exothermic reaction A reaction that releases energy; ΔH is negative for an exothermic reaction.

factor-label method (FLM) A technique for solving problems by using relationships among units.

fermentation The (anaerobic) oxidation of a sugar such as glucose to produce ethanol and carbon dioxide; the reaction is catalyzed by enzymes.

filtrate The part of a mixture that remains in solution after any residue is separated from it.

filtration The process in which a precipitate is recovered from a mixture.

first ionization energy The quantity of energy needed to remove the most loosely held electron from an isolated neutral atom.

first law of thermodynamics Energy is conserved in any process; $\Delta E = q + w$.

fission A nuclear reaction in which a heavy nuclide splits to form lighter nuclides and energy.

fission reactor A device for producing electrical energy by means of a controlled fission reaction.

formula mass The sum of the masses of the atoms in a formula; units are atomic mass units (u) or grams per mole (g/mol).

formula unit The notation that expresses the relative numbers of positive and negative ions in an ionic compound; the formula unit $CaCl_2$ indicates that calcium chloride contains two Cl^- ions for every Na^+ ion.

fractional distillation The separation of organic substances based on differences in their boiling points.

free-energy change (ΔG) A quantity that determines whether a reaction is spontaneous; $\Delta G = \Delta H - T\,\Delta S$; a negative value of ΔG indicates a spontaneous reaction.

freezing The transition from liquid to solid.

freezing point The temperature at which freezing occurs.

freezing point depression (lowering) The decrease in the freezing point of a solvent due to the presence of solute particles.

fuel rod The part of a nuclear reactor that contains the fissionable material.

functional group An atom or group of atoms that confers specific properties on an organic molecule.

fundamental forces The four basic forces in the universe: gravitational, electromagnetic, strong, weak; also called *fundamental interactions*.

fundamental interactions See **fundamental forces**.

fusion A synonym for *melting*; also, a nuclear process in which light nuclides join to form heavier nuclides and produce radiant energy.

fusion reactor An experimental device for producing a controlled fusion reaction and generating electrical energy from it.

gas The phase in which matter has neither definite shape nor definite volume.

glycol See **dihydroxy alcohol**.

Graham's law of effusion At constant temperature and pressure, the rate of effusion of a gas is inversely proportional to the square root of its molar mass; $\dfrac{r_1}{r_2} = \sqrt{\dfrac{M_2}{M_1}}$.

gram (g) A metric unit of mass.

gram-atomic mass The molar mass of an element expressed in grams per mole (g/mol).

gram-formula mass See **molar mass**.

gram-molecular mass The molar mass of a molecule.

ground state The electron configuration of an atom in the lowest energy state.

group The elements within a single vertical column of the Periodic Table.

half-cell The part of an electrochemical cell in which oxidation or reduction occurs.

half-cell potential See **electrode potential**.

half-life The time needed for a substance to decay to one-half its initial mass.

half-reaction The oxidation or reduction portion of a redox reaction.

halogen An element in Group 17 of the Periodic Table; F, Cl, Br, I, At.

heat The energy transferred between two objects when they are at different temperatures.

heat of formation ($\Delta H_{f}°$) The heat absorbed or released when 1 mole of a compound is formed from its elements under standard conditions.

heat of fusion (H_f) The heat absorbed when a unit mass of solid changes to liquid at its melting point; $H_{f(\text{ice})} = 333.6$ joules per gram.

heat of reaction (ΔH) The heat absorbed or released as a result of a chemical reaction.

heat of vaporization (H_v) The heat absorbed when a unit mass of liquid changes to gas at its boiling point; $H_{v(\text{water})} = 2259$ joules per gram.

heavy water A molecule of water in which the hydrogen atoms have a mass number of 2; deuterium oxide.

Hess's law The heat absorbed or released in a given reaction can be expressed as the sum of the heats associated with other reactions if these reactions can be "added" to produce the original reaction.

heterogeneous mixture A nonuniform mixture.

homogeneous mixture A mixture with a uniform distribution of particles; a solution is one example of a homogeneous mixture.

homologous series A group of organic compounds with related structures and properties; each successive member of the series differs from the one before it by a specific number of carbon and hydrogen atoms (usually CH_2).

hybrid orbital An orbital formed by the "mixing" of individual atomic orbitals.

hydrate A crystalline compound that has water molecules incorporated into its crystal structure; common examples include $CuSO_4 \cdot 5H_2O$ and $Na_2SO_4 \cdot 10H_2O$ [also written as $CuSO_4(H_2O)_5$ and $Na_2SO_4(H_2O)_{10}$].

hydration The association of water molecules with an ion or another molecule.

hydride A binary compound of an active metal and hydrogen; the H^- ion.

hydrocarbon An organic compound composed of carbon and hydrogen.

hydrogen bond An unusually strong intermolecular attraction that results when hydrogen is bonded to a small, highly electronegative atom such as F, O, or N.

hydrogen half-cell See **standard hydrogen half-cell**.

hydrolysis A reaction in which a water molecule breaks a chemical bond; the reaction between certain salts and water to produce an excess of hydronium or hydroxide ions.

hydronium ion H_3O^+; the conjugate acid of H_2O; responsible for acidic properties in water solutions.

hydroxide ion OH^-; the conjugate base of H_2O; responsible for basic properties in water solutions.

ideal gas A model of a gas in which the particles have no volume, do not attract or repel each other, and collide without loss of energy; real gases approximate ideal gas behavior under conditions of low pressure and high temperature.

ideal gas law The relationship obeyed by an ideal gas; $PV = nRT$, where n is the number of moles of gas and R is the universal gas constant.

indicator A substance that undergoes a color change to signal a change in chemical conditions; acid-base indicators change color over specified pH ranges.

inert gas An element in Group 18 of the Periodic Table; Ne, Ar, Kr, Xe, Rn. (He is also associated with Group 18.)

inorganic compound A compound that is not a hydrocarbon derivative.

internal energy (*E*) The total energy possessed by a system.

ion A particle in which the numbers of protons and electrons are not equal.

ion-dipole attraction The attractive force between an ion and the oppositely charged dipole of a neighboring polar molecule.

ionic bond The electrostatic attraction of positive and negative ions in an ionic compound; an electronegativity difference of 1.7 or greater indicates the presence of an ionic bond.

ionic compound A substance whose particles consist of positive and negative ions.

ionic radius The size of a monatomic ion; positive ions are generally smaller than their neutral atoms, while negative ions are generally larger.

ionization constant An equilibrium constant expression that describes an ionization reaction; a relatively large value indicates a more ionized substance; K_a, K_b, and K_w are examples of ionization constants.

ionization energy The quantity of energy needed to remove an electron from an atom or ion; see also **first ionization energy**.

isomers Different compounds that have the same molecular formula; see also **constitutional isomers**; **enantiomers**; **stereoisomers**.

isotopes Atoms having the same atomic number but different mass numbers; atoms of the same element with differing numbers of neutrons.

IUPAC International Union of Pure and Applied Chemistry; the scientific group responsible for all major policies in chemistry, including the naming of elements and compounds.

joule (J) The SI unit of work and energy.

Kelvin (K) The SI unit of temperature. A temperature *difference* of 1 K is equal to a temperature *difference* of 1 C°.

kernel The nucleus and inner (nonvalence) electrons of an atom.

ketone An organic compound containing a carbonyl group with no hydrogen atoms directly attached to the carbonyl carbon.

kilo- The metric prefix signifying 1000.

kilojoule (kJ) 1000 joules.

kilopascal (kPa) 1000 pascals.

kinetic energy The energy associated with the motion of an object.

kinetic-molecular theory (KMT) The theory that explains the structures and behaviors of idealized models of gases, liquids, and solids.

Le Châtelier's principle When a system at equilibrium is subjected to a stress, the system will shift in order to lessen the effects of the stress. Eventually, a new equilibrium point is established.

Lewis acid Any substance that can accept a pair of electrons.

Lewis base Any substance that can donate a pair of electrons.

Lewis structure A shorthand notation for illustrating the ground-state valence electron configuration of an atom.

limiting reactant Any reactant that is entirely consumed in a chemical reaction.

liquid The phase in which matter has a definite volume but an indefinite shape; a liquid takes the shape of its container.

liter (L) A unit of volume in the metric system; 1 liter = 1000 cubic centimeters; 1 liter = 1 cubic decimeter; 1 liter is approximately equal to 1 quart.

litmus An acid-base indicator that is red in acidic solutions and blue in basic solutions.

London dispersion forces See **dispersion forces**.

macromolecule A giant molecule formed by network bonding or by polymerization.

malleability The property by which a substance is able to be formed into various shapes; metallic substances possess malleability.

manometer A curved tube used to determine gas pressure by measuring the difference in the heights of the liquid that partially fills the tube. One end of the tube can be in contact with the atmosphere (*open-tube manometer*) or be isolated from it (*closed-tube manometer*).

mass defect The difference in mass between a nucleus and the nucleons that compose it; the energy equivalent of mass defect is known as *binding energy*.

mass number The number of nucleons in a nuclide.

melting The transition from solid to liquid.

melting point The temperature at which the vapor pressure of a solid equals the vapor pressure of the liquid; the temperature at which the solid and liquid phases of a substance are in equilibrium; see also **freezing point**.

meniscus The curved surface of a liquid, caused by the attraction of the particles of the liquid and the container holding the liquid (e.g., water in a graduated cylinder), or by the mutual attraction of the particles of the liquid (e.g., mercury).

metal A substance composed of atoms with low ionization energies and relatively vacant valence levels; Na, Fe, Ag, and Ba are metallic substances.

metallic bond The delocalization of the valence electrons among the kernels of the metal atoms; "mobile valence electrons immersed in a sea of positive ions."

metalloid An element that has both metallic and nonmetallic properties; examples of metalloids include B, Ge, Si, and Te.

methanal HCHO; formaldehyde; the simplest aldehyde.

methane CH_4; the parent of the alkane family of hydrocarbons.

methanoic acid HCOOH; formic acid; the simplest organic acid.

methanol CH_3OH; methyl (wood) alcohol; the simplest alcohol.

milli- The metric prefix signifying 1/1000.

milliliter (mL) 1/1000 liter; 1 liter = 1000 milliliters.

miscible Able to be mixed in a solution of liquids that is soluble in all proportions; ethanol and water are a miscible pair of liquids.

mixture A material consisting of two or more components and having a variable composition.

moderator A substance used to produce slow neutrons and promote nuclear fission; graphite and heavy water are used as moderators in fission reactors.

molal boiling point constant (k_b) The increase in the boiling point of a solvent when 1 mole of particles is dissolved in 1 kilogram of solvent; k_b for H_2O is 0.52 C° per molal.

molal freezing point constant (k_f) The decrease in the freezing point of a solvent when 1 mole of particles is dissolved in 1 kilogram of solvent; k_f for H_2O is 1.86 C° per molal.

molality The concentration of a solution, measured as the number of moles of solute per kilogram of solvent.

molar mass The mass of any atom, element, ion, or compound expressed in grams per mole (g/mol).

molarity The concentration of a solution, measured as the number of moles of solute per liter of solution.

molar volume The volume occupied by 1 mole of an ideal gas; 22.4 liters at STP.

mole The number of atoms contained in 12 grams of carbon-12; see also **Avogadro's number**.

molecular formula A chemical formula that lists the number of atoms present but does not show the arrangement of the atoms in space.

molecular mass The sum of the masses of the atoms in a molecule; units are atomic mass units (u) or grams per mole (g/mol).

molecule The smallest unit of a nonionic substance; Ar, Cl_2, and NH_3 are molecules.

mole fraction (X) A measure of concentration that expresses the ratio of the number of moles of a given substance to the total number of moles present;

$$X_i = \frac{n_i}{n_{\text{total}}}.$$

monatomic ion An ion containing only one atom; Ca^{2+} is a monatomic ion; see also **simple ion**.

monatomic molecule A molecule consisting of one atom; Xe and He are monatomic molecules.

monomer The basic unit of a polymer; the monomer of a protein is an amino acid; the monomer of starch is glucose.

monoprotic acid A Brønsted–Lowry acid that contains one ionizable hydrogen atom.

network solid A substance formed by a two- or three-dimensional web of covalent bonds to produce a macromolecule; diamond and SiO_2 are network solids.

neutralization The reaction of equivalent amounts of hydronium and hydroxide ions in aqueous solution; the principal product is water. When the water is evaporated, the spectator ions form a salt; see also **spectator ion**.

neutron A neutral nuclear particle with a mass comparable to that of a proton.

noble gas See **inert gas**.

nonelectrolyte A substance whose aqueous solution does not conduct electricity; glucose is a nonelectrolyte.

nonmetal A substance that does not have characteristic metallic properties; C and S are nonmetallic elements.

nonpolar bond A covalent bond in which the electron pair or pairs are shared equally by both atoms.

nonpolar molecule A molecule containing only nonpolar bonds, such as N_2, or a molecule with a symmetrical charge distribution, such as CCl_4 and CO_2.

normal boiling point The boiling temperature of a substance at a pressure of 1 atmosphere.

nuclear equation A shorthand listing of reactant and product nuclides in a nuclear reaction.

nucleon A constituent of an atomic nucleus; a proton or a neutron.

nucleus The portion of the atom that contains more than 99.9 percent of the atom's mass; the nucleus is small, dense, and positively charged.

orbital A designation that describes the orientation of the electron cloud in space; p_x, p_y, and p_z are orbital designations.

order of magnitude The power of 10 closest to the value of a measurement.

ore A native mineral from which a metal or metals can be extracted.

organic chemistry The study of the hydrocarbons and their derivatives; the chemistry of carbon.

organic compound A compound that is a hydrocarbon or a hydrocarbon derivative.

oxidation The loss or apparent loss of electrons in a chemical reaction.

oxidation number (state) The charge that an atom has or appears to have when certain arbitrary rules are applied; oxidation numbers are useful for identifying the atoms that are oxidized and reduced in a redox reaction.

oxidation potential An electrode potential for an oxidation half-reaction.

oxidizing agent The particle in a redox reaction that causes another particle to be oxidized; as a result, an oxidizing agent is reduced.

paraffin A common name for a mixture of solid alkanes; another name for paraffin is wax.

partial pressure The individual pressure due to each gas in a mixture of gases.

particle accelerator A device that uses electric and/or magnetic fields to accelerate charged subatomic particles in order to study them further.

parts per million (ppm) A unit of concentration that expresses the mass of a solute dissolved in 1 million parts of a very dilute solution;

$$\text{ppm} = \frac{\text{mass}_{solute}}{\text{mass}_{solution}} \times 10^6.$$

pascal (Pa) The SI unit of pressure.

percent composition by mass The number of grams of an element (or group of elements) present in 100 grams of an ion or compound.

percent concentration A unit of concentration that expresses the mass of a solute dissolved in 100 parts of a solution;

$$\text{percent} (\%) = \frac{\text{mass}_{solute}}{\text{mass}_{solution}} \times 10^2.$$

percent yield The ratio of the actual yield of a product of a chemical reaction to the predicted yield, multiplied by 100.

period One of the horizontal rows of the Periodic Table; the period number indicates the valence level of an element.

periodic law The properties of elements recur at regular intervals and depend on their nuclear charges; "The properties of elements are a periodic function of their atomic numbers."

peroxide The O_2^{2-} ion; H_2O_2 and BaO_2 are peroxides.

petroleum Crude oil containing a mixture of hydrocarbons.

pH A scale of acidity and basicity based on the hydronium ion concentration in aqueous solution; $pH = -\log [H_3O^+]$. At 298 K, a pH less than 7 indicates an acidic solution; a pH greater than 7, a basic solution; a pH of 7, a neutral solution.

phase equilibrium The state in which the rates of opposing phase changes (freezing–melting, boiling–condensation, sublimation–deposition) are equal.

phenolphthalein An acid–base indicator that is colorless in acidic solutions and pink in basic solutions.

photon A fundamental unit of radiant energy; a quantum of radiation.

physical change A change that occurs with no change in the composition of a substance. Phase changes are physical changes.

physical property A property that can be measured without changing the composition of a substance. Density is a physical property.

pOH A scale of acidity and basicity based on the hydroxide ion concentration in aqueous solution; $pOH = -\log [OH^-]$. At 298 K, a pOH less than 7 indicates a basic solution; a pOH greater than 7, an acidic solution; a pOH of 7, a neutral solution; $pOH = 14 - pH$.

polar bond A covalent bond in which the electron pair or pairs are shared unequally by two atoms; the atom with the larger electronegativity has more of the electron density surrounding it.

polar molecule A molecule with an unsymmetrical charge distribution, such as H_2O; a dipole.

polyatomic ion An ion composed of more than one atom; SO_4^{2-} is a polyatomic ion.

polymer A macromolecule consisting of a chain of simpler units; polyethylene is a polymer of ethene.

polymerization See **polymer**; **addition polymerization**; **condensation polymerization**.

polyprotic acid A Brønsted–Lowry acid that contains more than one ionizable hydrogen atom.

positron A positively charged electron; a particle of antimatter.

potential energy The energy associated with the position of an object; a "stored" form of energy.

precipitate A deposit formed by the appearance of an excess of solid solute in a saturated solution.

precipitation See **crystallization**.

precision The closeness of a series of measurements to one another; see also **accuracy**.

pressure The force exerted on an object divided by the surface area of the object; $P = \dfrac{F}{A}$.

primary alcohol An alcohol in which two or three hydrogen atoms are bonded to the carbon atom containing the hydroxyl group.

principal energy level An integer, beginning with 1, that describes the approximate distance of an electron from the nucleus of an atom.

product(s) The substance or substances that are formed in a chemical process; products are on the right side of a chemical equation.

propanone $(CH_3)_2CO$; acetone; the simplest ketone.

proton A nuclear particle with a positive charge equal to the negative charge on the electron; a nucleon.

pure substance An element or a compound.

quark One of the fundamental particles of matter, always found in combination with each other. Nuclear particles, such as protons and neutrons, are comprised of groups of three quarks each.

radiant energy Electromagnetic energy; visible light and X rays are examples of radiant energy.

radioactive dating The use of radioactive isotopes to measure the age of an object.

radioactive tracer A radioisotope used to indicate the path of an atom in a chemical reaction.

radioactivity The spontaneous breakdown of a radioactive nuclide.

radioisotope A radioactive isotope.

rate-limiting step The slowest of the elementary reactions in a complex reaction.

reactant(s) The substance or substances that react in a chemical process; reactants are on the left side of a chemical equation.

redox reaction A chemical reaction in which oxidation-reduction takes place.

reducing agent The particle in a redox reaction that causes another particle to be reduced; as a result, a reducing agent is oxidized.

reduction The gain or apparent gain of electrons in a chemical reaction.

reduction potential An electrode potential for a reduction half-reaction.

roasting The reaction of a metallic ore with oxygen in order to produce the metallic oxide from which the metal may be obtained.

salt The spectator-ion product of a neutralization reaction; see also **spectator ion**.

salt bridge A device for allowing the flow of ions in a voltaic cell.

saponification The reaction of an ester with a base to produce an alcohol and the sodium salt of an organic acid; soap is produced by saponifying fats with NaOH.

saturated hydrocarbon A hydrocarbon containing only single carbon-carbon bonds.

saturated solution A solution in which the pure solute is in equilibrium with the dissolved solute; a solution that contains the maximum amount of dissolved solute.

secondary alcohol An alcohol in which only one hydrogen atom is bonded to the carbon atom containing the hydroxyl group.

second law of thermodynamics A statement that relates spontaneous processes to the entropy (disorder) of the universe.

significant digit(s) [figure(s)] The number or numbers that are part of a measurement. If there are two or more, all but the last figure is known; the last figure is the experimenter's best estimate.

simple ion An ion containing only one atom; a monatomic ion; Na^+ is a simple ion.

single bond A covalent bond in which one pair of electrons is shared by two adjacent atoms.

solid The phase in which matter has both definite shape and definite volume.

solubility The amount of solute needed to produce a saturated solution with a given amount of solvent.

solubility product constant (K_{sp}) The equilibrium constant expression that provides a measure of the solubility of an ionic compound; a large K_{sp} value generally indicates a more soluble salt.

solute(s) The substance or substances dissolved in a solution.

solution A homogeneous mixture whose particles are extremely small.

solution equilibrium The state in which the undissolved and dissolved solutes are in dynamic equilibrium; in a solid-liquid solution, the rate of dissolving equals the rate of crystallization.

solvent The substance in which the solute is dissolved.

spectator ion An ion that does not take part in a chemical reaction; in the (acid-base) reaction between NaOH(aq) and HCl(aq), Na$^+$(aq) and Cl$^-$(aq) are spectator ions; in the (redox) reaction between Zn(s) and Cu(NO$_3$)$_2$(aq), NO$_3^-$(aq) is the spectator ion.

spontaneous reaction A reaction that can occur under a specified set of conditions without the application of external work.

standard atmosphere (atm) A unit of pressure; 1 standard atmosphere = 101.325 kilopascals (exactly); 760 millimeters Hg.

standard cell potential ($\mathscr{E}_{cell}°$) The maximum potential difference of an electrochemical cell, measured in volts, under standard conditions.

standard electrode (reduction) potential ($\mathscr{E}°$) An oxidation potential or a reduction potential measured under standard conditions.

standard hydrogen half-cell A hydrogen gas-hydrogen ion electrode whose electrode potential under standard conditions is set at 0 volt.

standard solution A solution whose concentration is accurately known; a standard solution is used for analyzing other substances.

standard state A set of conditions imposed on a pure substance that includes a pressure of 100.00 kilopascals and a specified temperature (usually 298.15 K).

standard-state pressure A pressure of 100.00 kilopascals (exactly).

standard temperature and pressure (STP) 273.15 Kelvin and 1 atmosphere.

stereoisomers Isomers in which the atoms are connected in the same order but are arranged differently in space. *Cis-trans* isomerism is an example of stereoisomerism.

Stock system A systematic method of naming chemical compounds in which the *positive* oxidation number is written as a Roman numeral in parentheses after the element. For example, the Stock name of the compound Fe$_2$O$_3$ is iron(III) oxide.

stoichiometry The study of quantitative relationships in substances and reactions; chemical mathematics.

strong acid In aqueous solution, a substance that ionizes almost completely to hydronium ion.

strong base In aqueous solution, a substance that ionizes or dissociates almost completely to hydroxide ion.

structural formula A chemical formula that illustrates the spatial arrangement of each atom.

sublevel The property of an electron that describes the shape of its cloud; the letter designations for sublevels are *s, p, d, f,* . . . .

sublimation The direct transition from solid to gas; CO$_2$(s) and I$_2$(s) sublime at atmospheric pressure.

supersaturated solution A solution that contains more dissolved solute than a saturated solution at the same temperature; supersaturation is an unstable condition.

surroundings In thermodynamics, the part of the universe that is *not* included in the system being studied.

system In thermodynamics, the part of the universe singled out for study.

temperature A measure of the average kinetic energy of the particles of a substance.

tertiary alcohol An alcohol in which no hydrogen atoms are bonded to the carbon atom containing the hydroxyl group.

thermal energy The part of internal energy that is associated with random molecular motions.

titration The addition of a known volume of a standard solution in order to determine the concentration of an unknown solution.

torr A unit of pressure equivalent to 1 millimeter of mercury (mm Hg).

transition element An element in Groups 3–11 of the Periodic Table.

transmutation The conversion of one element to another by a nuclear process.

trihydroxy alcohol An alcohol with three hydroxyl groups; glycerol (1,2,3-propanetriol) is a trihydroxy alcohol.

triple bond A covalent bond in which three pairs of electrons are shared by two adjacent atoms.

triple point The unique temperature and pressure at which the solid, liquid, and gaseous phases can coexist in equilibrium. The triple point of water is used to set the fixed point on the Kelvin temperature scale (273.16 K).

tritium The radioactive isotope of hydrogen with a mass number of 3.

unsaturated hydrocarbon A hydrocarbon containing double and/or triple carbon-carbon bonds.

unsaturated solution A solution that contains less dissolved solute than a saturated solution at the same temperature.

valence electron An electron in the outermost principle energy level.

valence shell electron-pair repulsion method (VSEPR) A technique for determining the shape of a molecule or polyatomic ion by considering the number of shared and unshared pairs of electrons around the central atom of the molecule or ion.

van der Waals forces All forces involving attractions of polar molecules and nonpolar molecules; see also **dipole-dipole attraction**; **dispersion forces**; **ion-dipole attraction**.

vapor pressure The pressure produced by a solid or a liquid when it is in equilibrium with its gas phase.

volt (V) The unit of electric potential; it measures the work output per unit of electric charge transferred.

voltaic cell A device that uses a spontaneous redox reaction to generate electrical energy.

water of hydration The water molecules that are part of the crystalline structure of certain compounds.

weak acid In aqueous solution, a substance that is poorly ionized and produces only a small concentration of hydronium ion.

weak base In aqueous solution, a substance that is poorly ionized or dissociated and produces only a small concentration of hydroxide ion.

494

NEW YORK STATE REGENTS
REFERENCE TABLES FOR CHEMISTRY

TABLE A
Standard Temperature and Pressure

Name	Value	Unit
Standard Pressure	101.3 kPa 1 atm	kilopascal atmosphere
Standard Temperature	273 K 0°C	kelvin degree Celsius

TABLE B
Physical Constants for Water

Heat of Fusion	334 J/g
Heat of Vaporization	2260 J/g
Specific Heat Capacity of $H_2O(\ell)$	4.18 J/g•K

TABLE C
Selected Prefixes

Factor	Prefix	Symbol
10^3	kilo-	k
10^{-1}	deci-	d
10^{-2}	centi-	c
10^{-3}	milli-	m
10^{-6}	micro-	μ
10^{-9}	nano-	n
10^{-12}	pico-	p

TABLE D
Selected Units

Symbol	Name	Quantity
m	meter	length
g	gram	mass
Pa	pascal	pressure
K	kelvin	temperature
mol	mole	amount of substance
J	joule	energy, work, quantity of heat
s	second	time
min	minute	time
h	hour	time
d	day	time
y	year	time
L	liter	volume
ppm	parts per million	concentration
M	molarity	solution concentration
u	atomic mass unit	atomic mass

TABLE E
Selected Polyatomic Ions

Formula	Name	Formula	Name
H_3O^+	hydronium	CrO_4^{2-}	chromate
Hg_2^{2+}	mercury(I)	$Cr_2O_7^{2-}$	dichromate
NH_4^+	ammonium	MnO_4^-	permanganate
$C_2H_3O_2^-$ CH_3COO^- } acetate		NO_2^-	nitrite
		NO_3^-	nitrate
CN^-	cyanide	O_2^{2-}	peroxide
CO_3^{2-}	carbonate	OH^-	hydroxide
HCO_3^-	hydrogen carbonate	PO_4^{3-}	phosphate
$C_2O_4^{2-}$	oxalate	SCN^-	thiocyanate
ClO^-	hypochlorite	SO_3^{2-}	sulfite
ClO_2^-	chlorite	SO_4^{2-}	sulfate
ClO_3^-	chlorate	HSO_4^-	hydrogen sulfate
ClO_4^-	perchlorate	$S_2O_3^{2-}$	thiosulfate

TABLE F
Solubility Guidelines for Aqueous Solutions

Ions That Form *Soluble* Compounds	Exceptions
Group 1 ions (Li$^+$, Na$^+$, etc.)	
ammonium (NH$_4^+$)	
nitrate (NO$_3^-$)	
acetate (C$_2$H$_3$O$_2^-$ or CH$_3$COO$^-$)	
hydrogen carbonate (HCO$_3^-$)	
chlorate (ClO$_3^-$)	
halides (Cl$^-$, Br$^-$, I$^-$)	when combined with Ag$^+$, Pb^{2+}, or Hg$_2^{2+}$
sulfates (SO$_4^{2-}$)	when combined with Ag$^+$, Ca^{2+}, Sr^{2+}, Ba^{2+}, or Pb^{2+}

Ions That Form *Insoluble* Compounds*	Exceptions
carbonate (CO$_3^{2-}$)	when combined with Group 1 ions or ammonium (NH$_4^+$)
chromate (CrO$_4^{2-}$)	when combined with Group 1 ions, Ca^{2+}, Mg^{2+}, or ammonium (NH$_4^+$)
phosphate (PO$_4^{3-}$)	when combined with Group 1 ions or ammonium (NH$_4^+$)
sulfide (S^{2-})	when combined with Group 1 ions or ammonium (NH$_4^+$)
hydroxide (OH$^-$)	when combined with Group 1 ions, Ca^{2+}, Ba^{2+}, Sr^{2+}, or ammonium (NH$_4^+$)

*compounds having very low solubility in H$_2$O

TABLE G
Solubility Curves at Standard Pressure

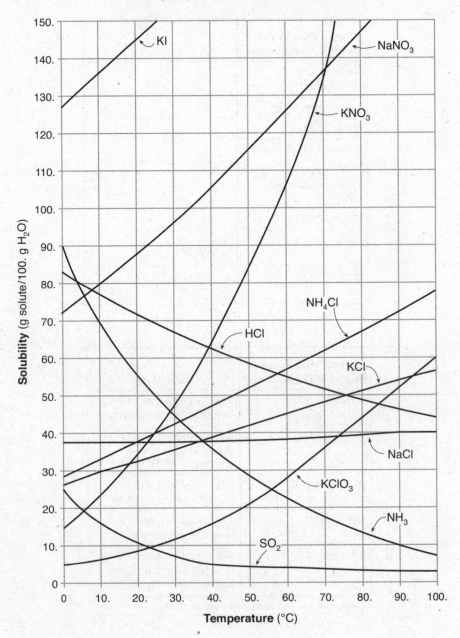

TABLE H
Vapor Pressure of Four Liquids

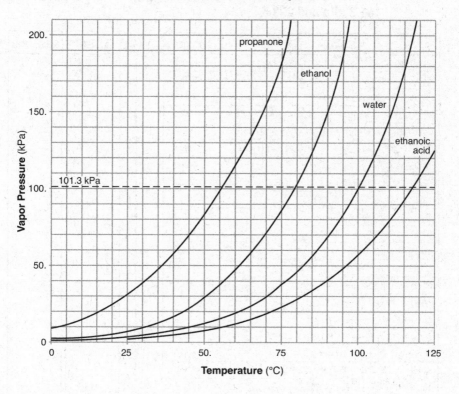

TABLE I
Heats of Reaction at
101.3 kPa and 298 K

Reaction	ΔH (kJ)*
$CH_4(g) + 2O_2(g) \longrightarrow CO_2(g) + 2H_2O(\ell)$	−890.4
$C_3H_8(g) + 5O_2(g) \longrightarrow 3CO_2(g) + 4H_2O(\ell)$	−2219.2
$2C_8H_{18}(\ell) + 25O_2(g) \longrightarrow 16CO_2(g) + 18H_2O(\ell)$	−10943
$2CH_3OH(\ell) + 3O_2(g) \longrightarrow 2CO_2(g) + 4H_2O(\ell)$	−1452
$C_2H_5OH(\ell) + 3O_2(g) \longrightarrow 2CO_2(g) + 3H_2O(\ell)$	−1367
$C_6H_{12}O_6(s) + 6O_2(g) \longrightarrow 6CO_2(g) + 6H_2O(\ell)$	−2804
$2CO(g) + O_2(g) \longrightarrow 2CO_2(g)$	−566.0
$C(s) + O_2(g) \longrightarrow CO_2(g)$	−393.5
$4Al(s) + 3O_2(g) \longrightarrow 2Al_2O_3(s)$	−3351
$N_2(g) + O_2(g) \longrightarrow 2NO(g)$	+182.6
$N_2(g) + 2O_2(g) \longrightarrow 2NO_2(g)$	+66.4
$2H_2(g) + O_2(g) \longrightarrow 2H_2O(g)$	−483.6
$2H_2(g) + O_2(g) \longrightarrow 2H_2O(\ell)$	−571.6
$N_2(g) + 3H_2(g) \longrightarrow 2NH_3(g)$	−91.8
$2C(s) + 3H_2(g) \longrightarrow C_2H_6(g)$	−84.0
$2C(s) + 2H_2(g) \longrightarrow C_2H_4(g)$	+52.4
$2C(s) + H_2(g) \longrightarrow C_2H_2(g)$	+227.4
$H_2(g) + I_2(g) \longrightarrow 2HI(g)$	+53.0
$KNO_3(s) \xrightarrow{H_2O} K^+(aq) + NO_3^-(aq)$	+34.89
$NaOH(s) \xrightarrow{H_2O} Na^+(aq) + OH^-(aq)$	−44.51
$NH_4Cl(s) \xrightarrow{H_2O} NH_4^+(aq) + Cl^-(aq)$	+14.78
$NH_4NO_3(s) \xrightarrow{H_2O} NH_4^+(aq) + NO_3^-(aq)$	+25.69
$NaCl(s) \xrightarrow{H_2O} Na^+(aq) + Cl^-(aq)$	+3.88
$LiBr(s) \xrightarrow{H_2O} Li^+(aq) + Br^-(aq)$	−48.83
$H^+(aq) + OH^-(aq) \longrightarrow H_2O(\ell)$	−55.8

*The ΔH values are based on molar quantities represented in the equations. A minus sign indicates an exothermic reaction.

TABLE J
Activity Series**

Most Active	Metals	Nonmetals	Most Active
	Li	F_2	
	Rb	Cl_2	
	K	Br_2	
	Cs	I_2	
	Ba		
	Sr		
	Ca		
	Na		
	Mg		
	Al		
	Ti		
	Mn		
	Zn		
	Cr		
	Fe		
	Co		
	Ni		
	Sn		
	Pb		
	H_2		
	Cu		
	Ag		
Least Active	Au		Least Active

**Activity Series is based on the hydrogen standard. H_2 is *not* a metal.

TABLE K
Common Acids

Formula	Name
HCl(aq)	hydrochloric acid
HNO_2(aq)	nitrous acid
HNO_3(aq)	nitric acid
H_2SO_3(aq)	sulfurous acid
H_2SO_4(aq)	sulfuric acid
H_3PO_4(aq)	phosphoric acid
H_2CO_3(aq) or CO_2(aq)	carbonic acid
CH_3COOH(aq) or $HC_2H_3O_2$(aq)	ethanoic acid (acetic acid)

TABLE L
Common Bases

Formula	Name
NaOH(aq)	sodium hydroxide
KOH(aq)	potassium hydroxide
$Ca(OH)_2$(aq)	calcium hydroxide
NH_3(aq)	aqueous ammonia

TABLE M
Common Acid–Base Indicators

Indicator	Approximate pH Range for Color Change	Color Change
methyl orange	3.1–4.4	red to yellow
bromthymol blue	6.0–7.6	yellow to blue
phenolphthalein	8–9	colorless to pink
litmus	4.5–8.3	red to blue
bromcresol green	3.8–5.4	yellow to blue
thymol blue	8.0–9.6	yellow to blue

Source: *The Merck Index*, 14th ed., 2006, Merck Publishing Group

TABLE N
Selected Radioisotopes

Nuclide	Half-Life	Decay Mode	Nuclide Name
^{198}Au	2.695 d	β^-	gold-198
^{14}C	5715 y	β^-	carbon-14
^{37}Ca	182 ms	β^+	calcium-37
^{60}Co	5.271 y	β^-	cobalt-60
^{137}Cs	30.2 y	β^-	cesium-137
^{53}Fe	8.51 min	β^+	iron-53
^{220}Fr	27.4 s	α	francium-220
^{3}H	12.31 y	β^-	hydrogen-3
^{131}I	8.021 d	β^-	iodine-131
^{37}K	1.23 s	β^+	potassium-37
^{42}K	12.36 h	β^-	potassium-42
^{85}Kr	10.73 y	β^-	krypton-85
^{16}N	7.13 s	β^-	nitrogen-16
^{19}Ne	17.22 s	β^+	neon-19
^{32}P	14.28 d	β^-	phosphorus-32
^{239}Pu	2.410×10^4 y	α	plutonium-239
^{226}Ra	1599 y	α	radium-226
^{222}Rn	3.823 d	α	radon-222
^{90}Sr	29.1 y	β^-	strontium-90
^{99}Tc	2.13×10^5 y	β^-	technetium-99
^{232}Th	1.40×10^{10} y	α	thorium-232
^{233}U	1.592×10^5 y	α	uranium-233
^{235}U	7.04×10^8 y	α	uranium-235
^{238}U	4.47×10^9 y	α	uranium-238

Source: *CRC Handbook of Chemistry and Physics*, 91st ed., 2010–2011, CRC Press

TABLE O
Symbols Used in Nuclear Chemistry

Name	Notation	Symbol
alpha particle	^4_2He or $^4_2\alpha$	α
beta particle	$^0_{-1}\text{e}$ or $^0_{-1}\beta$	β^-
gamma radiation	$^0_0\gamma$	γ
neutron	^1_0n	n
proton	^1_1H or ^1_1p	p
positron	$^0_{+1}\text{e}$ or $^0_{+1}\beta$	β^+

TABLE P
Organic Prefixes

Prefix	Number of Carbon Atoms
meth-	1
eth-	2
prop-	3
but-	4
pent-	5
hex-	6
hept-	7
oct-	8
non-	9
dec-	10

TABLE Q
Homologous Series of Hydrocarbons

Name	General Formula	Examples	
		Name	Structural Formula
alkanes	C_nH_{2n+2}	ethane	H—C—C—H (with H, H above and H, H below each carbon)
alkenes	C_nH_{2n}	ethene	C=C (with H and H attached to each carbon)
alkynes	C_nH_{2n-2}	ethyne	H—C≡C—H

Note: n = number of carbon atoms

TABLE R
Organic Functional Groups

Class of Compound	Functional Group	General Formula	Example
halide (halocarbon)	$-F$ (fluoro-) $-Cl$ (chloro-) $-Br$ (bromo-) $-I$ (iodo-)	$R-X$ (X represents any halogen)	$CH_3CHClCH_3$ 2-chloropropane
alcohol	$-OH$	$R-OH$	$CH_3CH_2CH_2OH$ 1-propanol
ether	$-O-$	$R-O-R'$	$CH_3OCH_2CH_3$ methyl ethyl ether
aldehyde	$-\overset{\overset{O}{\|\|}}{C}-H$	$R-\overset{\overset{O}{\|\|}}{C}-H$	$CH_3CH_2\overset{\overset{O}{\|\|}}{C}-H$ propanal
ketone	$-\overset{\overset{O}{\|\|}}{C}-$	$R-\overset{\overset{O}{\|\|}}{C}-R'$	$CH_3\overset{\overset{O}{\|\|}}{C}CH_2CH_2CH_3$ 2-pentanone
organic acid	$-\overset{\overset{O}{\|\|}}{C}-OH$	$R-\overset{\overset{O}{\|\|}}{C}-OH$	$CH_3CH_2\overset{\overset{O}{\|\|}}{C}-OH$ propanoic acid
ester	$-\overset{\overset{O}{\|\|}}{C}-O-$	$R-\overset{\overset{O}{\|\|}}{C}-O-R'$	$CH_3CH_2\overset{\overset{O}{\|\|}}{C}OCH_3$ methyl propanoate
amine	$-\overset{\overset{\|}{}}{N}-$	$R-\overset{\overset{R'}{\|}}{N}-R''$	$CH_3CH_2CH_2NH_2$ 1-propanamine
amide	$-\overset{\overset{O}{\|\|}}{C}-\overset{\overset{\|}{}}{N}H$	$R-\overset{\overset{O}{\|\|}}{C}-\overset{\overset{R'}{\|}}{N}H$	$CH_3CH_2\overset{\overset{O}{\|\|}}{C}-NH_2$ propanamide

Note: R represents a bonded atom or group of atoms.

Periodic Table of the Elements

Period		
	1	

1

1.00794	+1 −1
H	
1 1	

Group

1	2

2

6.941 +1	9.01218 +2
Li	**Be**
3 2-1	4 2-2

3

22.98977 +1	24.305 +2
Na	**Mg**
11 2-8-1	12 2-8-2

Group

3	4	5	6	7	8	9

4

39.0983 +1	40.08 +2	44.9559 +3	47.867 +2 +3 +4	50.9415 +2 +3 +4 +5	51.996 +2 +3 +6	54.9380 +2 +3 +4 +7	55.845 +2 +3	58.9332 +2 +3
K	**Ca**	**Sc**	**Ti**	**V**	**Cr**	**Mn**	**Fe**	**Co**
19 2-8-8-1	20 2-8-8-2	21 2-8-9-2	22 2-8-10-2	23 2-8-11-2	24 2-8-13-1	25 2-8-13-2	26 2-8-14-2	27 2-8-15-2

5

85.4678 +1	87.62 +2	88.9059 +3	91.224 +4	92.9064 +3 +5	95.94 +6	(98) +4 +6 +7	101.07 +3	102.906 +3
Rb	**Sr**	**Y**	**Zr**	**Nb**	**Mo**	**Tc**	**Ru**	**Rh**
37 2-8-18-8-1	38 2-8-18-8-2	39 2-8-18-9-2	40 2-8-18-10-2	41 2-8-18-12-1	42 2-8-18-13-1	43 2-8-18-13-2	44 2-8-18-15-1	45 2-8-18-16-1

6

132.905	137.33	138.9055 +3	178.49 +4	180.948 +5	183.84 +6	186.207 +4 +6 +7	190.23 +3 +4	192.217 +3
Cs	**Ba**	**La**	**Hf**	**Ta**	**W**	**Re**	**Os**	**Ir**
55 2-8-18-18-8-1	56 2-8-18-18-8-2	57 2-8-18-18-9-2	72 *18-32-10-2	73 -18-32-11-2	74 -18-32-12-2	75 -18-32-13-2	76 -18-32-14-2	77 -18-32-15-2

7

(223) +1	(226) +2	(227) +3	(261) +4	(262)	(266)	(272)	(277)	(276)
Fr	**Ra**	**Ac**	**Rf**	**Db**	**Sg**	**Bh**	**Hs**	**Mt**
87 -18-32-18-8-1	88 -18-32-18-8-2	89 -18-32-18-9-2	104	105	106	107	108	109

140.116 +3 +4	140.908 +3	144.24 +3	(145) +3	150.36 +2 +3
Ce	**Pr**	**Nd**	**Pm**	**Sm**
58	59	60	61	62

232.038 +4	231.036 +4 +5	238.029 +3 +4 +5 +6	(237) +3 +4 +5 +6	(244) +3 +4 +5 +6
Th	**Pa**	**U**	**Np**	**Pu**
90	91	92	93	94

*denotes the presence of (2-8-) for elements 72 and above

**The systematic names and symbols for elements of atomic numbers 113 and above will be used until the approval of trivial names by IUPAC.

Source: *CRC Handbook of Chemistry and Physics*, 91st ed., 2010–2011, CRC Press

KEY

Atomic Mass → 12.011

Selected Oxidation States → -4 +2 +4

Symbol → **C**

Relative atomic masses are based on $^{12}C = 12$ (exact)

Atomic Number → 6

Note: Numbers in parentheses are mass numbers of the most stable or common isotope.

Electron Configuration → 2-4

18

4.00260   0

He

2
2

Group

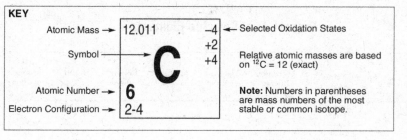

13	14	15	16	17	18
10.81 +3 **B** 5 2-3	12.011 -4 -2 +2 +4 **C** 6 2-4	14.0067 -3 -2 -1 +1 +2 +3 +4 +5 **N** 7 2-5	15.9994 -2 **O** 8 2-6	18.9984 -1 **F** 9 2-7	20.180 0 **Ne** 10 2-8
26.98154 +3 **Al** 13 2-8-3	28.0855 -4 +2 +4 **Si** 14 2-8-4	30.97376 -3 +3 +5 **P** 15 2-8-5	32.065 -2 +4 +6 **S** 16 2-8-6	35.453 -1 +1 +5 +7 **Cl** 17 2-8-7	39.948 0 **Ar** 18 2-8-8

10	11	12						
58.693 +2 +3 **Ni** 28 2-8-16-2	63.546 +1 +2 **Cu** 29 2-8-18-1	65.409 +2 **Zn** 30 2-8-18-2	69.723 +3 **Ga** 31 2-8-18-3	72.64 -4 +2 +4 **Ge** 32 2-8-18-4	74.9216 -3 +3 +5 **As** 33 2-8-18-5	78.96 -2 +4 +6 **Se** 34 2-8-18-6	79.904 -1 +1 +5 **Br** 35 2-8-18-7	83.798 0 +2 **Kr** 36 2-8-18-8
106.42 +2 +4 **Pd** 46 2-8-18-18	107.868 +1 **Ag** 47 2-8-18-18-1	112.41 +2 **Cd** 48 2-8-18-18-2	114.818 +3 **In** 49 2-8-18-18-3	118.71 +2 +4 **Sn** 50 2-8-18-18-4	121.760 -3 +3 +5 **Sb** 51 2-8-18-18-5	127.60 -2 +4 +6 **Te** 52 2-8-18-18-6	126.904 -1 +1 +5 +7 **I** 53 2-8-18-18-7	131.29 0 +2 +4 +6 **Xe** 54 2-8-18-18-8
195.08 +2 +4 **Pt** 78 -18-32-17-1	196.967 +1 +3 **Au** 79 -18-32-18-1	200.59 +1 +2 **Hg** 80 -18-32-18-2	204.383 +1 +3 **Tl** 81 -18-32-18-3	207.2 +2 +4 **Pb** 82 -18-32-18-4	208.980 +3 +5 **Bi** 83 -18-32-18-5	(209) +2 +4 **Po** 84 -18-32-18-6	(210) -1 **At** 85 -18-32-18-7	(222) 0 **Rn** 86 -18-32-18-8
(281) **Ds** 110	(280) **Rg** 111	(285) **Cn** 112	(284) **Uut** 113**	(289) **Uuq** 114	(288) **Uup** 115	(292) **Uuh** 116	(?) **Uus** 117	(294) **Uuo** 118

151.964 +2 +3 **Eu** 63	157.25 +3 **Gd** 64	158.925 +3 **Tb** 65	162.500 +3 **Dy** 66	164.930 +3 **Ho** 67	167.259 +3 **Er** 68	168.934 +3 **Tm** 69	173.04 +2 +3 **Yb** 70	174.9668 +3 **Lu** 71
(243) +3 +4 +5 +6 **Am** 95	(247) +3 **Cm** 96	(247) +3 +4 **Bk** 97	(251) +3 **Cf** 98	(252) +3 **Es** 99	(257) +3 **Fm** 100	(258) +2 +3 **Md** 101	(259) +2 +3 **No** 102	(262) +3 **Lr** 103

TABLE S
Properties of Selected Elements

Atomic Number	Symbol	Name	First Ionization Energy (kJ/mol)	Electro-negativity	Melting Point (K)	Boiling* Point (K)	Density** (g/cm^3)	Atomic Radius (pm)
1	H	hydrogen	1312	2.2	14	20.	0.000082	32
2	He	helium	2372	—	—	4	0.000164	37
3	Li	lithium	520.	1.0	454	1615	0.534	130.
4	Be	beryllium	900.	1.6	1560.	2744	1.85	99
5	B	boron	801	2.0	2348	4273	2.34	84
6	C	carbon	1086	2.6	—	—	—	75
7	N	nitrogen	1402	3.0	63	77	0.001145	71
8	O	oxygen	1314	3.4	54	90.	0.001308	64
9	F	fluorine	1681	4.0	53	85	0.001553	60.
10	Ne	neon	2081	—	24	27	0.000825	62
11	Na	sodium	496	0.9	371	1156	0.97	160.
12	Mg	magnesium	738	1.3	923	1363	1.74	140.
13	Al	aluminum	578	1.6	933	2792	2.70	124
14	Si	silicon	787	1.9	1687	3538	2.3296	114
15	P	phosphorus (white)	1012	2.2	317	554	1.823	109
16	S	sulfur (monoclinic)	1000.	2.6	388	718	2.00	104
17	Cl	chlorine	1251	3.2	172	239	0.002898	100.
18	Ar	argon	1521	—	84	87	0.001633	101
19	K	potassium	419	0.8	337	1032	0.89	200.
20	Ca	calcium	590.	1.0	1115	1757	1.54	174
21	Sc	scandium	633	1.4	1814	3109	2.99	159
22	Ti	titanium	659	1.5	1941	3560.	4.506	148
23	V	vanadium	651	1.6	2183	3680.	6.0	144
24	Cr	chromium	653	1.7	2180.	2944	7.15	130.
25	Mn	manganese	717	1.6	1519	2334	7.3	129
26	Fe	iron	762	1.8	1811	3134	7.87	124
27	Co	cobalt	760.	1.9	1768	3200.	8.86	118
28	Ni	nickel	737	1.9	1728	3186	8.90	117
29	Cu	copper	745	1.9	1358	2835	8.96	122
30	Zn	zinc	906	1.7	693	1180.	7.134	120.
31	Ga	gallium	579	1.8	303	2477	5.91	123
32	Ge	germanium	762	2.0	1211	3106	5.3234	120.
33	As	arsenic (gray)	944	2.2	1090.	—	5.75	120.
34	Se	selenium (gray)	941	2.6	494	958	4.809	118
35	Br	bromine	1140.	3.0	266	332	3.1028	117
36	Kr	krypton	1351	—	116	120.	0.003425	116
37	Rb	rubidium	403	0.8	312	961	1.53	215
38	Sr	strontium	549	1.0	1050.	1655	2.64	190.
39	Y	yttrium	600.	1.2	1795	3618	4.47	176
40	Zr	zirconium	640.	1.3	2128	4682	6.52	164

TABLE S
Properties of Selected Elements (Continued)

Atomic Number	Symbol	Name	First Ionization Energy (kJ/mol)	Electro-negativity	Melting Point (K)	Boiling* Point (K)	Density** (g/cm³)	Atomic Radius (pm)
41	Nb	niobium	652	1.6	2750.	5017	8.57	156
42	Mo	molybdenum	684	2.2	2896	4912	10.2	146
43	Tc	technetium	702	2.1	2430.	4538	11	138
44	Ru	ruthenium	710.	2.2	2606	4423	12.1	136
45	Rh	rhodium	720.	2.3	2237	3968	12.4	134
46	Pd	palladium	804	2.2	1828	3236	12.0	130.
47	Ag	silver	731	1.9	1235	2435	10.5	136
48	Cd	cadmium	868	1.7	594	1040.	8.69	140.
49	In	indium	558	1.8	430.	2345	7.31	142
50	Sn	tin (white)	709	2.0	505	2875	7.287	140.
51	Sb	antimony (gray)	831	2.1	904	1860.	6.68	140.
52	Te	tellurium	869	2.1	723	1261	6.232	137
53	I	iodine	1008	2.7	387	457	4.933	136
54	Xe	xenon	1170.	2.6	161	165	0.005366	136
55	Cs	cesium	376	0.8	302	944	1.873	238
56	Ba	barium	503	0.9	1000.	2170.	3.62	206
57	La	lanthanum	538	1.1	1193	3737	6.15	194
Elements 58–71 have been omitted.								
72	Hf	hafnium	659	1.3	2506	4876	13.3	164
73	Ta	tantalum	728	1.5	3290.	5731	16.4	158
74	W	tungsten	759	1.7	3695	5828	19.3	150.
75	Re	rhenium	756	1.9	3458	5869	20.8	141
76	Os	osmium	814	2.2	3306	5285	22.587	136
77	Ir	iridium	865	2.2	2719	4701	22.562	132
78	Pt	platinum	864	2.2	2041	4098	21.5	130.
79	Au	gold	890.	2.4	1337	3129	19.3	130.
80	Hg	mercury	1007	1.9	234	630.	13.5336	132
81	Tl	thallium	589	1.8	577	1746	11.8	144
82	Pb	lead	716	1.8	600.	2022	11.3	145
83	Bi	bismuth	703	1.9	544	1837	9.79	150.
84	Po	polonium	812	2.0	527	1235	9.20	142
85	At	astatine	—	2.2	575	—	—	148
86	Rn	radon	1037	—	202	211	0.009074	146
87	Fr	francium	393	0.7	300.	—	—	242
88	Ra	radium	509	0.9	969	—	5	211
89	Ac	actinium	499	1.1	1323	3471	10.	201
Elements 90 and above have been omitted.								

* boiling point at standard pressure
** density of solids and liquids at room temperature and density of gases at 298 K and 101.3 kPa
— no data available
Source: CRC Handbook for Chemistry and Physics, 91st ed., 2010–2011, CRC Press

TABLE T
Important Formulas and Equations

Density	$d = \dfrac{m}{V}$	d = density m = mass V = volume
Mole Calculations	number of moles = $\dfrac{\text{given mass}}{\text{gram-formula mass}}$	
Percent Error	% error = $\dfrac{\text{measured value} - \text{accepted value}}{\text{accepted value}} \times 100$	
Percent Composition	% composition by mass = $\dfrac{\text{mass of part}}{\text{mass of whole}} \times 100$	
Concentration	parts per million = $\dfrac{\text{mass of solute}}{\text{mass of solution}} \times 1\,000\,000$ molarity = $\dfrac{\text{moles of solute}}{\text{liter of solution}}$	
Combined Gas Law	$\dfrac{P_1 V_1}{T_1} = \dfrac{P_2 V_2}{T_2}$	P = pressure V = volume T = temperature
Titration	$M_A V_A = M_B V_B$	M_A = molarity of H^+ M_B = molarity of OH^- V_A = volume of acid V_B = volume of base
Heat	$q = mC\Delta T$ $q = mH_f$ $q = mH_v$	q = heat H_f = heat of fusion m = mass H_v = heat of vaporization C = specific heat capacity ΔT = change in temperature
Temperature	K = °C + 273	K = kelvin °C = degree Celsius

ADDITIONAL REFERENCE TABLES FOR CHEMISTRY

TABLE U
Selected Physical Constants

Name	Symbol or Abbreviation	Value
Avogadro constant	N_A	6.022×10^{23}/mol
Boltzmann constant	k	1.381×10^{-23} J/K
Elementary charge	e	1.602×10^{-19} C
Speed of light	c	2.998×10^{8} m/s
Planck constant	h	6.626×10^{-34} J·s
Molar gas constant	R	8.315 Pa·m³/mol·K 0.08206 L·atm/mol·K
Atomic mass unit	u (amu)	1.661×10^{-27} kg
Proton mass	m_p	1.673×10^{-27} kg 1.007 u
Neutron mass	m_n	1.675×10^{-27} kg 1.009 u
Electron mass	m_e	9.109×10^{-31} kg 0.0005486 u
Standard-state pressure	—	100.0 kPa
Standard atmosphere	atm	101.3 kPa 760.0 mm Hg 760.0 torr
Triple point of water	T_{tr}	273.16 K
Standard temperature	—	273.15 K
Standard temperature and pressure	STP	273.15 K and 1 atm
Molar volume of ideal gas at STP	V_m	22.41 L/mol

TABLE V
Elements of the Periodic Table
(in Alphabetical Order)

Name	Atomic Number	Symbol	Location in Periodic Table*	Electron Configuration†
Actinium	89	Ac	P7, G3	[Rn]$6d^17s^2$
Aluminum	13	Al	P3, G13	[Ne]$3s^23p^1$
Americium	95	Am	AS	[Rn]$5f^77s^2$
Antimony	51	Sb	P5, G15	[Kr]$4d^{10}5s^25p^3$
Argon	18	Ar	P3, G18	[Ne]$3s^23p^6$
Arsenic	33	As	P4, G15	[Ar]$3d^{10}4s^24p^3$
Astatine	85	At	P6, G17	[Xe]$4f^{14}5d^{10}6s^26p^5$
Barium	56	Ba	P6, G2	[Xe]$6s^2$
Berkelium	97	Bk	AS	[Rn]$5f^97s^2$
Beryllium	4	Be	P2, G2	[He]$2s^2$
Bismuth	83	Bi	P6, G14	[Xe]$4f^{14}5d^{10}6s^26p^3$
Bohrium	107	Bh	P7, G7	[Rn]$5f^{14}6d^57s^2$
Boron	5	B	P2, G13	[He]$2s^22p^1$
Bromine	35	Br	P4, G17	[Ar]$3d^{10}4s^24p^5$
Cadmium	48	Cd	P5, G12	[Kr]$4d^{10}5s^2$
Calcium	20	Ca	P4, G2	[Ar]$4s^2$
Californium	98	Cf	AS	[Rn]$5f^{10}7s^2$
Carbon	6	C	P2, G14	[He]$2s^22p^2$
Cerium	58	Ce	LS	[Xe]$4f^15d^16s^2$
Cesium	55	Cs	P6, G1	[Xe]$6s^1$
Chlorine	17	Cl	P3, G17	[Ne]$3s^23p^5$
Chromium	24	Cr	P4, G6	[Ar]$3d^54s^1$
Cobalt	27	Co	P4, G9	[Ar]$3d^74s^2$
Copernicium	112	Cn	P7, G12	[Rn]$5f^{14}6d^{10}7s^2$
Copper	29	Cu	P4, G11	[Ar]$3d^{10}4s^1$
Curium	96	Cm	AS	[Rn]$5f^76d^17s^2$
Darmstadtium	110	Ds	P7, G10	[Rn]$5f^{14}6d^97s^1$
Dubnium	105	Db	P7, G5	[Rn]$5f^{14}6d^37s^2$
Dysprosium	66	Dy	LS	[Xe]$4f^{10}6s^2$
Einsteinium	99	Es	AS	[Rn]$5f^{11}7s^2$
Erbium	68	Er	LS	[Xe]$4f^{12}6s^2$
Europium	63	Eu	LS	[Xe]$4f^76s^2$
Fermium	100	Fm	AS	[Rn]$5f^{12}7s^2$

Name	Atomic Number	Symbol	Location in Periodic Table*	Electron Configuration†
Flerovium	114	Fl	P7, G14	$[Rn]5f^{14}6d^{10}7s^27p^2$
Fluorine	9	F	P2, G17	$[He]2s^22p^5$
Francium	87	Fr	P7, G1	$[Rn]7s^1$
Gadolinium	64	Gd	LS	$[Xe]4f^75d^16s^2$
Gallium	31	Ga	P4, G13	$[Ar]3d^{10}4s^24p^1$
Germanium	32	Ge	P4, G14	$[Ar]3d^{10}4s^24p^2$
Gold	79	Au	P6, G11	$[Xe]4f^{14}5d^{10}6s^1$
Hafnium	72	Hf	P6, G4	$[Xe]4f^{14}5d^26s^2$
Hassium	108	Hs	P7, G8	$[Rn]5f^{14}6d^67s^2$
Helium	2	He	P1, G18	$1s^2$
Holmium	67	Ho	LS	$[Xe]4f^{11}6s^2$
Hydrogen	1	H	P1, G1	$1s^1$
Indium	49	In	P5, G13	$[Kr]4d^{10}5s^25p^1$
Iodine	53	I	P5, G17	$[Kr]4d^{10}5s^25p^5$
Iridium	77	Ir	P6, G9	$[Xe]4f^{14}5d^76s^2$
Iron	26	Fe	P4, G8	$[Ar]3d^64s^2$
Krypton	36	Kr	P4, G18	$[Ar]3d^{10}4s^24p^6$
Lanthanum	57	La	P6, G3	$[Xe]5d^16s^2$
Lawrencium	103	Lr	AS	$[Rn]5f^{14}6d^17s^2$
Lead	82	Pb	P6, G10	$[Xe]4f^{14}5d^{10}6s^26p^2$
Lithium	3	Li	P2, G1	$[He]2s^1$
Livermorium	116	Lv	P7, G16	$[Rn]5f^{14}6d^{10}7s^27p^4$
Lutetium	71	Lu	LS	$[Xe]4f^{14}5d^16s^2$
Magnesium	12	Mg	P3, G2	$[Ne]3s^2$
Manganese	25	Mn	P4, G7	$[Ar]3d^54s^2$
Meitnerium	109	Mt	P7, G9	$[Rn]5f^{14}6d^77s^2$
Mendelevium	101	Md	AS	$[Rn]5f^{13}7s^2$
Mercury	80	Hg	P6, G12	$[Xe]4f^{14}5d^{10}6s^2$
Molybdenum	42	Mo	P5, G6	$[Kr]4d^55s^1$
Moscovium	115	Mc	P7, G15	$[Rn]5f^{14}6d^{10}7s^27p^3$
Neodymium	60	Nd	LS	$[Xe]4f^46s^2$
Neon	10	Ne	P2, G18	$[He]2s^22p^6$
Neptunium	93	Np	AS	$[Rn]5f^46d^17s^2$
Nickel	28	Ni	P4, G10	$[Ar]3d^84s^2$
Nihonium	113	Nh	P7, G13	$[Rn]5f^{14}6d^{10}7s^27p^1$
Niobium	41	Nb	P5, G5	$[Kr]4d^45s^1$

Name	Atomic Number	Symbol	Location in Periodic Table*	Electron Configuration†
Nitrogen	7	N	P2, G15	$[He]2s^22p^3$
Nobelium	102	No	AS	$[Rn]5f^{14}7s^2$
Oganesson	118	Og	P7, G18	$[Rn]5f^{14}6d^{10}7s^27p^6$
Osmium	76	Os	P6, G8	$[Xe]4f^{14}5d^66s^2$
Oxygen	8	O	P2, G16	$[He]2s^22p^4$
Palladium	46	Pd	P5, G10	$[Kr]4d^{10}$
Phosphorus	15	P	P3, G15	$[Ne]3s^23p^3$
Platinum	78	Pt	P6, G10	$[Xe]4f^{14}5d^96s^1$
Plutonium	94	Pu	AS	$[Rn]5f^67s^2$
Polonium	84	Po	P6, G16	$[Xe]4f^{14}5d^{10}6s^26p^4$
Potassium	19	K	P4, G1	$[Ar]4s^1$
Praseodymium	59	Pr	LS	$[Xe]4f^36s^2$
Promethium	61	Pm	LS	$[Xe]4f^56s^2$
Protactinium	91	Pa	AS	$[Rn]5f^26d^17s^2$
Radium	88	Ra	P7, G2	$[Rn]7s^2$
Radon	86	Rn	P6, G18	$[Xe]4f^{14}5d^{10}6s^26p^6$
Rhenium	75	Re	P6, G7	$[Xe]4f^{14}5d^56s^2$
Rhodium	45	Rh	P5, G9	$[Kr]4d^85s^1$
Roentgenium	111	Rg	P7, G11	$[Rn]5f^{14}6d^{10}7s^1$
Rubidium	37	Rb	P5, G1	$[Kr]5s^1$
Ruthenium	44	Ru	P5, G8	$[Kr]4d^75s^1$
Rutherfordium	104	Rf	P7, G4	$[Rn]5f^{14}6d^27s^2$
Samarium	62	Sm	LS	$[Xe]4f^66s^2$
Scandium	21	Sc	P4, G3	$[Ar]3d^14s^2$
Seaborgium	106	Sg	P7, G6	$[Rn]5f^{14}6d^47s^2$
Selenium	34	Se	P4, G16	$[Ar]3d^{10}4s^24p^4$
Silicon	14	Si	P3, G14	$[Ne]3s^23p^2$
Silver	47	Ag	P5, G11	$[Kr]4d^{10}5s^1$
Sodium	11	Na	P3, G1	$[Ne]3s^1$
Strontium	38	Sr	P5, G2	$[Kr]5s^2$
Sulfur	16	S	P3, G16	$[Ne]3s^23p^4$
Tantalum	73	Ta	P6, G5	$[Xe]4f^{14}5d^36s^2$
Technitium	43	Tc	P5, G7	$[Kr]4d^55s^2$
Tellurium	52	Te	P5, G16	$[Kr]4d^{10}5s^25p^4$
Tennessine	117	Ts	P7, G17	$[Rn]5f^{14}6d^{10}7s^27p^5$
Terbium	65	Tb	LS	$[Xe]4f^96s^2$

Name	Atomic Number	Symbol	Location in Periodic Table*	Electron Configuration†
Thallium	81	Tl	P6, G13	[Xe]$4f^{14}5d^{10}6s^26p^1$
Thorium	90	Th	AS	[Rn]$6d^27s^2$
Thulium	69	Tm	LS	[Xe]$4f^{13}6s^2$
Tin	50	Sn	P5, G14	[Kr]$4d^{10}5s^25p^2$
Titanium	22	Ti	P4, G4	[Ar]$3d^24s^2$
Tungsten	74	W	P6, G6	[Xe]$4f^{14}5d^46s^2$
Uranium	92	U	AS	[Rn]$5f^36d^17s^2$
Vanadium	23	V	P4, G5	[Ar]$3d^34s^2$
Xenon	54	Xe	P5, G18	[Kr]$4d^{10}5s^25p^6$
Ytterbium	70	Yb	LS	[Xe]$4f^{14}6s^2$
Yttrium	39	Y	P5, G3	[Kr]$4d^15s^2$
Zinc	30	Zn	P4, G12	[Ar]$3d^{10}4s^2$
Zirconium	40	Zr	P5, G4	[Kr]$4d^25s^2$

*Refers to the Periodic Table in Appendix 1: P = period; G = group; LS = lanthanide series (elements 58–71); AS = actinide series (elements 90–103).

†Ground-state configuration

TABLE W-1
Specific Heats of Selected Substances and Mixtures*

Substance	C/(J/g · K)	Substance	C/(J/g · K)	Substance	C/(J/g · K)
Ag (s)	0.235	Cu (s)	0.385	Mg (s)	1.02
Al (s)	0.897	Fe (s)	0.449	Mn (s)	0.479
AlF$_3$ (s)	0.895	Glass (ℓ)	0.753	Na (s)	1.23
As (s)	0.329	He (g)	5.19	NaCl (s)	0.858
Au (s)	0.129	Hg (ℓ)	0.140	Pb (s)	0.129
BeO (s)	1.02	H$_2$O (s)	2.06	Si (s)	0.714
Ca (s)	0.647	H$_2$O (ℓ)	4.19	SiO$_2$ (s)	0.740
CaCO$_3$ (s)	0.920	H$_2$O (g)	2.02	Sn (s)	0.228
CaSO$_4$ (s)	0.732	In (s)	0.233	Steel (s)	0.449
C$_6$H$_6$ (ℓ)	1.74	In (ℓ)	0.216	Steel (ℓ)	0.719
C$_6$H$_5$CH$_3$ (ℓ)	1.80	K (s)	0.757	Ti (s)	0.523
CH$_3$CH$_2$OH (ℓ)	2.42	Kerosene (ℓ)	2.09	W (s)	0.132
CH$_3$COOH	2.05	Li (s)	3.58	Zn (s)	0.388

* (s) = solid phase, (ℓ) = liquid phase, (g) = gas phase. Note: 1 K = 1 C°.

TABLE W-2
Heats (Enthalpies) of Fusion and Vaporization
(Including Melting Points and Boiling Points)

Substance	M.P./°C	DH$_f$/(J/g)	B.P./°C	DH$_v$/(J/g)
Aluminum [Al]	660	397	2450	11400
Ammonia [NH$_3$]	−77.8	33.0	−33.4	137
Copper [Cu]	1083	134	1187	5060
Ethanol [C$_2$H$_5$OH]	−114	104	78	854
Gold [Au]	1063.00	64.4	2660	1580
Helium [He]	−269.65	5.23	−268.93	20.9
Iron [Fe]	1808	289	3023	6340
Lead [Pb]	327.3	24.5	1750	870
Nitrogen [N$_2$]	−209.97	25.5	−195.81	201
Oxygen [O$_2$]	−218.79	13.8	−182.97	213
Silver [Ag]	960.80	88.2	2193	2330
Sulfur [S]	119	38.1	444.60	326
Tungsten [W]	3410	184	5900	4800
Water [H$_2$O]	0.00	333	100.00	2260

TABLE W-3
Molal Freezing and Boiling Point Data*

Substance	F.P./°C	k$_f$/(C°·kg/mol)	B.P./°C	k$_b$/(C°·kg/mol)
Acetic acid [HC$_2$H$_3$O$_2$]	16.7	3.63	118	3.22
Benzene [C$_6$H$_6$]	5.53	5.07	80.1	2.64
Camphor [C$_{10}$H$_{16}$O]	178	37.8	207	—
Cyclohexane [C$_6$H$_{12}$]	6.54	20.8	80.7	2.92
Nitrobenzene [C$_6$H$_5$NO$_2$]	5.76	6.87	211	5.20
Phenol [C$_6$H$_5$OH]	40.9	6.84	182	3.54
Water [H$_2$O]	0.00	1.86	100.	0.513

* A dash (—) indicates that data are not available.

TABLE W-4
Standard Heats of Formation for Substances
(at 298.15 K and 100 kPa)

Substance	$DH°_f$/(kJ/mol)	Substance	$DH°_f$/(kJ/mol)
Aluminum		**Copper**	
Al(s)	0	Cu(s)	0
$AlCl_3$(s)	−704.2	CuCl(s)	−137.2
Al_2O_3(s)	−1675.7	$CuCl_2$(s)	−220.1
Barium		CuO(s)	−157.3
Ba(s)	0	$CuSO_4$(s)	−771.4
$BaCO_3$(s)	−1213.0	**Fluorine**	
$BaCl_2$(s)	−855.0	F_2(g)	0
$Ba(NO_3)_2$(s)	−988.0	F(g)	79.4
$BaSO_4$(s)	−1473.2	**Hydrogen**	
Bromine		H_2(g)	0
$Br_2(\ell)$	0	HBr(g)	−36.3
Br_2(g)	30.9	HCl(g)	−92.3
Br(g)	111.9	HCN(g)	135.1
Calcium		HF(g)	−273.3
Ca(s)	0	HI(g)	26.5
$CaCO_3$(s) [calcite]	−1207.6	$HNO_3(\ell)$	−174.1
$CaCl_2$(s)	−795.4	$H_2O(\ell)$	−285.8
CaF_2(s)	−1228.0	H_2O(g)	−241.8
$Ca(NO_3)_2$(s)	−938.2	$H_2O_2(\ell)$	−187.8
CaO(s)	−634.9	H_2O_2(g)	−136.3
$Ca(OH)_2$(s)	−985.2	H_2S(g)	−20.6
$Ca_3(PO_4)_2$(s)	−4120.8	$H_2SO_4(\ell)$	−814.0
$CaSO_4$(s)	−1434.5	**Iodine**	
Carbon		I_2(s)	0
C(s) [graphite]	0	I_2(g)	62.4
C(s) [diamond]	1.9	I(g)	106.8
CH_4(g)	−74.6	$ICl(\ell)$	−23.9
C_2H_2(g)	227.4	ICl(g)	17.8
C_2H_4(g)	52.4	**Iron**	
C_2H_6(g)	−84.0	Fe(s)	0
C_3H_8(g)	−103.8	$FeCl_2$(s)	−341.8
$C_6H_6(\ell)$	49.1	$FeCl_3$(s)	−399.5
$CH_3OH(\ell)$	−239.2	Fe_2O_3(s)	−824.2
$C_2H_5OH(\ell)$	−277.6	Fe_3O_4(s)	−1118.4
CO(g)	−110.5	**Lead**	
CO_2(g)	−393.5	Pb(s)	0
Chlorine		$PbCl_2$(s)	−359.4
Cl_2(g)	0	PbS(s)	−100.4
Cl(g)	121.3	$PbSO_4$(s)	−920.0
ClO_2(g)	102.5	**Lithium**	
Cl_2O(g)	80.3	Li(s)	0
Chromium		LiCl(s)	−408.6
Cr(s)	0	LiF(s)	−616.0
$CrCl_2$(s)	−395.4	$LiNO_3$(s)	−483.1
$CrCl_3$(s)	−556.5	LiOH(s)	−484.9

TABLE W-4: *(Continued)*

Substance	D$H°_f$/(kJ/mol)	Substance	D$H°_f$/(kJ/mol)
Magnesium		**Silver**	
Mg(s)	0	Ag(s)	0
MgCl$_2$(s)	−641.3	AgBr(s)	−100.4
MgF$_2$(s)	−1124.2	AgCl(s)	−127.0
MgO(s)	−601.6	AgI(s)	−61.8
Mg(OH)$_2$(s)	−924.5	AgNO$_3$(s)	−124
Manganese		**Sodium**	
Mn(s)	0	Na(s)	0
MnO$_2$(s)	−520.0	Na(g)	107.5
Nitrogen		NaBr(s)	−361.1
N$_2$(g)	0	NaCl(s)	−411.2
N(g)	472.7	Na$_2$CO$_3$(s)	−1130.7
NH$_3$(g)	−45.9	NaF(s)	−576.6
N$_2$H$_4$(ℓ)	50.6	NaH(s)	−56.3
NH$_4$Cl(s)	−314.4	NaOH(s)	−425.6
NH$_4$NO$_3$(s)	−365.6	Na$_2$SO$_4$(s)	−1387.1
NO(g)	91.3	**Sulfur**	
NO$_2$(g)	33.2	S(s) [rhombic]	0
N$_2$O(g)	81.6	SF$_6$(g)	−1220.5
N$_2$O$_4$(g)	11.1	SO$_2$(g)	−296.8
Oxygen		SO$_3$(g)	−395.7
O$_2$(g)	0	**Tin**	
O$_3$(g)	142.7	Sn(s) [white]	0
Potassium		Sn(s) [gray]	−2.1
K(s)	0	SnCl$_4$(ℓ)	−511.3
KBr(s)	−393.8	**Zinc**	
KCl(s)	−436.5	Zn(s)	0
KClO$_3$(s)	−397.7	ZnCl$_2$(s)	−415.1
KClO$_4$(s)	−432.8	ZnO(s)	−350.5
KF(s)	−567.3		
KMnO$_4$(s)	−837.2		
KOH(s)	−424.6		
K$_2$SO$_4$(s)	−1437.8		

TABLE X
Vapor Pressures of Water at Various Temperatures

t_c/°C	V.P. /mmHg	V.P. /kPa	t_c/°C	V.P. /mmHg	V.P. /kPa
0	4.585	0.6113	21	18.66	2.488
1	4.929	0.6572	22	19.84	2.645
2	5.296	0.7061	23	21.08	2.810
3	5.686	0.7581	24	22.39	2.985
4	6.102	0.8136	25	23.77	3.169
5	6.545	0.8726	26	25.22	3.363
6	7.016	0.9354	27	26.75	3.567
7	7.516	1.002	28	28.37	3.782
8	8.048	1.073	29	30.06	4.008
9	8.612	1.148	30	31.84	4.246
10	9.212	1.228	40	55.37	7.381
11	9.848	1.313	50	92.59	12.34
12	10.52	1.403	60	149.5	19.93
13	11.24	1.498	70	233.8	31.18
14	11.99	1.599	80	355.3	47.37
15	12.79	1.706	90	525.9	70.12
16	13.64	1.819	100	760.0	101.3
17	14.54	1.938	105	906.0	120.8
18	15.48	2.064	110	1074	143.2
19	16.48	2.198	115	1268	169.0
20	17.54	2.339	120	1489	198.5

TABLE Y-1
Solubility-Product Constants (at 298.15 K and 100 kPa)

Formula	K_{sp}	Formula	K_{sp}
Bromides		**Iodides**	
AgBr	5.35×10^{-13}	AgI	8.52×10^{-17}
CuBr	6.27×10^{-9}	CuI	1.27×10^{-12}
$PbBr_2$	6.60×10^{-6}	PbI_2	9.80×10^{-28}
Carbonates		**Phosphates**	
Ag_2CO_3	8.46×10^{-12}	Ag_3PO_4	8.89×10^{-17}
$BaCO_3$	2.58×10^{-9}	$Ca_3(PO_4)_2$	2.07×10^{-33}
$CaCO_3$	3.36×10^{-9}	$Cd_3(PO_4)_2$	2.53×10^{-33}
$MgCO_3$	6.82×10^{-6}	$Cu_3(PO_4)_2$	1.40×10^{-37}
$SrCO_3$	5.60×10^{-10}	$Mg_3(PO_4)_2$	1.04×10^{-24}
$ZnCO_3$	8×10^{-28}	$Ni_3(PO_4)_2$	4.74×10^{-32}
Chlorides		**Sulfates**	
AgCl	1.77×10^{-10}	Ag_2SO_4	1.20×10^{-5}
CuCl	1.72×10^{-7}	$BaSO_4$	1.08×10^{-10}
$PbCl_2$	1.70×10^{-5}	$CaSO_4$	4.93×10^{-5}
Chromate		$PbSO_4$	2.53×10^{-8}
Ag_2CrO_4	1.12×10^{-12}	**Sulfides**	
Fluorides		CdS	1.40×10^{-29}
BaF_2	1.84×10^{-6}	CuS	1.27×10^{-36}
CaF_2	3.45×10^{-11}	HgS	1.55×10^{-52}
MgF_2	5.16×10^{-11}	MnS	4.65×10^{-14}
PbF_2	3.30×10^{-8}	PbS	9.05×10^{-29}
Hydroxides		ZnS	2.93×10^{-25}
$Ca(OH)_2$	5.02×10^{-6}	**Sulfites**	
$Cd(OH)_2$	7.20×10^{-15}	Ag_2SO_3	1.50×10^{-14}
$Fe(OH)_2$	4.87×10^{-17}	$BaSO_3$	5.00×10^{-10}
$Fe(OH)_3$	2.79×10^{-39}		
$Mg(OH)_2$	5.61×10^{-12}		
$Ni(OH)_2$	5.48×10^{-16}		
$Zn(OH)_2$	3.00×10^{-17}		

TABLE Y-2
Ionization Constants of Weak Acids and Bases
(at 298.15 K and 100 kPa)

$$K_w = [H_3O^+] \cdot [OH^-] = 1.0 \times 10^{-14}$$

Monoprotic Acids (acids with one ionizable hydrogen atom)

Formula	Name	K_a
HIO_3	iodic acid	1.6×10^{-1}
HNO_2	nitrous acid	7.2×10^{-4}
HF	hydrofluoric acid	6.6×10^{-4}
$HCHO_2$	formic acid	1.8×10^{-4}
$HC_3H_5O_3$	lactic acid	1.4×10^{-4}
$HC_7H_5O_2$	benzoic acid	6.3×10^{-5}
$HC_4H_7O_2$	butanoic acid	1.5×10^{-5}
HN_3	hydrazoic acid	1.9×10^{-5}
$HC_2H_3O_2$	acetic acid	1.8×10^{-5}
$HC_3H_5O_2$	propanoic acid	1.3×10^{-5}
$HOCl$	hypochlorous acid	2.9×10^{-8}
HCN	hydrocyanic acid	6.2×10^{-10}
HC_6H_5O	phenol	1.0×10^{-10}
H_2O_2	hydrogen peroxide	2.2×10^{-12}

Diprotic Acids (acids with two ionizable hydrogen atoms)

Formula	Name	K_{a1}	K_{a2}	K_{a3}
H_2SO_4	sulfuric acid	Large	1.1×10^{-2}	
H_2CrO_4	chromic acid	5.0	1.5×10^{-6}	
$H_2C_2O_4$	oxalic acid	5.4×10^{-2}	5.3×10^{-5}	
H_3PO_3	phosphorous acid	3.7×10^{-2}	2.1×10^{-7}	
H_2SO_3	sulfurous acid	1.3×10^{-2}	6.2×10^{-8}	
H_2SeO_3	selenous acid	2.3×10^{-3}	5.4×10^{-9}	
$H_2C_3H_2O_4$	malonic acid	1.5×10^{-3}	2.0×10^{-6}	
$H_2C_8H_4O_4$	phthalic acid	1.1×10^{-3}	3.9×10^{-6}	
$H_2C_4H_4O_6$	tartaric acid	9.2×10^{-4}	4.3×10^{-5}	
H_2CO_3	carbonic acid	4.4×10^{-7}	4.7×10^{-11}	

Triprotic Acids (acids with three ionizable hydrogen atoms)

Formula	Name	K_{a1}	K_{a2}	K_{a3}
H_3PO_4	phosphoric acid	7.1×10^{-3}	6.3×10^{-8}	4.2×10^{-13}
H_3AsO_4	arsenic acid	6.0×10^{-3}	1.0×10^{-7}	3.2×10^{-12}
$H_3C_6H_5O_7$	citric acid	7.4×10^{-4}	1.7×10^{-5}	4.0×10^{-7}

Bases

Formula	Name	K_b
$(CH_3)_2NH$	dimethylamine	6.9×10^{-4}
CH_3NH_2	methylamine	4.2×10^{-4}
$CH_3CH_2NH_2$	ethylamine	4.3×10^{-4}
$(CH_3)_3N$	trimethylamine	6.3×10^{-5}
NH_3	ammonia	1.8×10^{-5}
N_2H_4	hydrazine	8.5×10^{-7}
C_5H_5N	pyridine	1.5×10^{-9}
$C_6H_5NH_2$	aniline	7.4×10^{-10}

TABLE Z
Standard Electrode Potentials
(at 298.15 K and 100 kPa)

Half-reaction	E°/V	Half-reaction	E°/V
$F_2 + 2e^- \rightarrow 2F^-$	2.87	$O_2 + 2H_2O + 4e^- \rightarrow 4OH^-$	0.40
$Ag^{2+} + e^- \rightarrow Ag^+$	1.99	$Cu^{2+} + 2e^- \rightarrow Cu$	0.34
$Co^{3+} + e^- \rightarrow Co^{2+}$	1.82	$Hg_2Cl_2 + 2e^- \rightarrow 2Hg + 2Cl^-$	0.34
$H_2O_2 + 2H^+ + 2e^- \rightarrow 2H_2O$	1.78	$AgCl + e^- \rightarrow Ag + Cl^-$	0.22
$Ce^{4+} + e^- \rightarrow Ce^{3+}$	1.70	$SO_4^{2-} + 4H^+ + 2e^-$	
$PbO_2 + 4H^+ + SO_4^{2-} + 2e^-$		$\rightarrow H_2SO_3 + H_2O$	0.20
$\rightarrow PbSO_4 + 2H_2O$	1.69	$Cu^{2+} + e^- \rightarrow Cu^+$	0.16
$MnO_4^- + 4H^+ + 3e^- \rightarrow MnO_2 + 2H_2O$	1.68	$\mathbf{2H^+ + 2e^- \rightarrow H_2}$	**0.00**
$2e^- + 2H^+ + IO_4^- \rightarrow IO_3^- + H_2O$	1.60	$Fe^{3+} + 3e^- \rightarrow Fe$	−0.036
$MnO_4^- + 8H^+ + 5e^- \rightarrow Mn^{2+} + 4H_2O$	1.51	$Pb^{2+} + 2e^- \rightarrow Pb$	−0.13
$Au^{3+} + 3e^- \rightarrow Au$	1.50	$Sn^{2+} + 2e^- \rightarrow Sn$	−0.14
$PbO_2 + 4H^+ + 2e^- \rightarrow Pb^{2+} + 2H_2O$	1.46	$Ni^{2+} + 2e^- \rightarrow Ni$	−0.23
$Cl_2 + 2e^- \rightarrow 2Cl^-$	1.36	$PbSO_4 + 2e^- \rightarrow Pb + SO_4^{2-}$	−0.35
$Cr_2O_7^{2-} + 14H^+ + 6e^- \rightarrow 2Cr^{3+} + 7H_2O$	1.33	$Cd^{2+} + 2e^- \rightarrow Cd$	−0.40
$O_2 + 4H^+ + 4e^- \rightarrow 2H_2O$	1.23	$Fe^{2+} + 2e^- \rightarrow Fe$	−0.44
$MnO_2 + 4H^+ + 2e^- \rightarrow Mn^{2+} + 2H_2O$	1.21	$Cr^{3+} + e^- \rightarrow Cr^{2+}$	−0.50
$IO_3^- + 6H^+ + 5e^- \rightarrow \frac{1}{2}I_2 + 3H_2O$	1.20	$Cr^{3+} + 3e^- \rightarrow Cr$	−0.73
$Br_2 + 2e^- \rightarrow 2Br^-$	1.09	$Zn^{2+} + 2e^- \rightarrow Zn$	−0.76
$VO_2^+ + 2H^+ + e^- \rightarrow VO^{2+} + H_2O$	1.00	$2H_2O + 2e^- \rightarrow H_2 + 2OH^-$	−0.83
$AuCl_4^- + 3e^- \rightarrow Au + 4Cl^-$	0.99	$Mn^{2+} + 2e^- \rightarrow Mn$	−1.18
$NO_3^- + 4H^+ + 3e^- \rightarrow NO + 2H_2O$	0.96	$Al^{3+} + 3e^- \rightarrow Al$	−1.66
$ClO_2 + e^- \rightarrow ClO_2^-$	0.95	$H_2 + 2e^- \rightarrow 2H^-$	−2.23
$2Hg^{2+} + 2e^- \rightarrow Hg_2^{2+}$	0.91	$Mg^{2+} + 2e^- \rightarrow Mg$	−2.37
$Ag^+ + e^- \rightarrow Ag$	0.80	$La^{3+} + 3e^- \rightarrow La$	−2.37
$Hg_2^{2+} + 2e^- \rightarrow 2Hg$	0.80	$Na^+ + e^- \rightarrow Na$	−2.71
$Fe^{3+} + e^- \rightarrow Fe^{2+}$	0.77	$Ca^{2+} + 2e^- \rightarrow Ca$	−2.76
$O_2 + 2H^+ + 2e^- \rightarrow H_2O_2$	0.68	$Ba^{2+} + 2e^- \rightarrow Ba$	−2.90
$MnO_4^- + e^- \rightarrow MnO_4^{2-}$	0.56	$K^+ + e^- \rightarrow K$	−2.92
$I_2 + 2e^- \rightarrow 2I^-$	0.54	$Li^+ + e^- \rightarrow Li$	−3.05
$Cu^+ + e^- \rightarrow Cu$	0.52		

ANSWERS TO END-OF-CHAPTER QUESTIONS

Chapter 1

1. (4)	**7.** (4)	**13.** (2)	**19.** (4)	**25.** (2)
2. (4)	**8.** (3)	**14.** (3)	**20.** (2)	**26.** (4)
3. (1)	**9.** (1)	**15.** (2)	**21.** (4)	**27.** (3)
4. (3)	**10.** (4)	**16.** (1)	**22.** (2)	**28.** (3)
5. (4)	**11.** (4)	**17.** (3)	**23.** (2)	
6. (1)	**12.** (2)	**18.** (1)	**24.** (3)	

Constructed-Response Questions

1. (a) 6.3477×10^4 (b) 2.30×10^{-4}

2. (a) 654,000   (b) 0.00555

3. $7.00 \ \cancel{\text{daddles}} \cdot \left(\dfrac{1 \text{ piddle}}{3.33 \ \cancel{\text{daddles}}} \right) = 2.10 \text{ piddles}$

4. 21 g

5. $f = \dfrac{cd}{abe}$

6. (a)

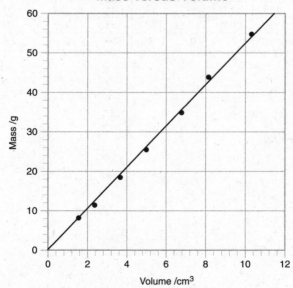

Mass versus Volume

(b) density (from slope) = 5.26 g/cm^3

7. (a) 4   (b) 5   (c) 2   (d) 3   (e) 4

8. (a) 343.2 m (b) 0.0165 kJ   (c) 2.2 cm^2 (d) 5.4 g/L

9. Sample 3

10. 2 molecules of compound y_2z

11. A compound must consist of two or more elements.

Chapter 2

1. (4)	8. (1)	15. (3)	22. (4)	29. (3)
2. (1)	9. (4)	16. (3)	23. (4)	30. (4)
3. (1)	10. (1)	17. (3)	24. (3)	31. (4)
4. (3)	11. (1)	18. (3)	25. (4)	32. (2)
5. (2)	12. (1)	19. (3)	26. (2)	33. (3)
6. (2)	13. (3)	20. (2)	27. (3)	34. (1)
7. (2)	14. (3)	21. (1)	28. (1)	

Constructed-Response Questions

1.

Symbol	^{39}K	^{51}V	^{197}Au	^{222}Rn	^{31}P^{3-}	^{206}Pb^{2+}	^{79}Se^{2-}	^{59}Ni^{2+}
Protons	19	23	79	86	15	82	34	28
Neutrons	20	28	118	136	16	124	45	31
Electrons	19	23	79	86	18	80	36	26
Atomic no.	19	23	79	86	15	82	34	28
Mass no.	39	51	197	222	31	206	79	59
Net charge	0	0	0	0	3−	2+	2−	2+

2.

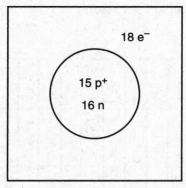

18 e⁻

15 p⁺

16 n

Model of $^{31}P^{3-}$ Ion

Chapter 3

1. (1)	**10.** (2)	**19.** (2)	**28.** (1)	**37.** (2)
2. (1)	**11.** (3)	**20.** (1)	**29.** (4)	**38.** (3)
3. (3)	**12.** (1)	**21.** (2)	**30.** (3)	**39.** (4)
4. (3)	**13.** (2)	**22.** (3)	**31.** (2)	**40.** (3)
5. (2)	**14.** (2)	**23.** (2)	**32.** (2)	**41.** (4)
6. (4)	**15.** (1)	**24.** (2)	**33.** (1)	
7. (4)	**16.** (2)	**25.** (1)	**34.** (1)	
8. (3)	**17.** (3)	**26.** (2)	**35.** (3)	
9. (2)	**18.** (1)	**27.** (2)	**36.** (4)	

Constructed-Response Questions

1.

	bromide Br^-	carbonate CO_3^{2-}	chlorate ClO_3^-	chloride Cl^-	chromate CrO_4^{2-}	nitrate NO_3^-	phosphate PO_4^{3-}	sulfate SO_4^{2-}	sulfide SO_4^{2-}
aluminum Al^{3+}	$AlBr_3$	$Al(CO_3)_3$	$Al(ClO_3)_3$	$AlCl_3$	$Al_2(CrO_4)_3$	$Al(NO_3)_3$	$AlPO_4$	$Al_2(SO_4)_3$	Al_2S_3
ammonium NH_4^+	NH_4Br	$(NH_4)_2CO_3$	NH_4ClO_3	NH_4Cl	$(NH_4)_2CrO_4$	NH_4NO_3	$(NH_4)_3PO_4$	$(NH_4)_2SO_4$	$(NH_4)_2S$
barium Ba^{2+}	$BaBr_2$	$BaCO_3$	$Ba(ClO_3)_2$	$BaCl_2$	$BaCrO_4$	$Ba(NO_3)_2$	$Ba_3(PO_4)_2$	$BaSO_4$	BaS
calcium Ca^{2+}	$CaBr_2$	$CaCO_3$	$Ca(ClO_3)_2$	$CaCl_2$	$CaCrO_4$	$Ca(NO_3)_2$	$Ca_3(PO_4)_2$	$CaSO_4$	CaS
copper (II) Cu^{2+}	$CuBr_2$	$CuCO_3$	$Cu(ClO_3)_2$	$CuCl_2$	$CuCrO_4$	$Cu(NO_3)_2$	$Cu_3(PO_4)_2$	$CuSO_4$	CuS
iron (II) Fe^{2+}	$FeBr_2$	$FeCO_3$	$Fe(ClO_3)_2$	$FeCl_2$	$FeCrO_4$	$Fe(NO_3)_2$	$Fe_3(PO_4)_2$	$FeSO_4$	FeS
iron (III) Fe^{3+}	$FeBr_3$	$Fe_2(CO_3)_3$	$Fe(ClO_3)_3$	$FeCl_3$	$Fe_2(CrO_4)_3$	$Fe(NO_3)_3$	$FePO_4$	$Fe_2(SO_4)_3$	Fe_2S_3
lead (II) Pb^{2+}	$PbBr_2$	$PbCO_3$	$Pb(ClO_3)_2$	$PbCl_2$	$PbCrO_4$	$Pb(NO_3)_2$	$Pb_3(PO_4)_2$	$PbSO_4$	PbS
lead (IV) Pb^{4+}	$PbBr_4$	$Pb(CO_3)_2$	$Pb(ClO_3)_4$	$PbCl_4$	$Pb(CrO_4)_2$	$Pb(NO_3)_4$	$Pb_3(PO_4)_4$	$Pb(SO_4)_2$	PbS_2
magnesium Mg^{2+}	$MgBr_2$	$MgCO_3$	$Mg(ClO_3)_2$	$MgCl_2$	$MgCrO_4$	$Mg(NO_3)_2$	$Mg_3(PO_4)_2$	$MgSO_4$	MgS
mercury (I) Hg_2^{2+}	Hg_2Br_2	Hg_2CO_3	$Hg_2(ClO_3)_2$	Hg_2Cl_2	Hg_2CrO_4	$Hg_2(NO_3)_2$	$(Hg_2)_3(PO_4)_2$	Hg_2SO_4	Hg_2S
mercury (II) Hg^{2+}	$HgBr_2$	$HgCO_3$	$Hg(ClO_3)_2$	$HgCl_2$	$HgCrO_4$	$(NO_3)_2$	$Hg_3(PO_4)_2$	$HgSO_4$	HgS
potassium K^+	KBr	K_2CO_3	$KClO_3$	KCl	K_2CrO_4	KNO_3	K_3PO_4	K_2SO_4	K_2S
silver Ag^+	$AgBr$	Ag_2CO_3	$AgClO_3$	$AgCl$	Ag_2CrO_4	$AgNO_3$	Ag_3PO_4	Ag_2SO_4	Ag_2S
sodium Na^+	$NaBr$	Na_2CO_3	$NaClO_3$	$NaCl$	Na_2CrO_4	$NaNO_3$	Na_3PO_4	Na_2SO_4	Na_2S
zinc Zn^{2+}	$ZnBr_2$	$ZnCO_3$	$Zn(ClO_3)_2$	$ZnCl_2$	$ZnCrO_4$	$Zn(NO_3)_2$	$Zn_3(PO_4)_2$	$ZnSO_4$	ZnS

2. $S = 4+, O = 2-$

3. $K = 1+, Mn = 7+, O = 2-$

4. (a) $3Zn(OH)_2 + 2H_3PO_4 \rightarrow 6H_2O + Zn_3(PO_4)_2$
   (b) $6CO_2 + 6H_2O \rightarrow C_6H_{12}O_6 + 6O_2$

5. (a) sodium iodide   (b) carbon(II) sulfide
   (c) iron(III) chloride   (d) phosphorus(V) oxide
   (e) chlorine(VII) oxide

6. $Mg + Br_2 \rightarrow MgBr_2$

7. $2FeO \rightarrow 2Fe + O_2$

8. $2Al + 3Zn(NO_3)_2 \rightarrow 3Zn + 2Al(NO_3)_3$

9. $2Na_3PO_4 + 3Pb(NO_3)_2 \rightarrow 6NaNO_3 + Pb_3(PO_4)_2$

   According to Reference Table F, $Pb_3(PO_4)_2$ is the precipitate.

10. decomposition

11. The total number of atoms (4 H and 2 O) on both sides of the equation are equal.

12. 4 molecules of O_2 are produced.

Chapter 4

1. (3)	11. (1)	21. (3)	31. (3)	41. (1)
2. (2)	12. (4)	22. (3)	32. (2)	42. (3)
3. (1)	13. (2)	23. (1)	33. (2)	43. (3)
4. (2)	14. (3)	24. (1)	34. (3)	44. (1)
5. (3)	15. (1)	25. (3)	35. (2)	45. (3)
6. (1)	16. (2)	26. (4)	36. (2)	46. (4)
7. (2)	17. (2)	27. (2)	37. (2)	47. (1)
8. (4)	18. (3)	28. (1)	38. (3)	48. (1)
9. (4)	19. (2)	29. (4)	39. (2)	49. (2)
10. (4)	20. (4)	30. (1)	40. (3)	

Constructed-Response Questions

Substance	Molar Mass ($\mathcal{M}$) (g/mol)	Mass of Substance (m) (g)	Moles of Substance (n) (mol)	Number of Particles of Substance (N)
O_3	48	24	0.50	3.0×10^{23}
NH_3	17	170	10.	6.0×10^{24}
F_2	38	38	1.0	6.0×10^{23}
CO_2	44	4.4	0.10	6.0×10^{22}
NO_2	46	9.2	0.20	1.2×10^{23}
Ne	20.	5.0	0.25	1.5×10^{23}
N_2O	44	88	2.0	1.2×10^{24}
NH_3	17	8.5	0.50	3.0×10^{23}

2. (a) 82 g/mol   (b) 314 g/mol

3. (a) 1.0 mol   (b) 6.0×10^{23} molecules

4. 154 g

5. $Na_2B_4O_7$

6. S_2Cl_2

7. (a) 3.93 g   (b) 47.2 g

8. 87.4 g

9. 26.6%

10. (a) 58.3 g/mol   (b) 0.144 mol

11. $(10.01)(0.1991) + (11.01)(0.8009) = 10.81$

12. $CaSO_4 = (40\text{ u}) + (32\text{ u}) + (64\text{ u}) = 136\text{ u}$

    $2H_2O = (4\text{ u}) + (32\text{ u}) = 36\text{ u}$

    $GFM = 136\text{ u} + 36\text{ u} = 172\text{ u}$

13. $\%H_2O = \left(\dfrac{36\text{ u}}{172\text{ u}} \right) \times 100\% = 21\%$

14. 71.6%

15. $(0.57)(12.90\text{ u}) + (0.43)(12.90\text{ u})$

16. $2Sb_2S_3(s) + 9O_2(g) \rightarrow 2Sb_2O_3(s) + 6SO_2(g)$

Chapter 5

1. (1)	**11.** (4)	**21.** (1)	**31.** (1)	**41.** (2)
2. (3)	**12.** (4)	**22.** (3)	**32.** (4)	**42.** (2)
3. (2)	**13.** (4)	**23.** (4)	**33.** (1)	**43.** (2)
4. (4)	**14.** (3)	**24.** (4)	**34.** (2)	**44.** (2)
5. (4)	**15.** (1)	**25.** (2)	**35.** (1)	
6. (4)	**16.** (2)	**26.** (2)	**36.** (3)	
7. (2)	**17.** (1)	**27.** (1)	**37.** (2)	
8. (1)	**18.** (2)	**28.** (2)	**38.** (1)	
9. (4)	**19.** (2)	**29.** (4)	**39.** (2)	
10. (1)	**20.** (4)	**30.** (1)	**40.** (1)	

Constructed-Response Questions

1. 6.026 kJ released

2. -179.0 kJ

3. -166.6 kJ/mol

4. D

5. C

6. Any substance that absorbs heat ($+\Delta H$) when it dissolves in water could be used. Examples include $KNO_3(s)$ and $NH_4NO_3(s)$.

7. The bracelet's temperature increased because heat flowed from the student's body to the copper bracelet.

8. 0.474 mol

9. $q = (30.1 \text{ g})(0.385 \text{ J/g} \cdot \text{K})(19°C – 33°C)$

Chapter 6

1. (2)	**12.** (2)	**23.** (4)	**34.** (4)	**45.** (3)
2. (2)	**13.** (3)	**24.** (1)	**35.** (1)	**46.** (4)
3. (3)	**14.** (3)	**25.** (3)	**36.** (3)	**47.** (2)
4. (2)	**15.** (1)	**26.** (1)	**37.** (4)	**48.** (2)
5. (1)	**16.** (1)	**27.** (1)	**38.** (3)	**49.** (3)
6. (4)	**17.** (1)	**28.** (2)	**39.** (2)	**50.** (2)
7. (4)	**18.** (3)	**29.** (4)	**40.** (4)	**51.** (3)
8. (3)	**19.** (1)	**30.** (4)	**41.** (4)	**52.** (2)
9. (4)	**20.** (4)	**31.** (1)	**42.** (2)	
10. (2)	**21.** (4)	**32.** (3)	**43.** (4)	
11. (2)	**22.** (2)	**33.** (2)	**44.** (3)	

Constructed-Response Questions

1. 1093 torr

2. 939°C (1212 K)

3. 98.000 kPa

4. 630. mL at STP

5. $\dfrac{\text{rate of } N_2}{\text{rate of } CO_2} = \dfrac{1.25}{1}$

6. 0.423 g/L

7. (a) 23 L (b) 250.0 L

8. Either graph is acceptable.

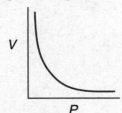 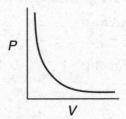

9. Use the combined gas law. Since the temperature is held constant, it is omitted from the equation.

   $(6.2 \text{ mL}) \bullet (1.4 \text{ atm}) = (3.1 \text{ mL}) \bullet P_2$

   $P_2 = 2.8 \text{ atm}$

10. The average kinetic energy decreases.

11. freezing (or solidification or change from liquid to solid)

12. 2.0 atm

13. The frequency of collisions between the gas molecules increases.

14. 710 (±10) mmHg

15. 114 (±2)°C

16. Liquid A will evaporate more rapidly. Compared with the vapor pressure of liquid B, the vapor pressure of liquid A is higher because liquid A has weaker intermolecular forces.

Chapter 7

1. (3)	**11.** (3)	**21.** (2)	**31.** (1)	**41.** (2)
2. (2)	**12.** (1)	**22.** (3)	**32.** (4)	**42.** (4)
3. (4)	**13.** (3)	**23.** (3)	**33.** (1)	**43.** (4)
4. (2)	**14.** (1)	**24.** (2)	**34.** (3)	**44.** (2)
5. (2)	**15.** (3)	**25.** (4)	**35.** (2)	
6. (2)	**16.** (2)	**26.** (1)	**36.** (3)	
7. (3)	**17.** (4)	**27.** (4)	**37.** (1)	
8. (4)	**18.** (2)	**28.** (3)	**38.** (4)	
9. (2)	**19.** (2)	**29.** (4)	**39.** (4)	
10. (3)	**20.** (4)	**30.** (1)	**40.** (4)	

Constructed-Response Questions

1. (a)

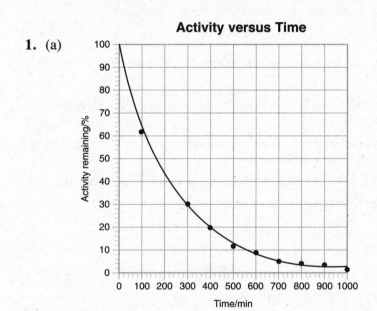

Activity versus Time

(b)

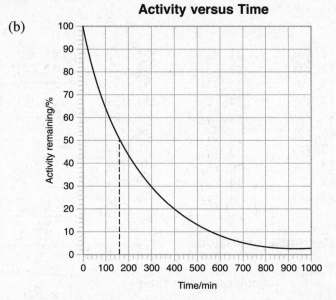

Activity versus Time

Half-life (as determined from the graph) = 160 min

2. ^4_2He

3. Alpha particles harm human tissue by altering genetic information in human cells.

4. Radium and calcium are located in the same periodic group and have similar chemical properties.

5. Zinc sulfide emits visible light in response to ionizing radiation.

6. The U-235 nucleus splits into smaller nuclei.

7. Cs-140

8. 5.00 min

Chapter 8

1. (1)	11. (2)	21. (3)	31. (2)	41. (1)
2. (4)	12. (2)	22. (2)	32. (4)	42. (3)
3. (4)	13. (4)	23. (2)	33. (4)	43. (1)
4. (3)	14. (3)	24. (1)	34. (2)	44. (1)
5. (2)	15. (1)	25. (2)	35. (1)	45. (3)
6. (4)	16. (3)	26. (2)	36. (3)	
7. (2)	17. (4)	27. (1)	37. (2)	
8. (3)	18. (2)	28. (3)	38. (4)	
9. (2)	19. (3)	29. (2)	39. (2)	
10. (2)	20. (4)	30. (3)	40. (3)	

Constructed-Response Questions

1.

Element	Symbol	Electron Configuration	Number of Valence Electrons	Lewis Structure
Argon	Ar	2-8-8	8	$:\overset{\cdot\cdot}{\underset{\cdot\cdot}{Ar}}:$
Barium	Ba	2-8-18-18-8-2	2	$\overset{\cdot\cdot}{Ba}$
Carbon	C	2-4	4	$\overset{\cdot\cdot}{\underset{\cdot}{C}}\cdot$
Chlorine	Cl	2-8-7	7	$\cdot\overset{\cdot\cdot}{\underset{\cdot\cdot}{Cl}}:$
Helium	He	2	2	$\overset{\cdot\cdot}{He}$
Nitrogen	N	2-5	5	$\cdot\overset{\cdot\cdot}{\underset{\cdot}{N}}\cdot$
Oxygen	O	2-6	6	$\cdot\overset{\cdot\cdot}{\underset{\cdot}{O}}:$
Potassium	K	2-8-8-1	1	$\overset{\cdot}{K}$
Strontium	Sr	2-8-18-8-2	2	$\overset{\cdot\cdot}{Sr}$
Sulfur	S	2-8-6	6	$\cdot\overset{\cdot\cdot}{\underset{\cdot}{S}}:$

2. 2 valence electrons

3. element *Y* (2-8-7-3)

4. H and He

5. As electrons transition from excited states to the ground state, photons are emitted. These emissions produce bright-line spectra.

6. Not all of the wavelengths of element *A* are shown in the spectrum of the mixture.

7. The wavelengths of the spectral lines for element *Z* are independent of the mass of the sample.

8. Different colors of light are produced when electrons return from higher energy states to lower energy states.

Chapter 9

1. (3)	11. (2)	21. (3)	31. (2)	41. (3)
2. (2)	12. (1)	22. (1)	32. (1)	42. (4)
3. (1)	13. (3)	23. (4)	33. (2)	43. (2)
4. (2)	14. (3)	24. (3)	34. (1)	44. (2)
5. (1)	15. (1)	25. (2)	35. (4)	
6. (1)	16. (1)	26. (2)	36. (4)	
7. (2)	17. (1)	27. (1)	37. (1)	
8. (1)	18. (2)	28. (2)	38. (4)	
9. (2)	19. (2)	29. (1)	39. (1)	
10. (4)	20. (4)	30. (2)	40. (2)	

Constructed-Response Questions

1. Na and Rb

2. They are located in the same group (Group 1) of the Periodic Table.

3. A K^+ ion has a total of 18 electrons.

4. A potassium atom has the electron configuration 2-8-8-1. A K^+ ion has the configuration 2-8-8. When a potassium atom forms an ion, the fourth shell is lost, thereby decreasing the radius.

5. A K^+ ion is electrically charged and is mobile within nerve fibers.

6. Na and K are located in the same group (Group 1) of the Periodic Table.

7. silicon (Si) or germanium (Ge)

8. The atomic radius of these elements increases down the group because each successive element has one more electron shell.

9. 4 valence electrons

10. The radius of a fluoride ion is larger than the radius of a fluorine atom.

11. A lithium atom loses its second-shell valence electron. So the lithium ion has only one shell of electrons.

12. As the elements of Period 2 are considered from left to right, the atomic radius generally decreases (with the exception of Ne).

Chapter 10

1. (1)	**11.** (3)	**21.** (1)	**31.** (3)	**41.** (2)
2. (4)	**12.** (2)	**22.** (2)	**32.** (4)	**42.** (2)
3. (1)	**13.** (2)	**23.** (3)	**33.** (4)	**43.** (1)
4. (4)	**14.** (4)	**24.** (2)	**34.** (3)	**44.** (1)
5. (1)	**15.** (3)	**25.** (3)	**35.** (2)	**45.** (4)
6. (2)	**16.** (2)	**26.** (2)	**36.** (3)	**46.** (3)
7. (2)	**17.** (2)	**27.** (2)	**37.** (1)	
8. (2)	**18.** (3)	**28.** (1)	**38.** (4)	
9. (4)	**19.** (3)	**29.** (1)	**39.** (3)	
10. (4)	**20.** (1)	**30.** (1)	**40.** (3)	

Constructed-Response Questions

1. (a) ionic   (b) covalent   (c) covalent   (d) ionic   (e) covalent

2.

3.

4. Although a CCl_4 molecule has polar bonds, its symmetry makes the entire molecule nonpolar.

5. The molecules of NH_3 are polar while the molecules of Cl_2 are not.

6. In molecules A, B, and C, electrons are shared. In KCl, the electrons are transferred from K to Cl.

7. Valence electrons are lost by potassium and gained by bromine.

8. 6

9. The charge is symmetrically distributed.

10. Either diagram is acceptable.

11. argon (Ar)

Chapter 11

1. (3)	11. (2)	21. (4)	31. (3)	41. (3)
2. (2)	12. (2)	22. (3)	32. (2)	42. (1)
3. (4)	13. (3)	23. (1)	33. (3)	
4. (1)	14. (2)	24. (3)	34. (4)	
5. (2)	15. (3)	25. (3)	35. (1)	
6. (4)	16. (2)	26. (3)	36. (4)	
7. (4)	17. (4)	27. (1)	37. (1)	
8. (1)	18. (4)	28. (3)	38. (3)	
9. (4)	19. (3)	29. (2)	39. (4)	
10. (3)	20. (1)	30. (1)	40. (3)	

Constructed-Response Questions

1. esterification

2. alcohol

3. Either diagram is acceptable.

$$
\begin{array}{c}
HOHH \\
||| \\
H-C-C-C-H \\
||| \\
HHH
\end{array}
$$

$$
\begin{array}{c}
HHH \\
||| \\
H-C-C-O-C-H \\
||| \\
HHH
\end{array}
$$

4. Two substances react to produce one substance.

5. halide (or alkyl halide or halogenalkane)

6. The compounds have the same molecular formulas but different structural formulas.

7. Any of the following diagrams is acceptable.

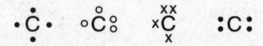

8. The molecule in Diagram *B* has only single carbon-to-carbon bonds.

9. The molecules have the same molecular formulas but different structural formulas.

Chapter 12

1. (1)	**10.** (2)	**19.** (3)	**28.** (1)	**37.** (2)
2. (3)	**11.** (3)	**20.** (2)	**29.** (1)	**38.** (1)
3. (2)	**12.** (1)	**21.** (2)	**30.** (3)	**39.** (1)
4. (2)	**13.** (4)	**22.** (2)	**31.** (3)	**40.** (3)
5. (3)	**14.** (3)	**23.** (1)	**32.** (3)	**41.** (2)
6. (2)	**15.** (3)	**24.** (4)	**33.** (1)	**42.** (4)
7. (1)	**16.** (3)	**25.** (4)	**34.** (2)	**43.** (3)
8. (4)	**17.** (2)	**26.** (2)	**35.** (1)	
9. (1)	**18.** (3)	**27.** (3)	**36.** (3)	

Constructed-Response Questions

1. 0.519 M

2. 0.0309 mol

3. 35.6 g

4. 0.728 M

5. Dilute 18.0 mL of the H_2SO_4 with enough water to make 325 mL of solution.

6. 2.26 kg

7. 744 g

8. $X = 0.0946$

9. 0.251 L

10.

Solute	Molar Mass of Solute ($\mathcal{M}$) (g/mol)	Moles of Solute Present (n) (mol)	Mass of Solute Present (m) (g)	Volume of Solution (V)	Molarity (M)
NH_3	17	2.00	34.0	250. mL	8.00
HNO_3	63	1.00	63.0	2.00 L	0.500
H_2SO_4	98	0.500	49.0	5.00 dm³	0.100
$MgCl_2$	95	1.00	95.0	0.500 L	2.00
$AlPO_4$	122	2.00	244	0.500 L	4.00
$C_6H_{12}O_6$	180	0.250	45.0	0.125 L	2.00
$HC_2H_3O_2$	60	2.00	120.	4000. cm³	0.500
KI	166	3.00	498	3.00 L	1.00
$Na_2Cr_2O_7$	262	3.00	786	0.333 L	9.00
$B(OH)_3$	62	4.00	248	12.0 L	0.333
H_2SO_4	98	3.00	294	2.00 dm³	1.50

11. $3.58°C$

12. $14.0 \ C°/m$

13. ionic bonds and polar covalent bonds

14. The solution is saturated because some $NH_4Cl(s)$ remained undissolved at the bottom of the test tube.

15. The dissolving of $NH_4Cl(s)$ is endothermic because the temperature of the solution at the end of the procedure is lower than the initial temperature of the water.

16. No solute remained in the bottom of the test tube.

Chapter 13

1. (2)	11. (4)	21. (3)	31. (2)	41. (1)
2. (2)	12. (4)	22. (3)	32. (2)	42. (1)
3. (3)	13. (4)	23. (1)	33. (4)	
4. (4)	14. (4)	24. (3)	34. (1)	
5. (1)	15. (4)	25. (4)	35. (3)	
6. (1)	16. (4)	26. (3)	36. (4)	
7. (3)	17. (1)	27. (1)	37. (2)	
8. (1)	18. (4)	28. (3)	38. (2)	
9. (1)	19. (3)	29. (4)	39. (3)	
10. (4)	20. (1)	30. (3)	40. (3)	

Constructed-Response Questions

1. (a) System is shifted to the right.
   (b) System is shifted to the right.
   (c) System is shifted to the left.
   (d) System is unaffected.
   (e) System is shifted to the left.

2. (a) $K_{eq} = \dfrac{[NO_2]^4[H_2O]^6}{[NH_3]^4[O_2]^7}$

   (b) $K = \dfrac{[H_2O]^3}{[H_2]^3}$

3. (a) $K_{eq} = \dfrac{[C]^3[D]^2}{[A][B]^2}$

   (b) $K_{eq} = 18$

4. $[H_2]_{eq} = 0.763 \ mol/L$, $[NH_3]_{eq} = 0.158 \ mol/L$

5. The forward reaction (the dissolving of KNO_3 in water) is endothermic. Raising the temperature favors the endothermic reaction.

6. The system is in equilibrium because the rate of dissolving is equal to the rate of crystallizing.

7. As the concentration of $CO_2(aq)$ decreases, the reaction shifts to the right, lowering the concentration of $H_2CO_3(aq)$.

Chapter 14

1. (4)	**11.** (1)	**21.** (1)	**31.** (3)	**41.** (1)
2. (4)	**12.** (3)	**22.** (2)	**32.** (2)	**42.** (4)
3. (1)	**13.** (4)	**23.** (4)	**33.** (3)	**43.** (3)
4. (2)	**14.** (3)	**24.** (2)	**34.** (3)	**44.** (4)
5. (1)	**15.** (1)	**25.** (3)	**35.** (4)	**45.** (2)
6. (3)	**16.** (4)	**26.** (2)	**36.** (3)	**46.** (4)
7. (2)	**17.** (1)	**27.** (4)	**37.** (4)	**47.** (2)
8. (2)	**18.** (1)	**28.** (1)	**38.** (2)	
9. (4)	**19.** (3)	**29.** (2)	**39.** (1)	
10. (2)	**20.** (1)	**30.** (2)	**40.** (1)	

Constructed-Response Questions

1. (a) $CH_3NH_2(aq) + H_2O(\ell) \rightleftharpoons CH_3NH_3^+(aq) + OH^-(aq)$

   (b) $K_b = \dfrac{[CH_3NH_3^+][OH^-]}{[CH_3NH_2]}$

2. (a) pH = 10.3   (b) pOH = 11.7

3. Amount of NaOH used: (35.2 mL – 23.2 mL) = 12.0 mL

4. Bromthymol blue. The indicator changes color around the end point of the titration (pH = 7.0).

5. $M_A V_A = M_B V_B$

   $MB = \dfrac{M_A V_A}{V_B} = \dfrac{(1.2 \text{ M}) \bullet (10.0 \text{ mL})}{(12.0 \text{ mL})} = 1.0 \text{ M}$

6. Multiple trials tend to cancel out experimental errors found in the individual trials.

7. titration

8. $HCl(aq) + NaOH(aq) \rightarrow H_2O(\ell) + NaCl(aq)$

9. 2

10. 0.12 M
11. yellow
12. hydroxide ion (OH^-)
13. 0.86 M

Chapter 15

1. (2)	**11.** (4)	**21.** (4)	**31.** (3)	**40.** (2)
2. (4)	**12.** (2)	**22.** (3)	**32.** (3)	**41.** (4)
3. (4)	**13.** (2)	**23.** (1)	**32.** (3)	
4. (4)	**14.** (4)	**24.** (4)	**33.** (1)	
5. (1)	**15.** (3)	**25.** (4)	**34.** (4)	
6. (4)	**16.** (2)	**26.** (4)	**35.** (3)	
7. (2)	**17.** (2)	**27.** (2)	**36.** (3)	
8. (3)	**18.** (1)	**28.** (2)	**37.** (1)	
9. (2)	**19.** (3)	**29.** (1)	**38.** (1)	
10. (1)	**20.** (1)	**30.** (2)	**39.** (4)	

Constructed-Response Questions

1. (a) $Zn(s) \rightarrow Zn^{2+}(aq) + 2e^-$

   $2H^+(aq) + 2e^- \rightarrow H_2(g)$

   (b) Zn is *higher* in Reference Table J than H_2; Cu is *lower* in Reference Table J than H_2.

   (c) $Cu^{2+}(aq) + H_2(g) \rightarrow Cu(s) + 2H^+(aq)$

2. (a) $7KOH + 4Zn + 6H_2O + KNO_3 \rightarrow 4K_2Zn(OH)_4 + NH_3$

   (b) $3H_2SO_4 + 6FeSO_4 + KClO_3 \rightarrow 3Fe_2(SO_4)_3 + KCl + 3H_2O$

   (c) $H_3AsO_4 + 4\,Zn + 8HNO_3 \rightarrow AsH_3 + 4Zn(NO_3)_2 + 4H_2O$

   (d) $3H_2O + 3LiOH + P_4 \rightarrow PH_3 + 3LiH_2PO_2$

   (e) $4Zn + 10H^+ + NO_3^- \rightarrow 4Zn^{2+} + NH_4^+ + 3H_2O$

   (f) $H_2O + 2MnO_4^- + 3CN^- \rightarrow 2MnO_2 + 2OH^- + 3CNO^-$

3. (a) $Cu^{2+}(aq) + 2e^- \rightarrow Cu(s)$

   (b) 2.000 moles of electrons are needed

   (c) $2.000 \; \cancel{\text{mole}} \cdot \left(\dfrac{96,470 \text{ C}}{1 \; \cancel{\text{mole}}} \right) = 1.929 \times 10^5 \text{C}$

   (d) $5.000 \; \cancel{\text{h}} \cdot \left(\dfrac{3600. \text{ s}}{1.000 \; \cancel{\text{h}}} \right) = 1.800 \times 10^4 \text{s}$

   $$\text{Current} = \frac{1.929 \times 10^5 \text{ C}}{1.800 \times 10^4 \text{s}} = 10.72 \text{ A}$$

4.

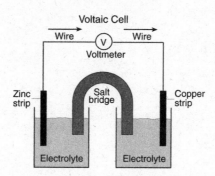

5. $Zn^0 \rightarrow Zn^{2+} + 2e^-$

6. The flow of ions through the salt bridge maintains the electrical neutrality of each half-cell.

7. The oxidation number changes from 0 to +2.

8. The number of moles of electrons lost is equal to the number of moles of electrons gained.

9. Risk: Mercury is toxic.

   Benefit: Mercury batteries are small in size.

10. The methane (CH_4) is combined with oxygen (O_2) to form two oxides (H_2O and CO_2).

11. The oxidation number changes from −4 to +4.

12. $Mg \rightarrow Mg^{2+} + 2e^-$

Chapter 16

1. (3)	**8.** (3)	**15.** (4)	**22.** (3)	**29.** (4)
2. (4)	**9.** (2)	**16.** (3)	**23.** (1)	**30.** (4)
3. (2)	**10.** (3)	**17.** (3)	**24.** (4)	**31.** (2)
4. (1)	**11.** (1)	**18.** (1)	**25.** (1)	**32.** (4)
5. (2)	**12.** (3)	**19.** (4)	**26.** (1)	**33.** (1)
6. (1)	**13.** (3)	**20.** (2)	**27.** (2)	**34.** (3)
7. (4)	**14.** (4)	**21.** (2)	**28.** (3)	**35.** (1)

Constructed-Response Questions

1. A
2. E
3. D
4. B

Appendix 4

ANSWERING CONSTRUCTED-RESPONSE QUESTIONS

A *constructed-response question* (also known as a *free-response question*) is an examination question that requires the test taker to do more than to choose among several responses or to fill in a blank. You may need to perform numerical calculations, draw or interpret a graph, draw a diagram, provide an extended written response to a question or problem, or write and balance a chemical equation. This appendix is designed to provide you with a number of general guidelines for answering constructed-response questions. *At this time, you should review the material in Chapter 1, pages 23–31.*

A. SOLVING PROBLEMS INVOLVING NUMERICAL CALCULATIONS

If you are presented with a numerical problem, you are advised to:

- Provide the appropriate equation(s).
- Substitute values and units into the equation(s).
- Display the answer, with appropriate units.

Chemistry courses use various types of metric units (SI and non-SI) to describe matter and energy, and you are expected to have some familiarity with them.

You may or may not be penalized if your answer to a calculation has an incorrect number of significant digits. It is always good practice, however, to pay attention to this detail.

You should write your solution in a clear, logical fashion. Many (if not all) teachers become more than a little annoyed if they have to jump visually around your paper in order to find your next calculation. It is also a good idea to identify your answer clearly, either by enclosing it in a box or by writing the word *answer* next to it.

A final caution: If you provide the correct answer but do not show any work, most teachers will award little or no credit for the problem!

Here are a sample problem and a model solution.

PROBLEM
A 5.00-gram object has a density of 4.00 grams per cubic centimeter. Calculate the volume of this object.

SOLUTION

$$d = \frac{m}{V}$$

Rearranging the equation gives

$$V = \frac{m}{d} = \frac{5.00 \text{ g}}{4.00 \text{ g/cm}^3}$$

$$\boxed{V = 1.25 \text{ cm}^3}$$

B. SOLVING PROBLEMS THAT REQUIRE THE GRAPHING OF EXPERIMENTAL DATA

If you are presented with a problem that requires you to graph experimental data, you are advised to:

- Label both axes with the appropriate variables and units.
- Divide the axes so that the given values occupy a significant portion of the graph. This allows a trend in the data to be observed.
- Plot all data points accurately.
- Depending on the instructions given, connect the data points with straight lines or draw a *best-fit* line or curve.
- If a part of the question requires that the slope of a line be calculated, calculate the slope *from the line*, not from individual data points.

Generally a graph should have a title, and the *independent variable* is usually displayed along the x-axis.

Here are a sample problem and a step-by-step model solution.

PROBLEM
A student attempts to estimate absolute zero in the following way: He subjects a sample of gas (at constant pressure and mass) to varying temperatures and measures the gas volume at each of the temperatures. The accompanying table contains his experimental data.

Temperature/°C	Volume/mL
−100	128
−60	148
−40	156
0	204
40	222
80	272
160	310

(a) Using axes that are appropriately labeled and scaled, draw a graph that accurately displays the student's data.

(b) Estimate the student's value for absolute zero by extending the graph to the Celsius temperature at which the volume of the gas is 0 milliliter.

SOLUTION

(a) The first step is to construct an appropriate set of axes if one is not provided. We will assume that you must start from scratch. Since temperature is the independent variable, you need to place it along the x-axis. Also, since absolute zero is −273.15°C, you must scale the axes properly. Here is one appropriate set of axes:

Volume versus Temperature

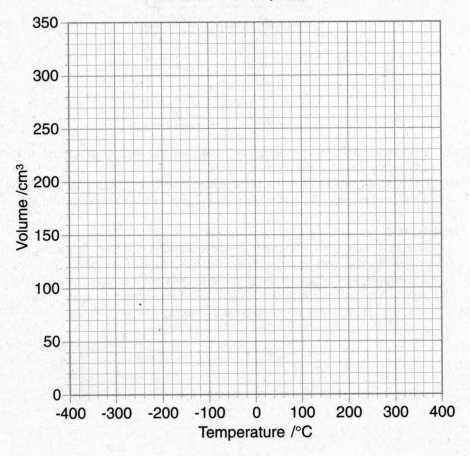

Second, you must plot the data points carefully on the axes as shown below:

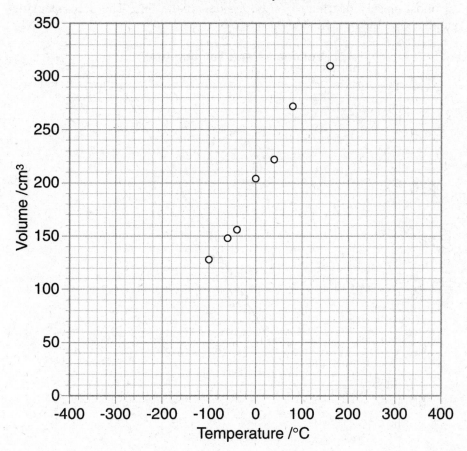

Volume versus Temperature

Your third task is to draw the graph. Examining *all* of the plotted points, you note that they fall approximately on a *straight line*. Therefore, the next step is to draw a *best-fit* straight-line graph. This is a graph in which the data points are most closely distributed on both sides of the line. The accompanying graph shows the best-fit straight line.

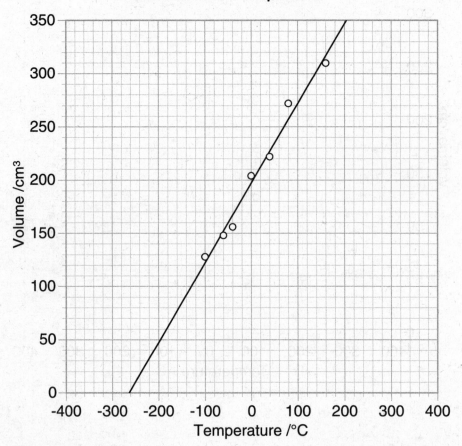

Note that the graph is extended beyond the data points. Such extensions are known as *extrapolated data*, and they are based on the assumption that the gas will continue to behave as it did within the experimental range for which the student collected data.

(b) Your final task is to inspect the graph closely and to estimate absolute zero. The calculated value is –263°C. (This corresponds to an experimental error of 3.7%.)

C. SOLVING PROBLEMS REQUIRING THE DRAWING OF A DIAGRAM

Since you may be presented with a problem requiring you to draw a diagram, you are advised to bring a straightedge with you. Then:

- Use the straightedge as needed.
- Draw the diagram neatly and accurately, and label it clearly.

The accompanying diagram represents a zinc–copper electrochemical cell containing an agar–KCl salt bridge. This is the type of diagram you may be asked to draw as part of an examination question.

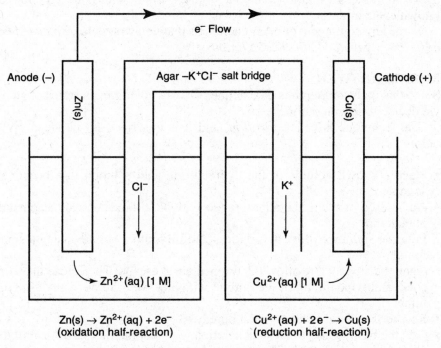

Zn(s) → Zn²⁺(aq) + 2e⁻ (oxidation half-reaction)

Cu²⁺(aq) + 2e⁻ → Cu(s) (reduction half-reaction)

Net reaction: Zn(s) + Cu²⁺(aq) → Cu(s) + Zn²⁺(aq)

Remember: your teacher has every right to expect you to draw an *accurate*, *complete*, and *neat* diagram!

D. WRITING AN EXTENDED RESPONSE TO AN EXAMINATION QUESTION

If the examination question requires an extended response, you are advised to:

* Use complete, clear sentences that make sense to the reader. You may decide to structure your response in outline or in paragraph form.
* Use correct chemistry in your explanations.

A sample question and its model answer are given below.

QUESTION

Describe in detail the technique for determining the concentration of a dilute hydrochloric acid solution using a dilute sodium hydroxide solution of known concentration.

Name any equipment or other chemicals that you would use. *You need not describe any mathematical calculations.*

MODEL ANSWER

Since this question requires an extended answer containing a series of steps, we decide to use an outline form.

The technique is called *titration*, and it is described in the steps given below.

* Place a known volume of the hydrochloric acid solution in a beaker of suitable size.
* Add a drop or two of an appropriatee acid–base indicator such as phenol-phthalein.
* Pour the sodium hydroxide solution carefully into a burette, using a small funnel.
* Open the stopcock to allow the trapped air to escape. Then close the stopcock, and wipe the tip of the burette with a tissue to remove any clinging liquid.
* Record the initial volume in the burette.
* Add the base slowly to the acid solution, with continuous stirring, until the phenolphthalein just changes from colorless to faint pink.
* Record the final volume in the burette.
* Repeat the experiment at least once.
* Rinse the apparatus with water to remove all traces of acid and base.
* Calculate the concentration of the hydrochloric acid solution.

E. WRITING AND BALANCING EQUATIONS IN A CHEMISTRY EXAMINATION

For all of the suggestions that follow, the reaction that occurs between aqueous solutions of sodium sulfate and barium nitrate is used as an example and is referred to as the given reaction.

- If you are asked to write a *word equation*, be certain to include the correct names of the reactants and products and their phases in the reaction.

For the given reaction, the word equation is

barium nitrate(aq) + sodium sulfate(aq) → barium sulfate(s) + sodium nitrate(aq)

- If you are asked to write a *balanced* equation, you are usually expected to balance using *smallest whole-number coefficients*.

For the given reaction, the balanced equation is

$$Ba(NO_3)_2(aq) + Na_2SO_4(aq) \rightarrow BaSO_4(s) + 2NaNO_3(aq)$$

- If you are asked to write an *ionic* equation occurring in aqueous solution, you must reduce everything to its component ions except *insoluble* compounds, such as $BaSO_4(s)$, and (of course!) *covalently* bonded substances, such as $H_2O(\ell)$.

For the given reaction, the ionic equation is

$$Ba^{2+}(aq) + 2NO_3^-(aq) + 2Na^+(aq) + SO_4^{2-}(aq) \rightarrow$$
$$BaSO_4(s) + 2Na^+(aq) + 2NO_3^-(aq)$$

- If you are asked to write a *net* ionic equation, you should omit all *spectator ions*, that is, all ions appearing *unchanged on both sides of the equation*.

For the given reaction, the net ionic equation is

$$Ba^{2+}(aq) + SO_4^{2-}(aq) \rightarrow BaSO_4(s)$$

THE NEW YORK STATE REGENTS EXAMINATION IN CHEMISTRY

The New York State Regents chemistry examination consists of questions that are based on the *New York State Chemistry Core*. The Core is divided into eleven areas as follows:

- Math Skills
- Atomic Concepts
- Periodic Table
- Moles/Stoichiometry
- Chemical Bonding
- Physical Behavior of Matter
- Kinetics/Equilibrium
- Organic Chemistry
- Oxidation-Reduction
- Acids, Bases, and Salts
- Nuclear Chemistry

The chemistry examination takes 3 hours and includes four parts: A, B-1, B-2, and C. You should be prepared to answer questions in multiple-choice format as well as questions that require more extended responses. Parts A, B-1, B-2, and C are divided as follows:

Part	Question Type(s)	Description of Questions
A	Multiple-choice	*Content-based* questions that will test your knowledge and understanding of the material contained in the NYS Chemistry Core.
B-1	Multiple-choice	*Content- and skills-based* questions that will test your ability to apply, analyze, and evaluate the NYS Chemistry Core material.
B-2	Constructed-response	*Content- and skills-based* questions that will test your ability to apply, analyze, and evaluate the NYS Chemistry Core material.
C	Constructed-response and/or *extended* constructed-response	*Content-based and application* questions that will test your ability to apply your knowledge of chemistry concepts and skills to "real-world" situations.

You may be required to graph data, complete a data table, label or draw diagrams, design experiments, make calculations, or write short or more extended responses. In addition you may be required to hypothesize, interpret, analyze, evaluate data, or apply your scientific knowledge and skills to real-world situations.

Some of the questions will require use of the New York State Regents Reference Tables for Chemistry in Appendix 1, pages 495–510.

Note that you will be required to answer ALL of the questions on the Regents examination.

Examination
June 2017
Chemistry
The Physical Setting

PART A

Answer all questions in this part.

Directions (1–30): For *each* statement or question, write in the answer space the *number* of the word or expression that, of those given, best completes the statement or answers the question. Some questions may require the use of the *2011 Edition Reference Tables for Physical Setting/Chemistry*.

1 Which statement describes the structure of an atom?

  (1) The nucleus contains positively charged electrons.

  (2) The nucleus contains negatively charged protons.

  (3) The nucleus has a positive charge and is surrounded by negatively charged electrons.

  (4) The nucleus has a negative charge and is surrounded by positively charged electrons.   1 _____

2 Which term is defined as the region in an atom where an electron is most likely to be located?

  (1) nucleus   (3) quanta

  (2) orbital   (4) spectra   2 _____

3 What is the number of electrons in an atom of scandium?

(1) 21   (3) 45

(2) 24   (4) 66   3 _____

4 Which particle has the *least* mass?

(1) a proton   (3) a helium atom

(2) an electron   (4) a hydrogen atom   4 _____

5 Which electron transition in an excited atom results in a release of energy?

(1) first shell to the third shell

(2) second shell to the fourth shell

(3) third shell to the fourth shell

(4) fourth shell to the second shell   5 _____

6 On the Periodic Table, the number of protons in an atom of an element is indicated by its

(1) atomic mass

(2) atomic number

(3) selected oxidation states

(4) number of valence electrons   6 _____

7 Which type of formula shows an element symbol for each atom and a line for each bond between atoms?

(1) ionic   (3) empirical

(2) structural   (4) molecular   7 _____

8 What is conserved during all chemical reactions?

(1) charge   (3) vapor pressure

(2) density   (4) melting point   8 _____

9 In which type of reaction can two compounds exchange ions to form two different compounds?

(1) synthesis
(2) decomposition
(3) single replacement
(4) double replacement 9 _____

10 At STP, two 5.0-gram solid samples of different ionic compounds have the same density. These solid samples could be differentiated by their

(1) mass   (3) temperature
(2) volume   (4) solubility in water   10 _____

11 What is the number of electrons shared between the atoms in an I_2 molecule?

(1) 7   (3) 8
(2) 2   (4) 4   11 _____

12 Which substance has nonpolar covalent bonds?

(1) Cl_2 (3) SiO_2
(2) SO_3 (4) CCl_4 12 _____

13 Compared to a potassium atom, a potassium ion has

(1) a smaller radius   (3) fewer protons
(2) a larger radius   (4) more protons   13 _____

14 Which form of energy is associated with the random motion of particles in a gas?

(1) chemical   (3) nuclear
(2) electrical   (4) thermal   14 _____

15 The average kinetic energy of water molecules *decreases* when

(1) $H_2O(\ell)$ at 337 K changes to $H_2O(\ell)$ at 300. K
(2) $H_2O(\ell)$ at 373 K changes to $H_2O(g)$ at 373 K
(3) $H_2O(s)$ at 200. K changes to $H_2O(s)$ at 237 K
(4) $H_2O(s)$ at 273 K changes to $H_2O(\ell)$ at 273 K   15 _____

16 The joule is a unit of

(1) concentration   (3) pressure
(2) energy   (4) volume   16 _____

17 Compared to a sample of helium at STP, the same sample of helium at a higher temperature and a lower pressure

(1) condenses to a liquid
(2) is more soluble in water
(3) forms diatomic molecules
(4) behaves more like an ideal gas   17 _____

18 A sample of a gas is in a sealed, rigid container that maintains a constant volume. Which changes occur between the gas particles when the sample is heated?

(1) The frequency of collisions increases, and the force of collisions decreases.
(2) The frequency of collisions increases, and the force of collisions increases.
(3) The frequency of collisions decreases, and the force of collisions decreases.
(4) The frequency of collisions decreases, and the force of collisions increases.   18 _____

19 At STP, which gaseous sample has the same number of molecules as 3.0 liters of $N_2(g)$?

    (1) 6.0 L of $F_2(g)$ (3) 3.0 L of $H_2(g)$

    (2) 4.5 L of $N_2(g)$ (4) 1.5 L of $Cl_2(g)$ 19 _____

20 Distillation of crude oil from various parts of the world yields different percentages of hydrocarbons. Which statement explains these different percentages?

    (1) Each component in a mixture has a different solubility in water.

    (2) Hydrocarbons are organic compounds.

    (3) The carbons in hydrocarbons may be bonded in chains or rings.

    (4) The proportions of components in a mixture can vary.   20 _____

21 In which 1.0-gram sample are the particles arranged in a crystal structure?

    (1) $CaCl_2(s)$ (3) $CH_3OH(\ell)$

    (2) $C_2H_6(g)$ (4) $CaI_2(aq)$ 21 _____

22 When a reversible reaction is at equilibrium, the concentration of products and the concentration of reactants must be

    (1) decreasing   (3) constant

    (2) increasing   (4) equal   22 _____

23 In chemical reactions, the difference between the potential energy of the products and the potential energy of the reactants is equal to the

    (1) activation energy

    (2) ionization energy

    (3) heat of reaction

    (4) heat of vaporization   23 _____

24 What occurs when a catalyst is added to a chemical reaction?

(1) an alternate reaction pathway with a lower activation energy

(2) an alternate reaction pathway with a higher activation energy

(3) the same reaction pathway with a lower activation energy

(4) the same reaction pathway with a higher activation energy

24 _____

25 What is the name of the compound with the formula $CH_3CH_2CH_2NH_2$?

(1) 1-propanol   (3) propanal

(2) 1-propanamine   (4) propanamide

25 _____

26 Which compound is an isomer of $C_2H_5OC_2H_5$?

(1) CH_3COOH (3) $C_3H_7COCH_3$

(2) $C_2H_5COOCH_3$ (4) C_4H_9OH

26 _____

27 Ethanoic acid and 1-butanol can react to produce water and a compound classified as an

(1) aldehyde   (3) ester

(2) amide   (4) ether

27 _____

28 During an oxidation-reduction reaction, the number of electrons gained is

(1) equal to the number of electrons lost

(2) equal to the number of protons gained

(3) less than the number of electrons lost

(4) less than the number of protons gained

28 _____

29 Which process requires energy for a nonspontaneous redox reaction to occur?

(1) deposition   (3) alpha decay

(2) electrolysis   (4) chromatography   29 _____

30 Which pair of compounds represents one Arrhenius acid and one Arrhenius base?

(1) CH_3OH and $NaOH$

(2) CH_3OH and HCl

(3) HNO_3 and $NaOH$

(4) HNO_3 and HCl 30 _____

PART B–1

Answer all questions in this part.

Directions (31–50): For *each* statement or question, write in the answer space the *number* of the word or expression that, of those given, best completes the statement or answers the question. Some questions may require the use of the *2011 Edition Reference Tables for Physical Setting/Chemistry.*

31 Which electron configuration represents the electrons of an atom of neon in an excited state?

    (1) 2-7   (3) 2-7-1

    (2) 2-8   (4) 2-8-1   31 _____

32 Some information about the two naturally occurring isotopes of gallium is given in the table below.

Natural Abundance of Two Gallium Isotopes

Isotope	Natural Abundance (%)	Atomic Mass (u)
Ga-69	60.11	68.926
Ga-71	39.89	70.925

Which numerical setup can be used to calculate the atomic mass of gallium?

    (1) $(0.6011)(68.926 \text{ u}) + (0.3989)(70.925 \text{ u})$

    (2) $(60.11)(68.926 \text{ u}) + (39.89)(70.925 \text{ u})$

    (3) $(0.6011)(70.925 \text{ u}) + (0.3989)(68.926 \text{ u})$

    (4) $(60.11)(70.925 \text{ u}) + (39.89)(68.926 \text{ u})$ 32 _____

33 A student measures the mass and volume of a sample of copper at room temperature and 101.3 kPa. The mass is 48.9 grams and the volume is 5.00 cubic centimeters. The student calculates the density of the sample. What is the percent error of the student's calculated density?

(1) 7.4%   (3) 9.2%

(2) 8.4%   (4) 10.2%   33 _____

34 What is the chemical formula for sodium sulfate?

(1) Na_2SO_4 (3) $NaSO_4$

(2) Na_2SO_3 (4) $NaSO_3$ 34 _____

35 Given the balanced equation representing a reaction:

$$2Na(s) + Cl_2(g) \rightarrow 2NaCl(s) + energy$$

If 46 grams of Na and 71 grams of Cl_2 react completely, what is the total mass of NaCl produced?

(1) 58.5 g   (3) 163 g

(2) 117 g   (4) 234 g   35 _____

36 Given the balanced equation representing a reaction:

$$2NO + O_2 \rightarrow 2NO_2 + energy$$

The mole ratio of NO to NO_2 is

(1) 1 to 1   (3) 3 to 2

(2) 2 to 1   (4) 5 to 2   36 _____

37 The particle diagram below represents a solid sample of silver.

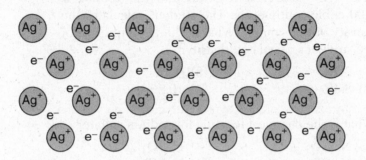

Which type of bonding is present when valence electrons move within the sample?

(1) metallic bonding   (3) covalent bonding

(2) hydrogen bonding   (4) ionic bonding   37 _____

38 Given the formula representing a molecule:

$$\begin{array}{ccc} H & H & H \\ | & | & | \\ H-C-C-C-H \\ | & | & | \\ H & H & H \end{array}$$

Which statement explains why the molecule is nonpolar?

(1) Electrons are shared between the carbon atoms and the hydrogen atoms.

(2) Electrons are transferred from the carbon atoms to the hydrogen atoms.

(3) The distribution of charge in the molecule is symmetrical.

(4) The distribution of charge in the molecule is asymmetrical.   38 _____

39 A solid sample of a compound and a liquid sample of the same compound are each tested for electrical conductivity. Which test conclusion indicates that the compound is ionic?

(1) Both the solid and the liquid are good conductors.

(2) Both the solid and the liquid are poor conductors.

(3) The solid is a good conductor, and the liquid is a poor conductor.

(4) The solid is a poor conductor, and the liquid is a good conductor. 39 _____

40 Which statement explains why 10.0 mL of a 0.50 M H_2SO_4(aq) solution exactly neutralizes 5.0 mL of a 2.0 M NaOH(aq) solution?

(1) The moles of H^+(aq) equal the moles of OH^-(aq).

(2) The moles of H_2SO_4(aq) equal the moles of NaOH(aq).

(3) The moles of H_2SO_4(aq) are greater than the moles of NaOH(aq).

(4) The moles of H^+(aq) are greater than the moles of OH^-(aq). 40 _____

41 Which particle diagram represents *one* substance in the gas phase?

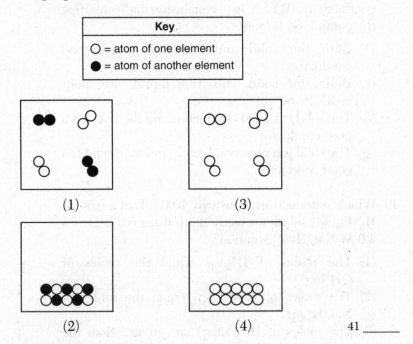

Key
O = atom of one element
● = atom of another element

(1)

(3)

(2)

(4)

41 _____

42 Given the equation representing a chemical reaction at equilibrium in a sealed, rigid container:

$$H_2(g) + I_2(g) + energy \rightleftharpoons 2HI(g)$$

When the concentration of $H_2(g)$ is increased by adding more hydrogen gas to the container at constant temperature, the equilibrium shifts

(1) to the right, and the concentration of $HI(g)$ decreases

(2) to the right, and the concentration of $HI(g)$ increases

(3) to the left, and the concentration of $HI(g)$ decreases

(4) to the left, and the concentration of $HI(g)$ increases

42 _____

43 Which diagram represents the potential energy changes during an exothermic reaction?

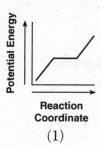

Reaction
Coordinate

(1)

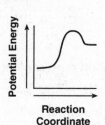

Reaction
Coordinate

(3)

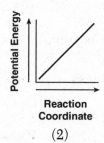

Reaction
Coordinate

(2)

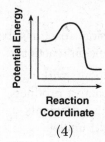

Reaction
Coordinate

(4)

43 _____

44 Which compound is classified as an ether?

(1) CH_3CHO (3) CH_3COCH_3

(2) CH_3OCH_3 (4) CH_3COOCH_3

44 _____

45 Given the equation representing a reversible reaction:

$$HCO_3^-(aq) + H_2O(\ell) \rightleftharpoons H_2CO_3(aq) + OH^-(aq)$$

Which formula represents the H^+ acceptor in the forward reaction?

(1) $HCO_3^-(aq)$ (3) $H_2CO_3(aq)$

(2) $H_2O(\ell)$ (4) $OH^-(aq)$

45 _____

46 What is the mass of an original 5.60-gram sample of iron-53 that remains unchanged after 25.53 minutes?

(1) 0.35 g   (3) 1.40 g
(2) 0.70 g   (4) 2.80 g   46 _____

47 Given the equation representing a nuclear reaction:

$$_1^1H + X \rightarrow {}_3^6Li + {}_2^4He$$

The particle represented by X is

(1) $_4^9Li$ (3) $_5^{10}Be$
(2) $_4^9Be$ (4) $_6^{10}C$ 47 _____

48 Fission and fusion reactions both release energy. However, only fusion reactions

(1) require elements with large atomic numbers
(2) create radioactive products
(3) use radioactive reactants
(4) combine light nuclei   48 _____

49 The chart below shows the crystal shapes and melting points of two forms of solid phosphorus.

Two Forms of Phosphorus

Form of Phosphorus	Crystal Shape	Melting Point (°C)
white	cubic 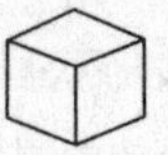	44
black	orthorhombic 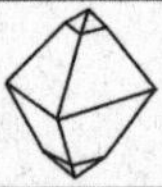	610

Which phrase describes the two forms of phosphorus?

(1) same crystal structure and same properties
(2) same crystal structure and different properties
(3) different crystal structures and different properties
(4) different crystal structures and same properties   49 _____

568

50 Which graph shows the relationship between pressure and Kelvin temperature for an ideal gas at constant volume?

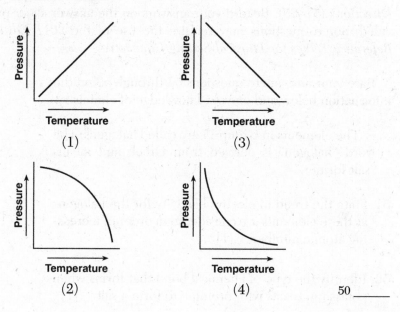

(1)

(3)

(2)

(4)

50 _____

PART B-2

Answer all questions in this part.

Directions (51–65): Record your answers on the answer sheet provided. Some questions may require the use of the *2011 Edition Reference Tables for Physical Setting/Chemistry*.

Base your answers to questions 51 through 53 on the information below and on your knowledge of chemistry.

The elements in Group 17 are called halogens. The word "halogen" is derived from Greek and means "salt former."

51 State the trend in electronegativity for the halogens as these elements are considered in order of increasing atomic number.   [1]

52 Identify the type of chemical bond that forms when potassium reacts with bromine to form a salt.   [1]

53 Based on Table *F*, identify *one* ion that reacts with iodide ions in an aqueous solution to form an insoluble compound.   [1]

Base your answers to questions 54 through 57 on the information below and on your knowledge of chemistry.

The diagrams below represent four different atomic nuclei.

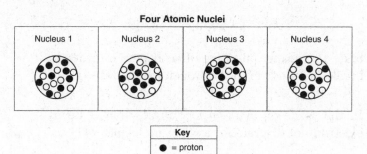

Four Atomic Nuclei

54 Identify the element that has atomic nuclei represented by nucleus 1.   [1]

55 Determine the mass number of the nuclide represented by nucleus 2.   [1]

56 Explain why nucleus 2 and nucleus 4 represent the nuclei of two different isotopes of the same element.   [1]

57 Identify the nucleus above that is found in an atom that has a stable valence electron configuration.   [1]

Base your answers to questions 58 through 60 on the information below and on your knowledge of chemistry.

The equation below represents a chemical reaction at 1 atm and 298 K.

$$2H_2(g) + O_2(g) \rightarrow 2H_2O(g)$$

58 State the change in energy that occurs in order to break the bonds in the hydrogen molecules. [1]

59 In the space *on your answer sheet*, draw a Lewis electron-dot diagram for a water molecule. [1]

60 Compare the strength of attraction for electrons by a hydrogen atom to the strength of attraction for electrons by an oxygen atom within a water molecule. [1]

Base your answers to questions 61 through 63 on the information below and on your knowledge of chemistry.

- A test tube contains a sample of solid stearic acid, an organic acid.
- Both the sample and the test tube have a temperature of 22.0°C.
- The stearic acid melts after the test tube is placed in a beaker with 320. grams of water at 98.0°C.
- The temperature of the liquid stearic acid and water in the beaker reaches 74.0°C.

61 Identify the element in stearic acid that makes it an organic compound.   [1]

62 State the direction of heat transfer between the test tube and the water when the test tube was placed in the water.   [1]

63 Show a numerical setup for calculating the amount of thermal energy change for the water in the beaker.   [1]

Base your answers to questions 64 and 65 on the information below and on your knowledge of chemistry.

A nuclear reaction is represented by the equation below.

$$_1^3H \rightarrow _2^3He + _{-1}^0e$$

64 Identify the decay mode of hydrogen-3.   [1]

65 Explain why the equation represents a transmutation.   [1]

PART C

Answer all questions in this part.

Directions (66–85): Record your answers on the answer sheet provided. Some questions may require the use of the *2011 Edition Reference Tables for Physical Setting/Chemistry.*

Base your answers to questions 66 through 68 on the information below and on your knowledge of chemistry.

A technician recorded data for two properties of Period 3 elements. The data are shown in the table below.

Two Properties of Period 3 Elements

Element	Na	Mg	Al	Si	P	S	Cl	Ar
Ionic Radius (pm)	95	66	51	41	212	184	181	—
Reaction with Cold Water	reacts vigorously	reacts very slowly	no observable reaction	no observable reaction	no observable reaction	no observable reaction	reacts slowly	no observable reaction

66 Identify the element in this table that is classified as a metalloid. [1]

67 State the phase of chlorine at 281 K and 101.3 kPa. [1]

68 State evidence from the technician's data which indicates that sodium is more active than aluminum. [1]

Base your answers to questions 69 through 71 on the information below and on your knowledge of chemistry.

Ammonia, $NH_3(g)$, can be used as a substitute for fossil fuels in some internal combustion engines. The reaction between ammonia and oxygen in an engine is represented by the unbalanced equation below.

$$NH_3(g) + O_2(g) \rightarrow N_2(g) + H_2O(g) + energy$$

69 Balance the equation *on the answer sheet* for the reaction of ammonia and oxygen, using the smallest whole-number coefficients. [1]

70 Show a numerical setup for calculating the mass, in grams, of a 4.2-mole sample of O_2. Use 32 g/mol as the gram-formula mass of O_2. [1]

71 Determine the new pressure of a 6.40-L sample of oxygen gas at 300. K and 100. kPa after the gas is compressed to 2.40 L at 900. K. [1]

Base your answers to questions 72 through 76 on the information below and on your knowledge of chemistry.

Fruit growers in Florida protect oranges when the temperature is near freezing by spraying water on them. It is the freezing of the water that protects the oranges from frost damage. When $H_2O(\ell)$ at 0°C changes to $H_2O(s)$ at 0°C, heat energy is released. This energy helps to prevent the temperature inside the orange from dropping below freezing, which could damage the fruit. After harvesting, oranges can be exposed to ethene gas, C_2H_4, to improve their color.

72 Write the empirical formula for ethene.   [1]

73 Explain, in terms of bonding, why the hydrocarbon ethene is classified as unsaturated.   [1]

74 Determine the gram-formula mass of ethene.   [1]

75 Explain, in terms of particle arrangement, why the entropy of the water *decreases* when the water freezes.   [1]

76 Determine the quantity of heat released when 2.00 grams of $H_2O(\ell)$ freezes at 0°C.   [1]

Base your answers to questions 77 through 80 on the information below and on your knowledge of chemistry.

A student constructs an electrochemical cell during a laboratory investigation. When the switch is closed, electrons flow through the external circuit. The diagram and ionic equation below represent this cell and the reaction that occurs.

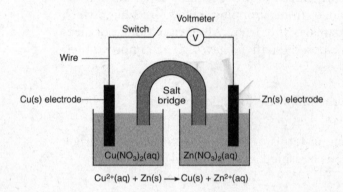

$$Cu^{2+}(aq) + Zn(s) \longrightarrow Cu(s) + Zn^{2+}(aq)$$

77 State the form of energy that is converted to electrical energy in the operating cell. [1]

78 State, in terms of the Cu(s) electrode and the Zn(s) electrode, the direction of electron flow in the external circuit when the cell operates. [1]

79 Write a balanced equation for the half-reaction that occurs in the Cu half-cell when the cell operates. [1]

80 State what happens to the mass of the Cu electrode and the mass of the Zn electrode in the operating cell. [1]

Base your answers to questions 81 and 82 on the information below and on your knowledge of chemistry.

A solution is made by dissolving 70.0 grams of $KNO_3(s)$ in 100. grams of water at 50.°C and standard pressure.

81 Show a numerical setup for calculating the percent by mass of KNO_3 in the solution.   [1]

82 Determine the number of additional grams of KNO_3 that must dissolve to make this solution saturated.   [1]

Base your answers to questions 83 through 85 on the information below and on your knowledge of chemistry.

Vinegar is a commercial form of acetic acid, $HC_2H_3O_2(aq)$. One sample of vinegar has a pH value of 2.4.

83 Explain, in terms of particles, why $HC_2H_3O_2(aq)$ can conduct an electric current.   [1]

84 State the color of bromthymol blue indicator in a sample of the commercial vinegar.   [1]

85 State the pH value of a sample that has ten times *fewer* hydronium ions than an equal volume of a vinegar sample with a pH value of 2.4.   [1]

Answer Sheet
June 2017
Chemistry
The Physical Setting

PART B–2

51 _____

_____

_____

52 _____

53 _____

54 _____

55 _____

56 _____

_____

_____

57 _____

58 _____

_____

_____

59

60 _____

_____

_____

61 _____

62 _____

_____

_____

63

64 _____

65 _____

_____

_____

PART C

66 _____

67 _____

68 _____

_____

_____

69 ___ $NH_3(g)$ + ___ $O_2(g)$ → ___ $N_2(g)$ + ___ $H_2O(g)$ + energy

70

71 _____ kPa

72 _____

73 _____

_____

_____

74 _____ **g/mol**

75 _____

_____

_____

76 _____ **J**

77 _____

78 _____

_____

_____

79 _____

80 Cu electrode: _____

Zn electrode: _____

81

82 _____ g

83 _____

_____

_____

84 _____

85 _____

Answers
June 2017

Chemistry
The Physical Setting

Answer Key

PART A

1. 3	7. 2	13. 1	19. 3	25. 2
2. 2	8. 1	14. 4	20. 4	26. 4
3. 1	9. 4	15. 1	21. 1	27. 3
4. 2	10. 4	16. 2	22. 3	28. 1
5. 4	11. 2	17. 4	23. 3	29. 2
6. 2	12. 1	18. 2	24. 1	30. 3

PART B–1

31. 3	36. 1	41. 3	46. 2
32. 1	37. 1	42. 2	47. 2
33. 3	38. 3	43. 4	48. 4
34. 1	39. 4	44. 2	49. 3
35. 2	40. 1	45. 1	50. 1

INDEX